The Path to Sustainable Corporate Management

D1618962

Ralf T. Kreutzer

The Path to Sustainable Corporate Management

How to Take Responsibility for People, the Environment and the Economy

Ralf T. Kreutzer
Hochschule für Wirtschaft und Recht
Berlin, Germany

ISBN 978-3-658-43973-6 ISBN 978-3-658-43974-3 (eBook)
https://doi.org/10.1007/978-3-658-43974-3

Translation from the German language edition: "Der Weg zur nachhaltigen Unternehmensführung" by Ralf T. Kreutzer, © Der/die Herausgeber bzw. der/die Autor(en), exklusiv lizenziert an Springer Fachmedien Wiesbaden GmbH, ein Teil von Springer Nature 2023. Published by Springer Fachmedien Wiesbaden. All Rights Reserved.

This book is a translation of the original German edition "Der Weg zur nachhaltigen Unternehmensführung" by Ralf T. Kreutzer, published by Springer Fachmedien Wiesbaden GmbH in 2023. The translation was done with the help of an artificial intelligence machine translation tool. A subsequent human revision was done primarily in terms of content, so that the book will read stylistically differently from a conventional translation. Springer Nature works continuously to further the development of tools for the production of books and on the related technologies to support the authors.

© The Editor(s) (if applicable) and The Author(s), under exclusive license to Springer Fachmedien Wiesbaden GmbH, part of Springer Nature 2024

This work is subject to copyright. All rights are solely and exclusively licensed by the Publisher, whether the whole or part of the material is concerned, specifically the rights of translation, reprinting, reuse of illustrations, recitation, broadcasting, reproduction on microfilms or in any other physical way, and transmission or information storage and retrieval, electronic adaptation, computer software, or by similar or dissimilar methodology now known or hereafter developed.
The use of general descriptive names, registered names, trademarks, service marks, etc. in this publication does not imply, even in the absence of a specific statement, that such names are exempt from the relevant protective laws and regulations and therefore free for general use.
The publisher, the authors, and the editors are safe to assume that the advice and information in this book are believed to be true and accurate at the date of publication. Neither the publisher nor the authors or the editors give a warranty, expressed or implied, with respect to the material contained herein or for any errors or omissions that may have been made. The publisher remains neutral with regard to jurisdictional claims in published maps and institutional affiliations.

This Springer imprint is published by the registered company Springer Fachmedien Wiesbaden GmbH, part of Springer Nature.
The registered company address is: Abraham-Lincoln-Str. 46, 65189 Wiesbaden, Germany

Paper in this product is recyclable.

We need a legal, economic, and societal framework in which those who deal particularly responsibly with people, natural resources, and the environment as a whole can be more successful than those who do not.

We need not only a social, but also a sustainable market economy, to meet the great challenges of today and tomorrow.

Finally, the General Assembly of the ***United Nations*** *adopted a resolution on July 28, 2022, recognizing the right to access a clean, healthy, and sustainable environment as a human right. However, this resolution is not binding, but a so-called Loft Law—a law with political symbolism.*

It is up to all of us to derive powerful action from this symbol!

After all, we did not inherit the earth from our parents, but borrowed it from our children!

Ralf T. Kreutzer

Preface

Dear reader,

One thing is becoming visible to all of us:

We must not go into a **collective early retirement** in terms of climate and environmental protection, because the changes here—so far—will rather show in decades than in years. Especially in recent years, it has become clear that **climate change** is no longer a **calculable threat scenario**—with the option of "sitting it out". The climate change that is already visible today demands its **tribute** daily—somewhere in the world.

We are approaching various **climate tipping points.** Such tipping points occur when changes in large parts of the climate system continue autonomously beyond a certain warming threshold. Then, collapsing ice sheets not only lead to a significant rise in sea levels and further global warming, but also to the death of species-rich biomes in the rainforests and warm water corals. Due to increasing global warming, more carbon is released by thawing permafrost soils. This further drives the warming. As a result, a **negative spiral** sets in, leading to further **warming dynamics**—a true vicious circle!

The **best time to act** would have been a few decades ago, when the **limits of growth** were first reported in the early 1970s. However, it is said: **The second best time to act is now!**

The motto is: **Time for change!**

Today we are faced with an **equation with a multitude of unknowns**:

- What are the most important challenges in each case?
- And what are the most convincing solutions?

In this case, it is necessary to overcome a partially encountered **disappointment prevention**. If such a thing exists, then one refrains from any kind of expectations, because these could remain unfulfilled and therefore disappointments would be preprogrammed.

What is it called so aptly?

Every expectation is a trap!

We must prevent too many people from giving up out of fear of failing to cope with the climate crisis and not taking any action at all. At the same time, society must not slide into a **climate change tinnitus**. Tinnitus is the Latin word for ringing. The term climate change tinnitus refers to such a disturbing noise that one constantly hears but tries to ignore in order to continue one's normal life as unchanged as possible. Ignorance and unwillingness to see the climate-related changes are expressions of this. So how can a climate change tinnitus be avoided, whose disturbing whistling, humming, buzzing, or knocking is simply filtered out?

Questions upon questions, for which there are no simple and often no clear answers. What do we as a society and economy need in this situation?

We still need a **relaxed soul,** to be able to tackle the challenges powerfully. In addition, a **tense mind** and **a lot of willpower** are also needed to take the necessary steps against resistance. After all, a lot of energy is needed for every change process because every change naturally encounters resistance. Therefore, we must activate the existing resources at all levels to master the challenges ahead of us.

In economy and society, it is becoming increasingly important to drive an ecological transformation parallel to the digital transformation. Therefore, companies today need not only a **Chief Digital Officer,** to significantly advance the digital transformation of a company. They also need a **Chief Sustainability Officer,** who trims the entirety of the company's activities towards sustainability.

In this context, a quote from ***Ludwig Erhard*** from the year 1957 takes on a whole new relevance:

> "We will indeed reach a point where it is entirely appropriate to question whether it is still correct and beneficial to produce more goods, more material wealth, or whether it would not be more sensible to gain more leisure time, more contemplation, more leisure and more relaxation by renouncing this 'progress'."[1]

Germany is the **land of TÜV** and the **land of tinkerers.** In this combination, many important things can and will succeed. I can also say: It must succeed!

I would like to thank my guest authors *Dirk Ziems, Thomas Ebenfeld, Rochus Winkler* and *Werner Detering* from *concept m research + strategies* and *flying elephant marketing consultancy, Ingo Lies,* the founder and owner of *Chameleon,* as well as *Andreas Huthwelker* and *Christian Schlimok* from *Novamondo* for their valuable contributions.

I would like to express my heartfelt thanks to my editorial team *Barbara Roscher* and *Angela Meffert,* who have also accompanied this work with their esteemed professionalism. I thank my student assistant *Joleen Grogoll,* who has significantly supported my work through her research and the creation of illustrations.

---

[1] Erhard, L. (1957). Prosperity for all. Düsseldorf: Econ.

I wish my esteemed readers an inspiring study of this work and many important impulses to change something in their own lives as well as in their professional work towards sustainability. Good luck with that!

> We must be the change we want to see in the world.
> Mahatma Gandhi

Königswinter, Berlin
April 2023

Ralf T. Kreutzer

Abstract

The world is in upheaval—politically, socially, and in terms of natural resources. More and more people and institutions are basing their **purchasing decisions** on criteria that contribute in a broader sense to **sustainability.** To meet the requirements of the **economy, ecology, and social sector (Triple Bottom Line)**, it makes sense to implement the **three-pillar model of sustainability** in the company. At the same time, further requirements are defined, among other things, with the **supply chain law** and demanding expectations are also formulated by customers and other stakeholders towards companies with the **ESG criteria** (Environment, Social, Governance). This work shows how companies can use the **sustainability trend** in a **value-creating** way in the long term, by having those responsible at all corporate levels, and not just in **marketing** and **brand management**, more strongly address these challenges and deal with the associated implications for sustainable **corporate management**.

Contents

About the Author

Prof. Dr. Ralf T. Kreutzer has been a Professor of Marketing at the University of Economics and Law/Berlin School of Economics and Law since 2005. In parallel, he works as a trainer, coach, and marketing and management consultant. He spent 15 years in various leadership positions at Bertelsmann (last position Director of the International Division of a subsidiary), Volkswagen (Managing Director of a subsidiary), and Deutsche Post (Managing Director of a subsidiary) before he was appointed Professor of Marketing in 2005.

Prof. Kreutzer has provided significant impulses on various topics around marketing, dialogue marketing, CRM/customer loyalty systems, database marketing, online marketing, social media marketing, digital Darwinism, digital branding, dematerialization, change management, digital transformation, artificial intelligence, agile management, sustainability, strategic and international marketing through regular publications and keynote speeches (including in Germany, Austria, Switzerland, France, Belgium, Singapore, India, Japan, Russia, and the USA). He has advised a multitude of companies both domestically and abroad on these subject areas. In addition, Prof. Kreutzer is active as a trainer and coach.

Prof. Kreutzer is a founding member of the consulting firm Green Elephant Consulting (www.green-elephant.info), which specializes in making sustainability achievable for companies.

His most recent book publications are "Toolbox for Marketing and Management" (2019), "Practice-oriented Marketing" (2023), "Toolbox for Digital Business" (2022).

Glossary of Abbreviations

AEPW	Alliance to End Plastic Waste
BAFA	Federal Office for Economic Affairs and Export Control (Bundesamt für Wirtschaft und Außenkontrolle)
BIM	Building Information Modeling
BMAS	Federal Ministry for Labour and Social Affairs (Bundesministerium für Arbeit und Soziales)
BMWSB	Federal Ministry for Housing, Urban Development and Building (Bundesministerium für Wohnen, Stadtentwicklung und Bauwesen)
BVL	Federal Office for Consumer Protection and Food Safety (Bundesamt für Verbraucherschutz und Lebensmittelsicherheit)
C2M	Consumer to Manufacturer
CCS	Carbon Capture and Storage
CEO	Chief Executive Officer
CMO	Chief Marketing Officer
CSRD	Corporate Sustainability Reporting Directive
CSuO	Chief Sustainability Officer
D2C	Direct-to-Customer
DGNB	German Society for Sustainable Building (Deutsche Gesellschaft für Nachhaltiges Bauen)
DJSI	Dow Jones Sustainability Indices
ECORE	ESG-Circle of Real Estate
EDR	Electrodermal response
EMEA	Europe, Middle East, Africa
ESG	Environment(al), Social, Governance
ETS	Emissions Trading System
EVPG	Energy Consumption Relevant Products Act (Energieverbrauchsrelevante-Produkte-Gesetz)
EZB	European Central Bank (Europäische Zentralbank)
FSC	Forest Stewardship Council
GHG	Greenhouse Gas

GOTS	Global Organic Textile Standard
GRI	Global Reporting Initiative
GRP	Gross Rating Point
GVM	Society for Packaging Market Research (Gesellschaft für Verpackungsmarktforschung)
GWW	Association of the Promotional Products Industry (Gesamtverband der Werbeartikel-Wirtschaft)
HR	Human Resources
IFC	Industry Foundation Classes
IPCC	Intergovernmental Panel on Climate Change (Climate Council)
KMU	Small and medium-sized enterprises (Kleine und mittlere Unternehmen)
KrWG	Circular Economy Act (Kreislaufwirtschaftsgesetz)
LCA	Life Cycle Assessment
LEH	Retail Food (Lebensmittel-Einzelhandel)
LkSG	Supply Chain Law (Lieferkettengesetz)
NFRD	Non-Financial Reporting Directive
NGO	Non-Governmental Organization
PaaS	Product as a Service
PEF	Product Environmental Footprint
PGR	Psychogalvanic response
QNG	Sustainable Building Quality Seal (Qualitätssiegel Nachhaltige Gebäude)
SASB	Sustainability Accounting Standards Board
SAF	Sustainable Aviation Fuel
SBTi	Science-Based Target Initiative
SFDR	Sustainable Finance Disclosure Regulation
TCFD	Task Force on Climate-Related Financial Disclosures
VerpackG	Packaging Act (Verpackungsgesetz)
VO	Regulation (Verordnung)
W2E	Waste-to-Energy

1 Sustainability in Corporate Governance and on the Customer Side

Nature is relentless and unchangeable, and it is indifferent whether the hidden reasons and ways of its actions are understandable to man or not.

Galileo Galilei

Abstract

Sustainability is a term that is intensively discussed today, which is further explained here. In addition, the differences between linear and circular economy are elaborated. The openness of customers and companies to the topic of sustainability is worked out based on various studies. Additionally, terms such as the attitude behavior gap, mental accounting, as well as external effects and external costs are deepened in their significance for a sustainable economy.

1.1 What Does "Sustainability" Mean?

"Sustainability" is now thought of much more broadly than it was in the past. The original core of sustainability was a **principle of action for the use of natural resources.** This principle was intended to ensure a lasting satisfaction of (human) needs by ensuring a natural regenerative capacity of the entire ecosystem. This is also illustrated by the English term "to sustain", which can be translated as "to maintain". Sustainable action is intended to ensure that the extent of resource use does not endanger the entire ecosystem in the long term. Consequently, only as many resources may be taken from this system as can be replaced within the same period. This is about ecological sustainability. Often this is also referred to as "green".

© The Author(s), under exclusive license to Springer Fachmedien Wiesbaden GmbH, part of Springer Nature 2024

R. T. Kreutzer, *The Path to Sustainable Corporate Management*,
https://doi.org/10.1007/978-3-658-43974-3_1

Ecological sustainability means that raw materials and other materials are used in the broadest sense in an "environmentally friendly" way. This is intended to protect nature and secure natural resources for future generations. This refers to organic products that meet the requirements of organic farming. The use of plant protection products and artificial fertilizers is avoided. In addition, a "more species-appropriate" animal husbandry is implemented. This is intended to ensure that healthy food can still be produced in the future. Many organic labels visualize such ecological sustainability, although this is not always adhered to (see Sect. 5.3.5).

However, humanity is increasingly failing to meet the requirements of such ecological sustainability. This is evident on the so-called Earth Overshoot Day. This is also referred to as **Earth Overshoot Day**. The *Global Footprint Network,* a non-profit organization, annually determines the **global ecological footprint.** This describes the demand of people for biological resources. This can determine from which day of a current year the human demand for renewable raw materials is higher than the supply that can be generated by reproduction within one year. This supply is referred to as **global biocapacity**. Earth Overshoot Day illustrates from which day people—but only with renewable raw materials—live beyond their means. The consumption of non-renewable raw materials (such as oil, gas, sand, iron, etc.) is not taken into account here.

▶ **Food for Thought** The **Earth Overshoot Day** was on **May 15, 1962** in the year **1961**—and thus in the following year! The global ecosystem could still recover well here, because only 73% of the renewable resources were consumed within one year. In 1961, the global annual resource consumption therefore still left reserves. From 1970 onwards, the annual consumption of these resources exceeds the annually globally renewable supply. In the year **2022**, Earth Overshoot Day fell on **July 28, 2022.** In this year, 1.75 Earths would be needed to balance the global consumption of humanity's renewable resources (see Earth Overshoot Day, 2022). However, as is well known, humanity only has one Earth!

▶ **Note Box** Since 1970, humanity has been living significantly beyond its means—and thus at the expense of future generations.

How the **resource consumption** has changed over the last decades is shown in Fig. 1.1 (see Global Footprint Network, 2022; see also Land, 2018). With this development and in view of the growing world population, it is not foreseeable when a balance of supply (**biocapacity**) and demand (**ecological footprint**) of renewable raw materials could be achieved again.

Therefore, the achievement of ecological sustainability should not and must not be delegated solely to politics and the economy. Rather, every single citizen is called upon to act in a resource-conserving manner. This starts with the use of recyclable products

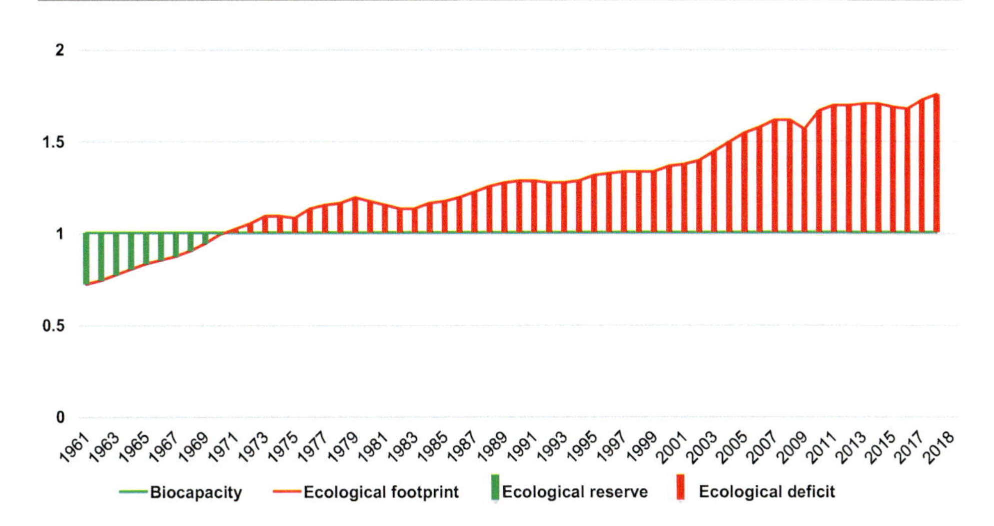

Fig. 1.1 Development of the ecological footprint. (Data source: Global Footprint Network, 2022)

and extends to the use of renewable energies and the avoidance of the introduction of toxins into the environment (e.g., through the use of pesticides in the home garden).

> Note-Box With **ecological sustainability**, it's about ensuring that our **planet** can survive.

However, today a focus on ecological sustainability is no longer sufficient. Therefore, sustainability has been supplemented by two additional dimensions. With **social sustainability**, it's about the major issues of **human rights,** equal opportunities, meeting basic needs and thus **quality of life.** In the economic environment, this is intended to ensure that the stakeholders involved in the value chain are treated "fairly"—regardless of the country. This includes a "fair" remuneration of suppliers, as promised in fair trade concepts. The renunciation of child labor, adequate protection of the stakeholders in the production process, and the preservation of human rights are also part of social sustainability.

Social sustainability can simply be translated as **"humanity"**. Who would want to adorn themselves with a piece of clothing that was made under inhumane working conditions in a sweatshop in Pakistan—perhaps even in twelve-hour shifts? Provided, one knows about it! Knowledge or suspicion of such poor working conditions can cause feelings of guilt in the buyer—and possibly prevent a purchase.

> Note-Box With **social sustainability**, it's about ensuring that our **people** can survive (well).

Under **economic sustainability**, it is understood that companies are able to successfully implement a business model over a longer period of time due to their profit-making. After all, no society, no economy, and no employee benefits if companies only have a minimal life expectancy and then have to cease their business operations, lay off their employees, and no longer pay taxes.

> **Note-Box** With **economic sustainability**, it's about **profit,** so that our companies can survive.

The interplay of the **requirements of people, planet and profit** can be found in Fig. 1.2.

The requirements of people and profit stand here **equally** side by side. One does not succeed without the other. Human activities must be **bearable** for people and the planet—both must be able to exist in the long term. In addition, the measures must be designed in such a way that the planet and companies remain **viable**. Where all fields overlap, lies the desired sustainability!

In the *Brundtland* report "Our common future" of the United Nations (1987, p. 15), the **term sustainability** was defined as follows:

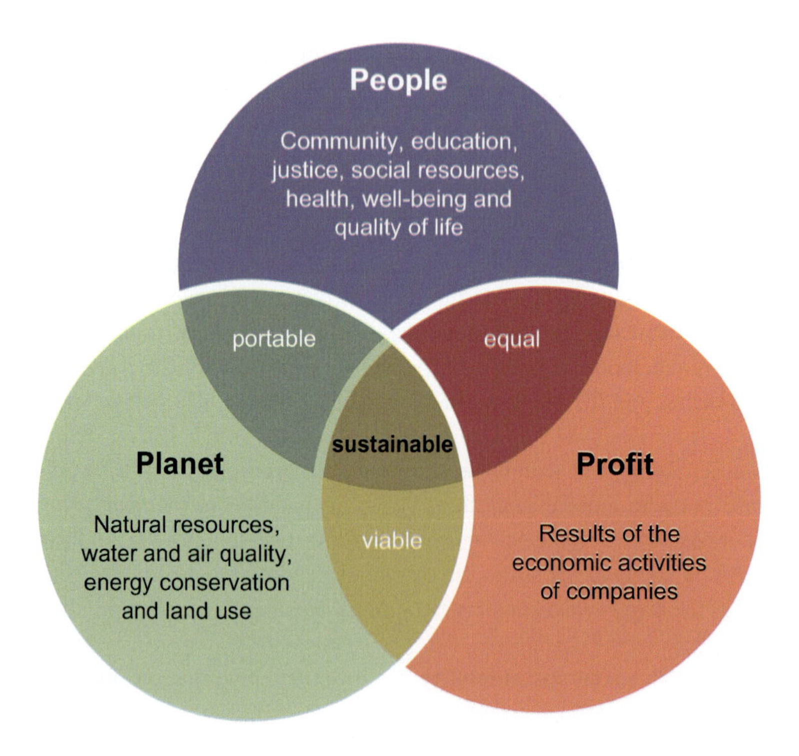

Fig. 1.2 People—Planet—Profit

"Humanity has the ability to make development sustainable to ensure that it meets the needs of the present without compromising the ability of future generations to meet their own needs."

To anchor these three dimensions of sustainability in the company, the **Triple-Bottom-Line concept** was developed. This concept, also known as the **Three-Pillar Model**, provides a framework. It is intended to help companies incorporate not only economic, but also social and ecological goals into their decisions and activities. This concept assumes that companies should be evaluated not only by the profit they generate, but also by their impact on society and the environment. Based on the Triple-Bottom-Line concept, companies measure and report their performance in three areas:

- **Economic area**
  This is about the financial performance of the company, such as profit, sales, and market shares.
- **Social area**
  This area examines the company's impact on society. This includes the creation of jobs, the promotion of education, and the support of communities.
- **Ecological area**
  This area looks at the company's impact on the environment. This includes the use of resources, the emission of greenhouse gases, and waste management.

▶ **Remember Box** The **Triple-Bottom-Line concept** brings together the three performance fields of People, Planet, and Profit. The implementation of this concept "forces" companies to strive for a permanent balance between ecological, social, and economic sustainability. The Bottom Line "Profit" is supplemented by criteria for measuring the impact on Planet and People.

The Triple-Bottom-Line concept advocates that companies improve their performance in all these areas and are oriented towards sustainable development. It represents an extension of traditional financial goals and emphasizes that companies also have responsibility for their impact on society and the environment.

These forms of sustainability need to be placed in larger contexts. As early as 2015, the *United Nations* adopted the **Agenda 2030**. It includes **17 overarching goals**—the so-called **Sustainable Development Goals**—and **169 sub-goals,** to secure a livable future worldwide (see Fig. 1.3; for more detail, United Nations, 2022).

The 17 goals of this **Agenda 2030** are addressed to everyone. This includes individual states, their respective civil societies, the economy, science, and every individual. To document progress with regard to the Agenda 2030, a report on **"Sustainable Development in Germany"** is published every two years.

Fig. 1.3 (The content of this publication has not been approved by the United Nations and does not reflect the views of the United Nations or its officials or Member States.) The 17 Sustainable Development Goals of the United Nations. *Source* (United Nations, 2020, p. 40)

The wish of many—or even all?—is for **green growth** or **Green Growth.** Because without such economic growth, the goals described cannot be achieved. Because where else but in the economy should the resources necessary for sustainable action be generated? Green Growth describes an alternative growth path. Green growth is based on a **sustainable use of natural resources,** in order to at least maintain the basis of quality of life in the long term. Requirements of **social sustainability** are also taken into account in Green Growth.

> **Food for Thought** Of course, it would be even better if we could pass on the world in a better state to the next generation than we received it. The fact that this—at least in parts—does not have to remain a utopia is shown by many developments in Germany: You can swim in the Rhine again, the Ruhr is no longer a stinking cesspool, the air quality in Bitterfeld is much better than before, child labor no longer exists, and the rights and health of working people are much more comprehensively protected than before. So it is possible!

A study conducted in Germany determined the influence of various actors on sustainability. The result is shown in Fig. 1.4. One thing becomes clear: The **civil society** and thus the **consumers** can exert the greatest influence towards a sustainable society with their behavior (29%)—closely followed by **profit-oriented companies** with 27%. **International and national political actors** can intervene with 22% and 9% respectively. The **non-profit organizations** are in between with 10% (cf. Simon-Kucher & Partner, 2022).

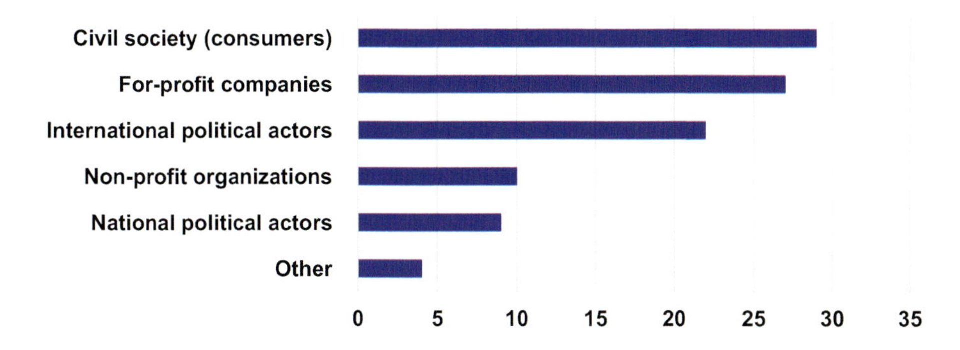

Fig. 1.4 Actors with a significant influence on sustainability in Germany

Green growth requires—in addition to more sustainable consumption—a restructuring of the economy aimed at higher resource efficiency. By developing new (green) markets and developing ecological products and services, competitive advantages are to be achieved. Green growth is intended to reconcile economic development with the necessities of sustainable corporate governance. However, it is also necessary to involve society on the path to sustainability, because not everything can stay as it is.

> **Food for Thought** In sustainable corporate governance, the thinking and actions of those responsible go far beyond their own immediate area of responsibility. In order to meet the demands of sustainability, the value chains must be examined in their entire length. Today and in the future, companies will be much more comprehensively forced by laws and regulations to define corporate responsibility in this broader sense.

1.2 Circular Economy

Due to its great importance for sustainable corporate governance, the basics of the circular economy will be deepened here. The circular economy is the counter-draft to the **linear economy (Linear Economy).** The linear economy is also referred to as **throw-away economy**. The motto is **Cradle to Grave**—"from the cradle to the grave" or also "from the cradle to the landfill for eternity". Here, the use of materials, processing, consumption and disposal take place without regard for the permanent preservation of resources.

The dominant **pattern of the linear economy** is: take, manufacture, consume/use, throw away—and don't think about it! This model assumes an inexhaustible amount of cheap, easily accessible resources. This approach was the dominant economic system for more than a century. In such an economic system, the processed raw materials or the products produced with them are mostly either deposited or burned after their use (cf. Fig. 1.5).

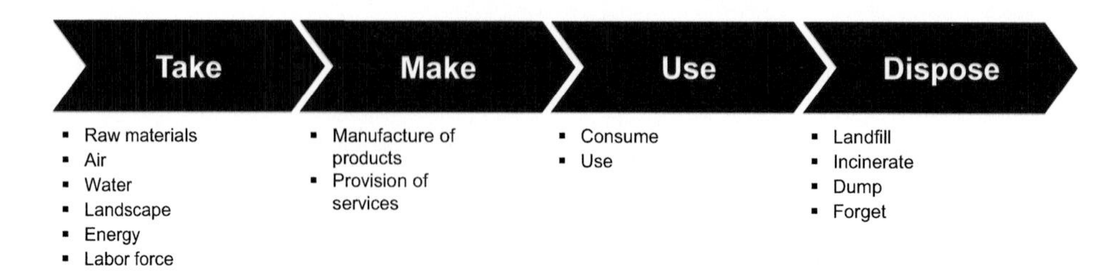

Fig. 1.5 Concept of the linear economy—Linear Economy

> **Note Box** The **linear economy** corresponds to the course of a river. It starts with the source and ends in the sea. In this case, the water only knows one direction of flow—towards the end.
>
> The **motto of the linear economy** is: **Cradle to Grave.**

In a **circular economy** or in the **Circular Economy** on the other hand, a renewal-oriented approach—a regenerative system—is aimed for. This is a **resource-saving development, production, consumption/use and recovery model.** Sustainability criteria must already be taken into account in the design. A "design for further and recycling" (including the conception for disassembly) is important in order to reduce the negative environmental impacts of the products at the end of their life cycle. In addition, this can reduce the costs for recycling. A recycling-friendly design facilitates the transition to further cycles (cf. fundamentally Stahel, 2019).

Additionally, it is about ending a **built-in obsolescence of products** (keyword Built-in Obsolescence). Through artificial obsolescence, companies try to limit the technical lifespan of products. This is intended to motivate customers to make a new purchase earlier than technically necessary. This endeavor does not fit with the circular economy. Therefore, a **"right to repair"** should be created for customers.

The circular economy aims to align production and distribution as resource-efficient as possible and to reuse a high proportion of renewable resources or recovered raw materials. It also aims to reduce waste in production and consumption. In addition, the pollutant load during the production and use of the marketed products (e.g., emissions of a vehicle) should be reduced.

> **Note Box** The circular economy leads to a decoupling of growth and resource consumption.

The concept of the circular economy is called **Cradle to Cradle**—"from cradle to cradle". In contrast to the Cradle-to-Grave approach, the **Cradle-to-Cradle concept** aims to align the products and their packaging from the first idea to a biological and/or technical cycle. The desired closed material cycles are achieved when the materials used allow a safe and complete return to the biosphere—or when these materials can be recovered

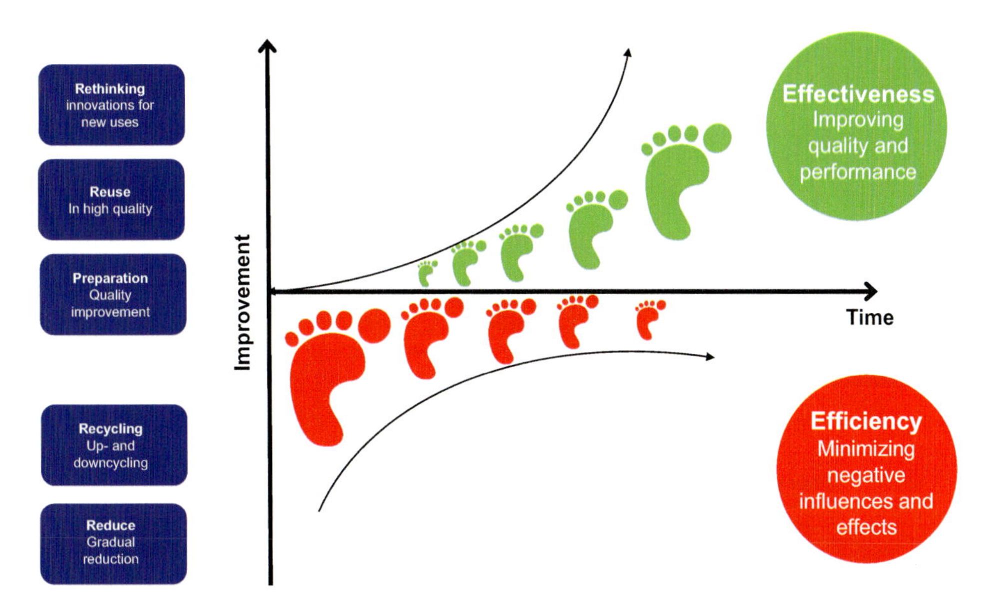

Fig. 1.6 Eco-effectiveness instead of efficiency. *Source* Based on (EPEA, 2023)

or reused as high-quality as possible. The core idea of the Cradle-to-Cradle concept—a focus on eco-effectiveness instead of efficiency—is illustrated in Fig. 1.6 (see EPEA, 2023).

> **Note Box** The **circular economy** is more like a lake. Here, life arises and passes away in one place. An eternal cycle that supports and amplifies itself—a system that can last in the long term.
>
> The **motto of the circular economy** is: **Cradle to Cradle.**

How the individual **phases of the circular economy** interlock is shown in Fig. 1.7 (see EPEA, 2023). Ideally, non-renewable energy is avoided—and no waste is produced.

> **Food for Thought When it comes to sustainability, the 5-R rule should be applied:**
>
> - **Refuse:** for example, by refusing plastic bags and single-use products
> - **Reduce:** by reducing the use of resources along the entire value chain
>
> At the **center of the circular economy** are also the following activities:
>
> - **Reuse:** further use of the products
> - **Repurpose:** assigning products to a new field of application
> - **Recycle:** recovery of the resources used

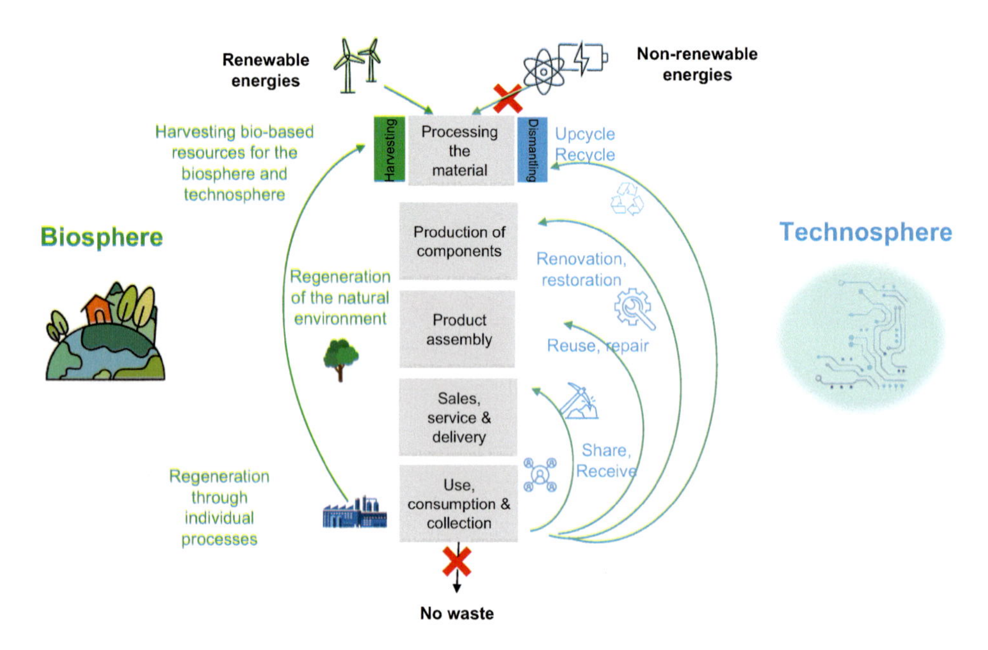

Fig. 1.7 Cradle-to-Cradle concept. *Source* Based on (EPEA, 2023)

In the circular economy, products and resources should be used as long as possible, reused or recycled, repaired or overhauled/maintained or refurbished. Only when these possibilities are exhausted and the lifespan can no longer be (economically) extended, should recycling take place. The core idea is to close the **resource, production, usage, and disposal cycles** (see Fig. 1.8).

The concept of the circular economy uses the following terms:

- **Reuse and repurposing**
  In reuse and repurposing, a material or product is used again for its original purpose after use. Such products are often referred to as "used", "second hand" or "pre-used". In this case, the original product usually remains unchanged when used again.
  Such second-hand use occurs with used cars, cameras, computers, mobile phones, and increasingly also with clothing. This does not deplete the cycle of resources, as the **usage period of the original product** is extended. **No primary resources are consumed.** Also, **no waste** is generated. The **value of the original product** is maintained at a high level.
- **Refurbishing** ("refurbish" stands for renovate, overhaul, revise)
  Refurbishing involves a **quality-assured overhaul and maintenance** of products or entire systems. These are then intended to be reused or remarketed.
  Refurbishing is mainly used for IT devices (including PCs, laptops, monitors, cartridges for ink and toner) and for car components (e.g., tires). Accordingly prepared

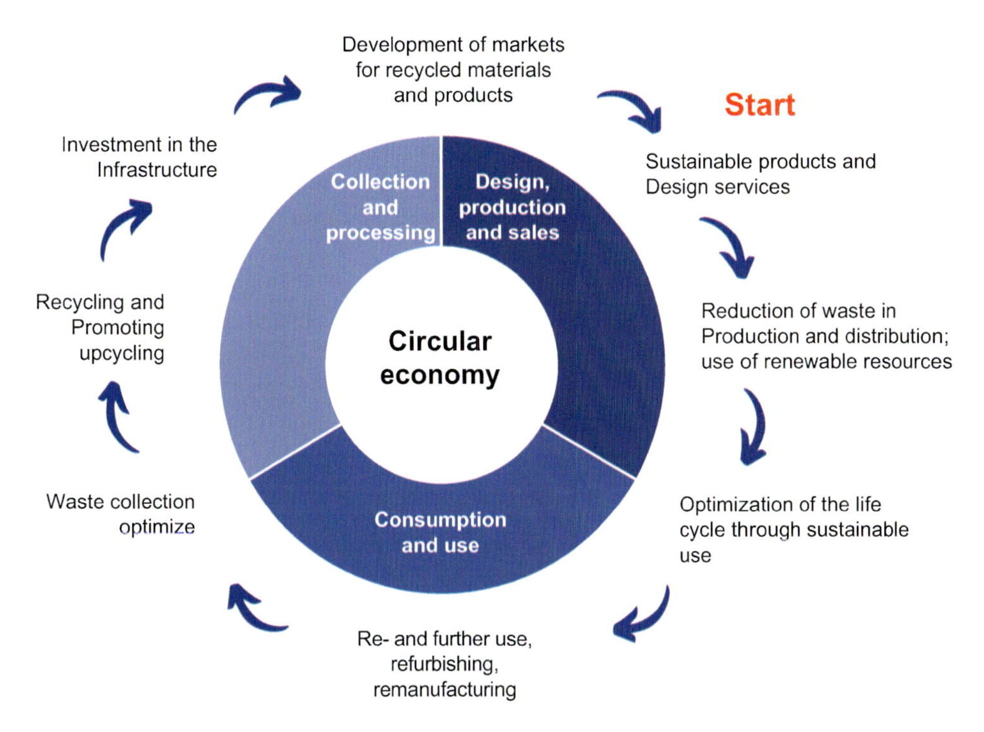

Fig. 1.8 Concept of the circular economy—Circular Economy

products are reintroduced into the usage cycle. Also, production facilities are reused by companies after an overhaul and maintenance. In this way, refurbishing contributes to the **conservation of primary resources** and the **avoidance of waste**. The **value of the original product** is largely preserved.

An overhaul is also necessary when a product is returned to the dealer or manufacturer due to the applicable right of withdrawal or due to a warranty case. Even today, such products are often destroyed instead of being reintroduced into the cycle. This is the case, for example, with *Amazon* (see Goebel, 2021). Such destruction deprives the circular economy of important resources. Instead, an examination and possibly an overhaul and/or maintenance should be carried out to be able to resell the products. This is often done with the note "refurbished"—associated with price reductions.

- **Refabrication** or **Remanufacturing**

  In refabrication or remanufacturing, used devices and systems are brought up to the quality standard of a new device. This often requires the product to be disassembled. This allows individual components and assemblies to be checked, refurbished, and possibly replaced. In this process, outdated components can also be replaced with new ones. The result is a product that is very close to the condition of a new product.

Estimates suggest that refabrication could save up to 90% of raw materials and over 50% of energy worldwide each year—compared to the production of new products. The "restoration" of products also results in fewer CO_2 emissions.
Today, refabrication is used, for example, in engines, pumps, robots, railway cars, and many machines. In this way, it contributes to the **conservation of primary resources** and the **avoidance of waste**. The **value of the original product** is largely preserved through the refurbishment.

- **Recycling**
  Recycling describes the process of reprocessing valuable materials. In this process, the original product is destroyed, leading to the partial destruction of primary resources. This results in a—often considerable—loss of the original product value. Therefore, recycling should only take place when the lifespan—for example, through refurbishing or remanufacturing—can no longer be (economically) extended. The result of recycling are **secondary raw materials** and **reusable parts.** In total, this **conserves primary resources.**
  In the case of a **decomposition** (i.e., a disassembly) of a smartphone, the battery is first removed. Often, it must still be disposed of professionally today if recycling is not possible. Other modules of the smartphone can be reintroduced into the cycle unchanged (e.g., power supplies). Decomposition also occurs in old cars and decommissioned airplanes, which essentially become "spare parts warehouses". These components can be reused and retain part of their original value.
  Other elements of a smartphone are crushed and sorted according to their respective components (plastics, metal mixtures, gold). Metals and glass are recovered through **melting processes**. This involves a phase transition from a solid substance or a solid mixture into a liquid state. Heat and/or pressure are applied for this. This allows the starting material to be broken down into different substances. Estimates suggest that nearly 10 g of copper, 150 mg of silver, and 25 mg of gold can be recovered per smartphone. In addition, smaller amounts of platinum and palladium are produced. Melting processes are also used in the recycling of aluminum cans, plastic bottles, and many other food packaging. The recovery of raw materials reduces the need for primary resources in new production.
  In the recycling of waste paper, used cardboard, and paperboard, a **dissolution process** is used. Here, the phase transition from the solid starting material to a liquid state is brought about by the addition of water. In the course of this process, printer ink in the pulp is removed using chemicals. Here too, primary resources are conserved.
  In the course of recycling, **secondary raw materials** are obtained. The end product from a material processing process of plastic is called **recyclate**. This is an umbrella term for the following results of a recycling process (see Kunststoffe.de, 2023):
  - **Grind**
    Grind is obtained by grinding plastic. Grind has different and irregular particle sizes from 2 mm to 5 mm. It can also contain dust particles.
  - **Regranulate**

Regranulate is obtained from grind through a melting process as granulate. The regranulate has a uniform grain size and is free of dust particles.

- **Regenerate**

  Regenerate is also obtained through a melting process. The use of additives can improve the properties of the regenerate. Regenerate has a uniform grain size and no dust particles. In addition, it is characterized by other properties.

The secondary raw materials obtained in this way can be reused, for example, for the production of new products. This often presents a particular challenge: The recyclate is, for example, often more expensive and less hygienic than the new plastic obtained from oil.

To enable recycling, the—preferably pure—**collection of waste** needs to be optimized. In addition, further, often **highly complex recycling plants,** are required to promote genuine recycling. Estimates suggest that today in Germany only about 5% of waste is—truly—recycled.

- **Upcycling**

  Upcycling refers to the reuse of a product or a recyclable material—but in a different way than before. This is essentially about recycling—but "upwards" ("up"). Waste products are transformed into other products.

  There are many examples of this type of **valuable reuses** through upcycling. In some cases, the original products are destroyed in the process of upcycling. Then, "new" hats and bags are made from discarded jeans. Car tires are turned into soles for flip-flops, truck tarps into handbags, and screw-top jars into flower vases. In Fig. 1.9, you can see the **penguin sculpture** of the expedition ship *Hanseatic Spirit*. This sculpture is called *Mother Penguin and Chick*. It was made from steel and recycled flip-flops by the artist cooperative *Ocean Sole* from Kenya. There, flip-flops were collected on the beaches and processed into this work of art. At the same time, the issue of marine plastic is addressed in an appealing way (see Fig. 1.9).

  However, there are also reuses that leave the original product intact but put it to a different use. For example, wooden railway sleepers find a wide range of new uses—including in gardening. Whole car tires are turned into swings or other playground equipment. An example is shown in Fig. 1.9 on a playground in Greenland. At go-kart tracks, car tires are used as "shock-absorbing" track boundaries.

Note Box Through upcyclingthe value of waste products increases—instead of decreasing.

Compared to downcycling, upcycling thus leads to a material upgrade of waste. The reuse or different reuse of already used material or used products in other contexts also reduces the consumption of primary raw materials.

To advance the circular economy, the **European Parliament** adopted a resolution on the new **Action Plan for the Circular Economy** in February 2021. It calls for additional measures. The EU's goal is to achieve a carbon-neutral, ecologically sustainable, toxin-

Fig. 1.9 Examples of upcycling—playground and sculpture. (Images: Private photos)

free, and fully circular economy by 2050. To this end, stricter recycling regulations and more binding targets for the use and consumption of materials are to be defined by 2030 (see European Parliament, 2021).

In light of these EU goals, it is worth taking a look at the **infrastructure of the circular economy in Germany**. Figure 1.10 shows the status for the years 2009 and 2020 (see Statista, 2021g, p. 17). At first glance, it is clear that the infrastructure in Germany has not been significantly expanded, but the number of individual facilities has even decreased. This is particularly the case with **dismantling companies for end-of-life vehicles** as well as **biological** and **chemical-physical treatment plants**. The number of **dismantling facilities for waste electrical and electronic equipment** has slightly increased. The fact that the number of **shredder plants** has significantly increased is not a testament to a successful circular economy. However, the number of facilities here and in all other cases alone says nothing about the volume they can handle.

Figure 1.10 shows that the number of **combustion plants** has decreased. However, **waste incineration** is not part of the circular economy. Sometimes, this is euphemistically referred to as **“thermal waste treatment”“** and **Waste-to-Energy(W2E)**. Waste incineration involves burning waste or subjecting it to other thermal processes. Depending on the material used, this can generate electricity and/or heat. The high temperatures used in the combustion process break down the molecules, irreversibly destroying the value of the original materials. This can also release harmful emissions.

▶ **Food for Thought** How far Germany is from a circular economy is shown by the handling of discarded—but sometimes still functional—household electronics (such as lamps, irons, toasters, screens, keyboards, computers). Often these are dropped off at municipal—so-called—recycling yards. There they are thrown into

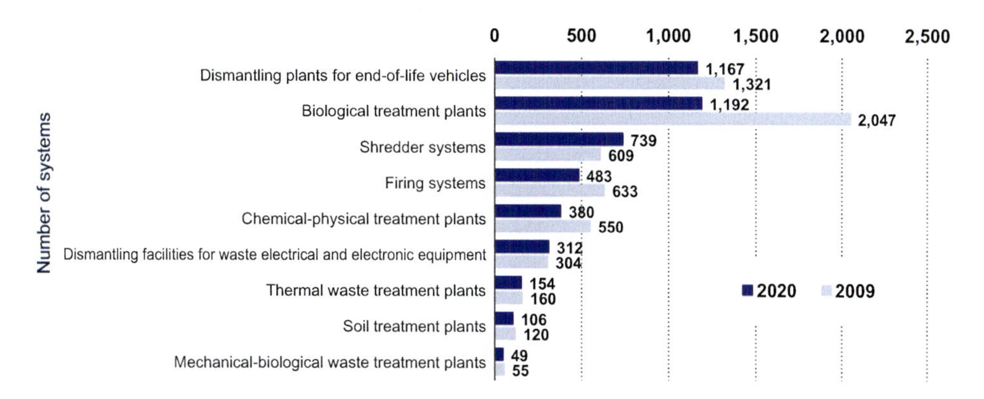

Fig. 1.10 Waste disposal facilities—Number by type of facility in Germany 2009 and 2020. (Data source: Statista, 2021g, p. 17)

large rubble bins—without a thought to the fact that the technical and economic lifespan of these products may not yet have expired. The throw into such bins abruptly ends the product's life cycle and destroys its original value and further raw materials if no further recycling steps follow.

In waste management, there is a lack—even in the industrial nation of Germany—not only of the necessary infrastructure for a circular economy, but also of societal awareness for dealing with finite resources.

How long do we want to tolerate such an approach, which is at the expense of the environment and thus also of future generations?

The long way still to go is shown by Fig. 1.11 (see Statista, 2021g, p. 19). In one of the richest industrial nations in the world, the majority of waste still ends up in **landfills**—stored for eternity! The **landfilling** is also not part of the circular economy. In the course of this waste disposal, the waste material is first piled up and then covered with a layer of earth or other materials. Due to the lack of air circulation and sunlight, a very slow decomposition process begins. This is also referred to as mummification. No value or raw materials are recovered through landfilling. An exception are landfills where methane gas is extracted for energy production.

What does Fig. 1.11 tell us? 41,434,000 t were deposited in Germany alone in 2020:

▶ **Waste has no value at all in a landfill.**

On the contrary: Landfills not only destroy landscapes. They can also release toxic substances into the air, soil, and water. This is especially the case in many developing countries, where waste—including from Western industrial nations—is openly stored and often burned in landfills.

Food for Thought Imagine a **freight train** that would have to transport all the waste that was deposited in Germany in 2020. Each hopper car could hold 25 t of waste. This train would have to have 1,656,000 cars. If each car is 10 m long, this train would be 16,560 km long. That's the straight-line distance from Hamburg to Sydney! For a single year—and for a "small" country like Germany!

The **sorting plants** found in second place at least offer the chance of recycling. The extent to which this is successful depends on what happens to the material after sorting. **Combustion plants** and the **thermal waste treatment** are almost identical. If you add up the figures for the use of combustion plants and thermal waste treatment in Fig. 1.11, the total amount is 39,432 t. This corresponds to second place in this ranking—ahead of the sorting plants. The aim of incineration is to reduce the volume of waste and at the same time generate energy and/or heat. The residues are usually deposited afterwards.

Shredding plants and **scrap shears** also do not stand for high-quality recycling. In Fig. 1.11, it also becomes visible what small proportion of waste still ends up in **dismantling facilities** and **demolition companies**.

Food for Thought Germany, one of the richest industrial nations, a country of technological expertise, primarily relies on landfilling and incineration for waste disposal—as well as on the export of waste!

An important question regarding the circular economy is to what extent recycling actually takes place. The values shown in Fig. 1.11 do not really correspond with the displayed **recycling rates of household waste** in Fig. 1.12. This also includes commercial waste similar to household waste, which is collected via public waste disposal (cf. Statista, 2021g, p. 36). Looking at Fig. 1.12, one could actually sit back complacently.

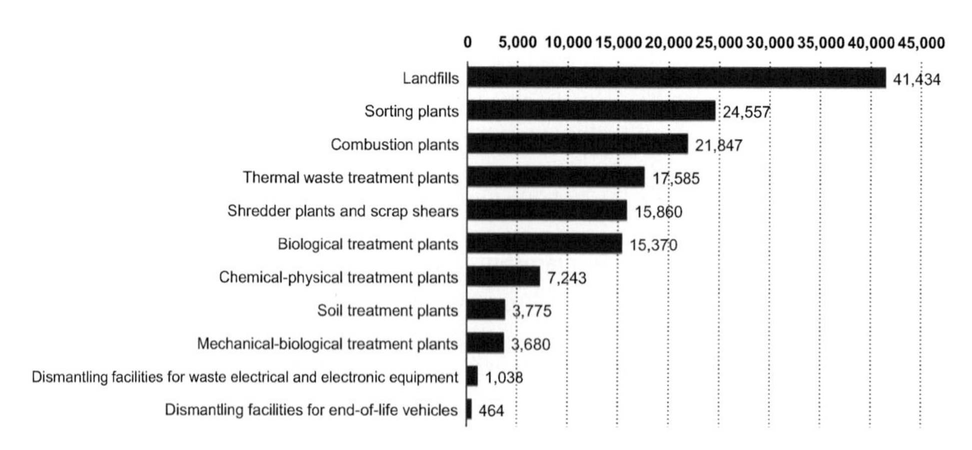

Fig. 1.11 Waste volumes in waste disposal facilities in Germany by type of facility—2020 (in 1000 t). (Data source: Statista, 2021g, p. 19)

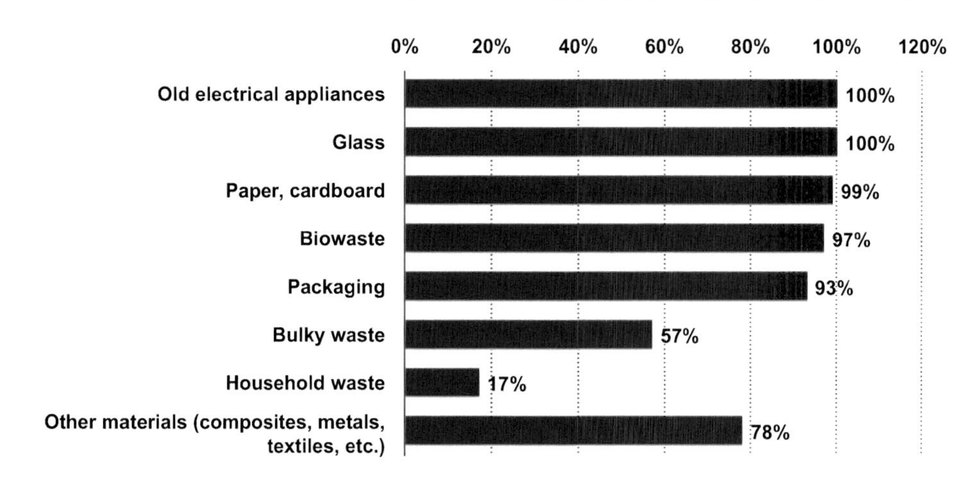

Fig. 1.12 Recycling rates of household waste in Germany by type of waste—2020. (Data source: Statista, 2021g, p. 36)

However, it should be taken into account that the shipping of waste from Germany is also partly credited to the recycling rates. When plastic waste from Germany is exported to certified recycling plants abroad, the corresponding amount of waste is included in the calculation of the German recycling rates. However, the verification and control systems as well as the recycling infrastructure in the target countries are often inadequate. Therefore, only a part of the waste is actually recycled there. Nevertheless, this export is included in the recycling rates in Germany, even if our waste is burned in the open, stored in landfills or dumped on the beaches in the recipient countries!

Note Box We must not believe the official statistics that present us with extremely high recycling rates. Many waste exports to certified recycling facilities abroad are credited to these recycling rates—regardless of the actual waste management in the target country. What happens to the waste there is usually of no interest to anyone in the country of origin of the waste. Motto: Out of sight, out of mind!

In addition, when evaluating recycling rates, it should be taken into account that only the quantities that actually arrive in the recycling system are included in their calculation. Waste that ends up in the landscape, in the sea, in rivers or directly in the landfill is not recorded in the determination of recycling rates due to lack of data. Recycling rates therefore always only refer to the quantities recorded in the system.

Food for Thought It remains quite "simple". The best way to deal with waste is to consistently reduce or avoid it. This applies to both individuals and companies.

Regardless of this, the rule is: **We have to close the circle!**

1.3 Customers' Openness to the Topic of "Sustainability"

Each and every one of us is called upon to reduce our own resource consumption and thus also our own negative impacts on the planet and its finite resources. After all, many of the **measures of the circular economy** are only successful if the following offers also meet with acceptance among customers. These include:

- Purchase of used products
- Purchase of durable products—possibly at higher prices
- Purchase of sustainably produced products—possibly at higher prices
- Purchase of products that contain recycled materials
- Purchase of products with less negative impact on the environment
- No purchases of disposable products
- And in general: fewer purchases

Only if consumers vote with their wallets for sustainably produced solutions can these also be offered by companies in the long term.

1.3.1 Study Results on Customers' Orientation Towards Sustainability

The **"achievement of sustainability"** in itself is already a value. However, companies are in competition and therefore have to strive to take their customers along on the path towards "sustainability". But what about the customers? How ready are customers already for sustainable consumption? And what do they accept for or with a green brand leadership?

The relevance of **sustainability as a purchasing criterion** in Germany in various sectors is shown in Fig. 1.13 (cf. Statista, 2021b). To determine this by sector, 1022 people (18 years and older) were interviewed. The question was: "How important is sustainability to you in your purchasing decisions for the following products?".

In total, 58% say that **sustainability as a purchasing criterion** is "rather important" or "very important" to them. However, the evaluation varies greatly by sector. The greatest importance is attributed to this criterion in the **energy/supply** sector (75%). This is followed at some distance by **construction** (64%), **travel and tourism** (63%) and **consumer goods** (62%). Sustainability is less important as a purchasing criterion in the **automobile** (54%) and **financial services** sectors (36%; each based on the number of those who indicated "rather important" or "very important"). The values suggest a relevance of "sustainability" in purchasing. However, whether these attitudes are also reflected in actual purchasing behavior remains to be seen.

Interesting results also emerge when the topic of **sustainability** is broken down to one's own **dietary style**. For this purpose, 1184 flexitarians and 280 vegetarians in Ger-

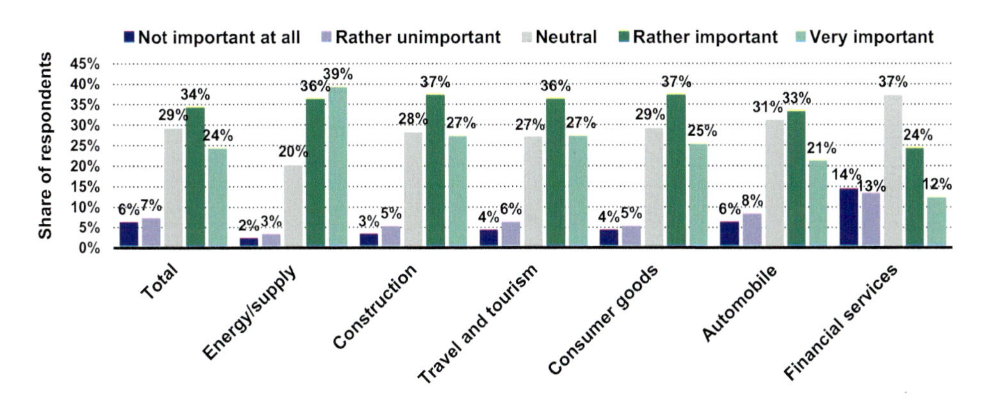

Fig. 1.13 Relevance of sustainability as a purchasing criterion in Germany by sector—2021. (Data source: Statista, 2021b)

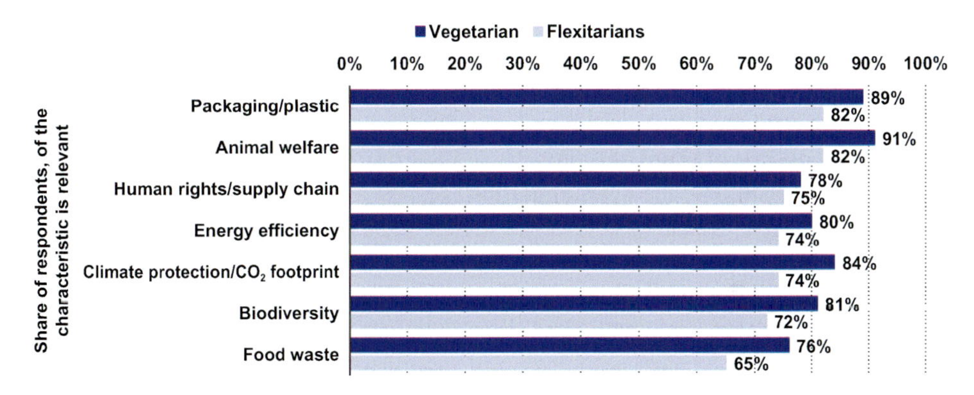

Fig. 1.14 Relevant attributes on the topic of sustainability of dietary style—2019. (Data source: Statista, 2019)

many were surveyed in 2019. The respondents were between 16 and 69 years old. To determine which attributes of sustainability are important in their own dietary style, the question was asked: "Which current issues on the topic of sustainability do you consider relevant?" Multiple answers were possible. Figure 1.14 shows the results of the statements "very relevant" and "rather relevant" (cf. Statista, 2019).

Here too, it is again evident that sustainability criteria are given great relevance. Among vegetarians, the importance is usually much more pronounced across all criteria. Both among vegetarians and flexitarians, the criteria **packaging/plastic** and **animal welfare** are each given the highest weight—followed by **human rights/supply chain.**

To achieve the **balance between desire and reality**, creative terms are created at *Rügenwalder* such as *"Vegan Teewurst", "Vegan Ham Spicker", "Vegan Mill Bratwurst"* or *"Vegan Mill Salami"*. How has *Tim Raue,* a top-class German chef, formulated about such **semantic scarecrow** (Raue, 2022, p. 18):

> "Because people actually want to eat meat. They want to eat a chicken—but no meat. Market research has clearly shown us: If we call it 'pea patty in chicken breast format', it doesn't pull."

> **Food for Thought** Is the following too bold a thought? It is becoming increasingly difficult to feed the population, which has grown to eight billion since November 2022. Why do we primarily approach the problem from the supply side and less from the demand side? One aspect in particular should stimulate thought and action: In most industrialized nations and in emerging developing countries, a large proportion of people are overweight. Most are not overweight due to illness, but simply because they eat and drink too much.
>
> If overweight people could be motivated to consume less, they would not only be doing themselves and the society in which they live a favor. Such a change in behavior would also counteract the food shortage. After all, many affluent countries buy up the food produced worldwide at high prices, motivate the burning down of rainforests to gain agricultural land—and promote the increasingly industrial production of plants and animals.
>
> Because not only the **overpopulation** is a problem, but also the **overconsumption** of the few! Reducing overweight to relieve the earth: a win-win-win solution!

What about **sustainability in clothing, fashion, and shoes**? In 2021, a representative survey of 10,281 people aged 18 and over was conducted in 17 countries. The question was: "How important is sustainability to you when making purchasing decisions for clothing, fashion, and shoes?" The results are shown in Fig. 1.15 (see Statista, 2021c).

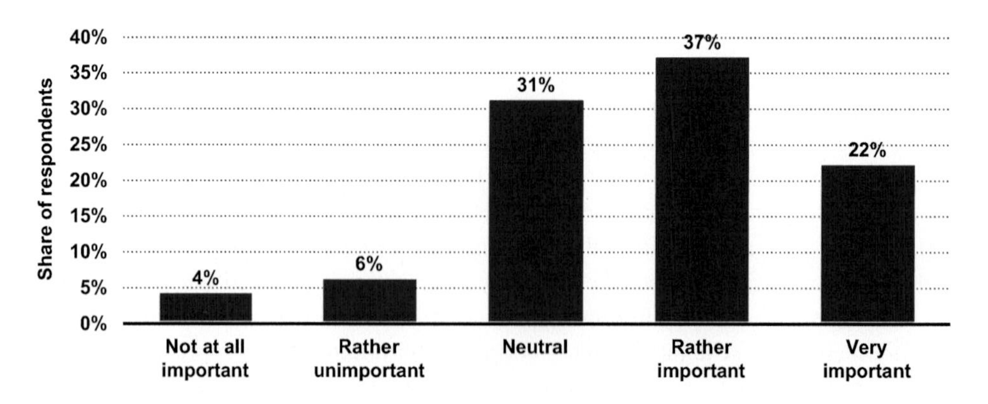

Fig. 1.15 Importance of sustainability as a purchasing criterion for clothing worldwide—2021. (Data source: Statista, 2021c)

Here too, the majority—59%—state that **sustainability** in clothing is a "rather important" or "very important" purchasing criterion. It is also necessary to check to what extent such an attitude is actually relevant to behavior.

What criteria are used when **purchasing household and electrical appliances** in Germany? Figure 1.16 shows the results of a representative population study in which 1002 people aged 18 and over were surveyed (see Statista, 2021g, p. 42). The first-mentioned criteria **durability** and **quality** can already contribute to **sustainability**, which is only mentioned as an independent criterion in fifth place here.

This raises the question of whether there are significant differences between different age groups with regard to these purchase criteria. It is noticeable that the older groups of people (34 to 54 and 55+) place considerably more value on **energy efficiency** and **durability** than the younger target group (see Fig. 1.17). There are no clear differences between the age cohorts for two options: This is the case once when **buying devices with environmental labels**. Such labels are based on CO_2-neutral production or the use of recycled materials for household and electrical appliances, for example. The values are also at a comparable level when **buying used, but as-new refurbished and overhauled devices** (keywords "refurbished" or "renewed"). The youngest consumer group shows significantly different behavior when **buying used devices without additional refurbishment.** The younger generation is much more open to this (see Statista, 2021g, p. 43).

In order to delve deeper into the issue of **sustainability during shopping**, McKinsey (2021, pp. 2 f.) surveyed German, Austrian, and Swiss consumers aged over 18 years. The target group was people who had purchased food, household, personal care, or health products in the last two months. The following findings were obtained:

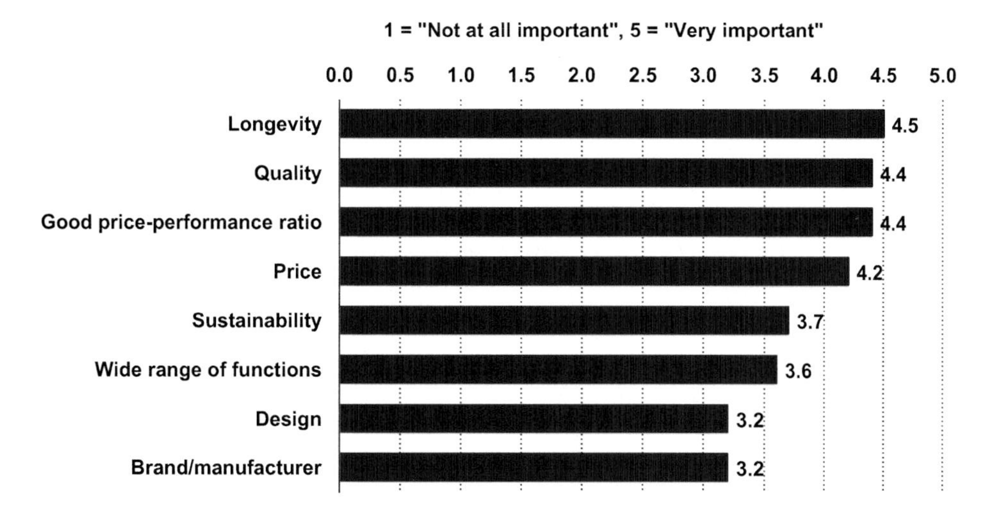

Fig. 1.16 Purchasing criteria for household and electrical appliances in Germany—20. (Data source: Statista, 2021g, p. 42)

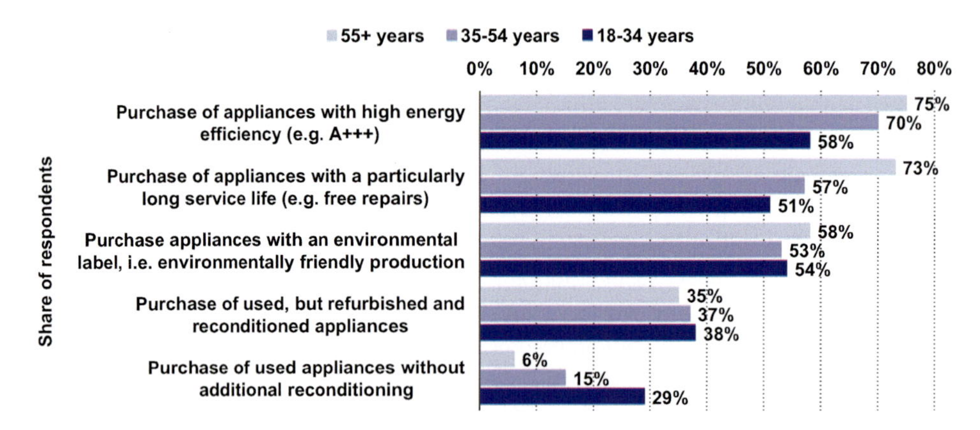

Fig. 1.17 Sustainability when buying household and electrical appliances in Germany—2021. (Data source: Statista, 2021g, p. 43)

- Half of the consumers were willing to pay up to 20% **more for sustainable products** during the pandemic than before. After the pandemic, an increase was expected. However, the other half of consumers are not willing to pay more for sustainable products.
- During the pandemic, consumers consciously strived to reduce their **impact on the environment and society**. About a quarter of consumers also switched more frequently to products or brands if they were labeled as sustainable.
- There are clear **differences between different consumer groups** in terms of their willingness to pay a premium for sustainable products. In general, **women, Generation Z consumers, people with high incomes** and **online shoppers** are more willing to pay a premium for sustainable products. Generation Z consumers are also more willing to accept higher prices for sustainable snacks, ready meals, and personal care products.
- Consumers are most willing to pay a **premium for sustainable fresh food and skin care products**. This does not apply to the same extent for sustainable household care products.
- The willingness to pay a premium for sustainable products is primarily determined by **ecological and social aspects**. **Aspects of personal health** are less important according to the respondents.
- A **fair remuneration of employees** is the most important **driver for the willingness to pay for sustainable products.**
- The most important **drivers for more environmentally friendly consumption** are **labels such as "free from…"** Also, labels such as **low greenhouse gas emissions, resource conservation** and **avoidance of packaging** promote more environmentally friendly consumption.
- The **importance of specific sustainability aspects** varies from category to category. For example, the avoidance of packaging is more important for fruits and vegetables

as well as for personal care and household care. This aspect is given less importance for meat, fish, and dairy products.

From these results, the **recommendation for manufacturers of consumer goods and for retailers** is derived to focus even more on sustainability issues. This also includes bold changes in the range, pricing, distribution channel, and in the areas of marketing, production, and logistics. Only in this way can sustainable corporate management and green brand management be achieved.

After these overarching results, it is now exciting to delve more deeply into the role of the **brand** in sustainable corporate management. For this purpose, 12,000 consumers in twelve countries (18 years and older, who made an online purchase in the last year) were asked whether they would be more likely to buy from a **brand with a clear commitment to sustainability**. Of these, 44% indicated that they would be more likely to buy from a brand with a clear commitment to sustainability in 2021 (see Fig. 1.18). In the **EMEA region** (Europe, Middle East, Africa), it was particularly important that brands take measures this year that align with their values. The fact that the large regions of Europe, the Middle East, and Africa were evaluated together in one group affects the power of insight. The orientation towards sustainability is least pronounced in North America at 40%.

The following presents study results separated by generations. The **names of the generations** refer to the birth years of the respondents. The classifications vary depending on the source, the most common are depicted here.

- **GenerationZ**
  - Individuals who were twelve to 25 years old at the time of the survey or were born in the years 1996 to 2009 inclusive.

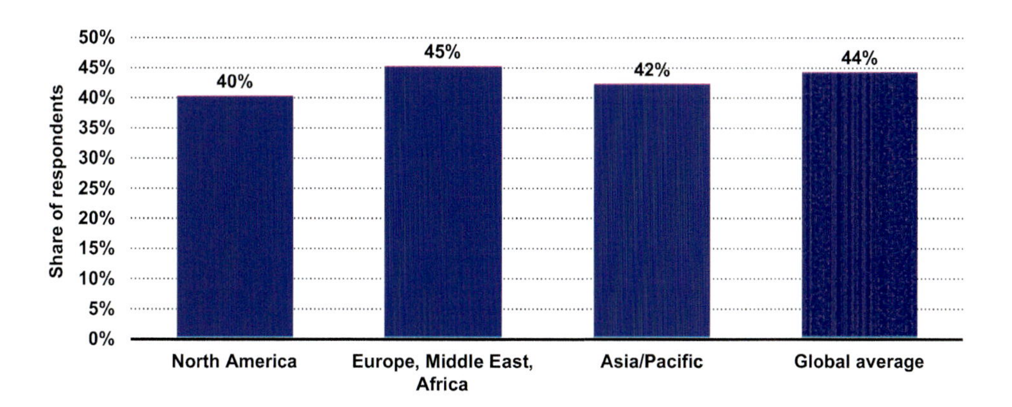

Fig. 1.18 Proportion of consumers who prefer to buy from sustainable brands—globally 2021. (Data source: Statista, 2022b)

  - Extrapolation to 9.91 million people (the number of twelve to 25-year-olds in unified Germany has never been smaller than today).
- **Generation Y**
  - Individuals who were 26 to 40 years old at the time of the survey or were born in the years 1981 to 1995 inclusive. This generation is often also referred to as the **Millennials**.
  - Extrapolation to 15.82 million people.
- **Generation X**
  - Individuals who were 41 to 55 years old at the time of the survey or were born in the years 1966 to 1980 inclusive.
  - Extrapolation to 16.93 million people.

Figure 1.19 shows to what extent different generations of the German-speaking population agreed in 2021 with selected **statements about shopping/consumption and branded goods** (see Statista, 2021d). For this purpose, a total of 23,299 people aged 14 to 55 were surveyed. The results obtained were extrapolated to 70.54 million people. Since the underlying *Consumer and Media Analysis* only includes data from the population aged 14 and over, the age spectrum of Generation Z cannot be fully depicted.

There are some clear differences between the generations. **Brand loyalty** is slightly less pronounced in Generation Z than in the other generations: 78.3% compared to 79.2% or 79.8%. This corresponds with the higher willingness of Generation Z to **try new things** (78.2% compared to 74.3% or 71.7%). At the same time, Generation Z associates **known branded goods** with higher quality (65.8% compared to 60.1% or 61%). It is not surprising that Generation Z has less **financial means** to satisfy their own needs. Nevertheless, the majority of Generation Z members confirm this with 54.5%—compared to 64.1% of Generation Y and 69.6% of Generation X.

	Generation Z (14-25 years)	Generation Y (26-40 years)	Generation X (41-55 years)
If I'm happy with a brand, I'll stick with it	78.3 %	79.2 %	79.8 %
I like to try out new products	78.2 %	74.3 %	71.7 %
When you buy well-known brand products, you can be sure that you are getting good quality	65.8 %	60.1 %	61 %
I always know what's on trend	60.8 %	48.4 %	39.8 %
Sharing things or services makes an active contribution to environmental protection and sustainability	59.4 %	53.4 %	51.6 %
I buy products from the region whenever possible	59.3 %	66.3 %	72 %
When buying products, it is important to me that the company in question acts in a socially and ecologically responsible manner	57.2 %	56.7 %	58.4 %
All in all, I have enough money to satisfy my needs	54.5 %	64.1 %	69.6 %

Fig. 1.19 Consumer behavior and brand importance by generation—2021. (Data source: Statista, 2021d)

What differences are there in the statements aimed at **sustainability**? Generation Z agrees to a much greater extent with the **Sharing Economy**—the sharing of things and services, with 59.4%. The approval rate is significantly lower for generations Y and X, at 53.4% and 51.6% respectively. Interestingly, the **regionality of products** is much less important for Generation Z at 59.3% than for the other generations (Y: 66.3% and X: 72%). There are no major differences in the statement on the **social and ecological orientation of the company.** Here, the results are very close to each other (Z: 57.2%, Y: 56.7%, X: 58.4%).

It is also interesting to ask how the number of people in Germany who pay attention to **fair trade products (Fair Trade)** when shopping has changed over time. For this purpose, the German-speaking population aged 14 and over was asked in personal interviews. The question was: "Here are various characteristics. Could you please look through them and tell me the ones where you would say: This applies to me, this is true for me?—I pay attention when shopping that the products come from fair trade (Fair Trade), i.e. that the producers in the Third World countries receive a fair price for their products." The number of people in Germany who affirm this question is shown in Fig. 1.20 for the years 2013 to 2021 (cf. Statista, 2021e).

To put these numbers into perspective, the number of people under 14 years of age (almost eleven million) must be subtracted from the total population in Germany in 2021 (almost 84 million). Then it becomes clear that in 2021, with almost 16 million people, only 22% of the total population over 14 years of age (73 million) value Fair Trade in shopping. In addition, Fig. 1.20 reveals that this value has not shown any significant growth since 2019. Rather, there seems to be a certain stagnation at a low level.

Also, always keep in mind: **Saying is not doing!**

What about **sustainable behavior patterns in online shopping**? For this purpose, a representative survey of consumers aged 18 to 90 was conducted. In Austria, 1054 people were surveyed, in Germany 1047, and in Switzerland 1054. It was determined what

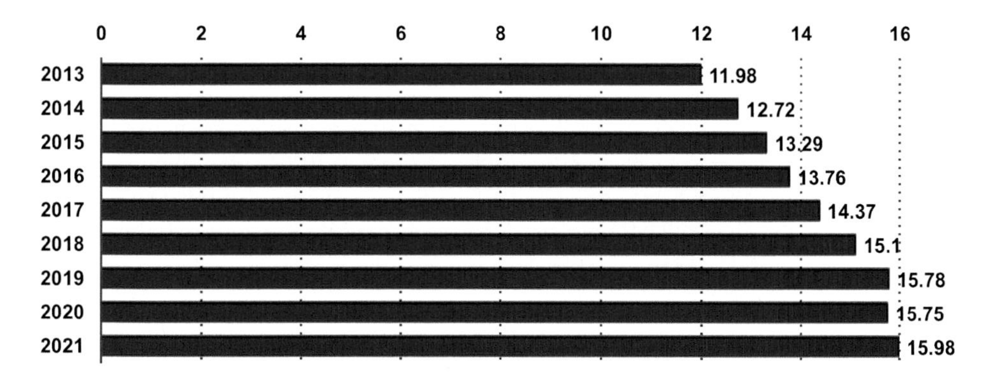

Fig. 1.20 Number of people in Germany who pay attention to fair trade products (Fair Trade) when shopping, in millions—2013 to 2021. (Data source: Statista, 2021e)

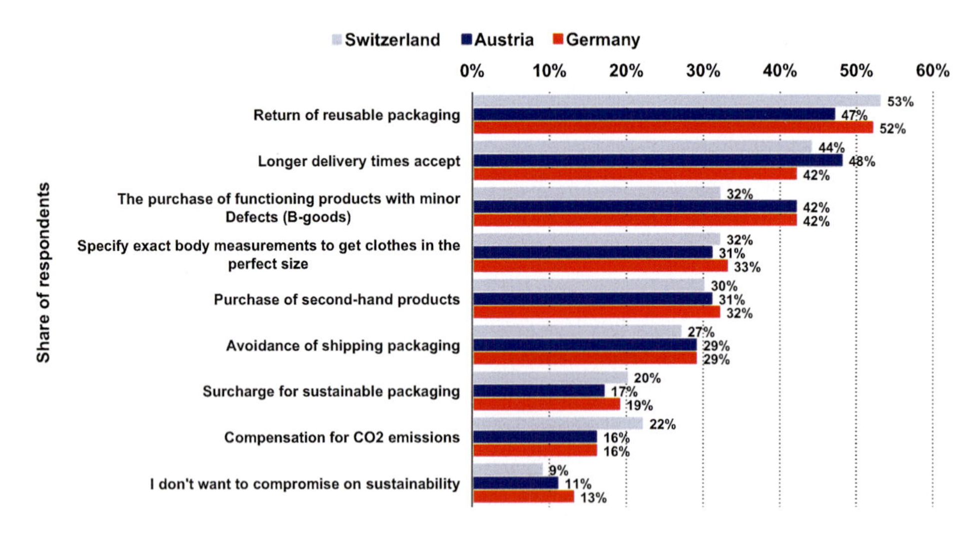

Fig. 1.21 Sustainable online shopping habits in German-speaking countries—2021. (Data source: Statista, 2021f)

proportion of consumers in German-speaking countries adopt sustainable online shopping habits. Figure 1.21 shows what percentage of respondents agree with each of the options (cf. Statista, 2021f).

In sum, the openness to sustainable behavior patterns documented in Fig. 1.20 is clearly limited. With the exception of **returning reusable packaging**, the majority of respondents do not agree to act more sustainably. **Longer delivery times** are much less accepted. Overall, the same assessments are shown in the three countries—with one exception: The purchase of functioning products with minor defects (B-goods) meets with significantly less approval in Switzerland than in Austria and Germany. Around 10% of respondents do not want to make **any compromises for sustainability**.

Which aspects are relevant for consumers in Germany in terms of **order processing** is shown in Fig. 1.22. For this, 1047 people aged between 18 and 90 were asked about their shopping behavior (see Statista, 2021h).

A look at the responses "very important" and "important" shows a great willingness to **consolidate orders** (83%). However, most online shoppers will have noticed that a consolidated order does not necessarily mean a consolidated delivery. Often, a single larger order triggers three, four, or five delivery processes—with correspondingly harmful effects on people and the environment. Also, 83% indicate that the **return policy**—here resale or disposal—is important to them. Despite this high value, two aspects must be assumed:

- Consumers usually have no or insufficient information about whether a returned item is put back on sale or destroyed.

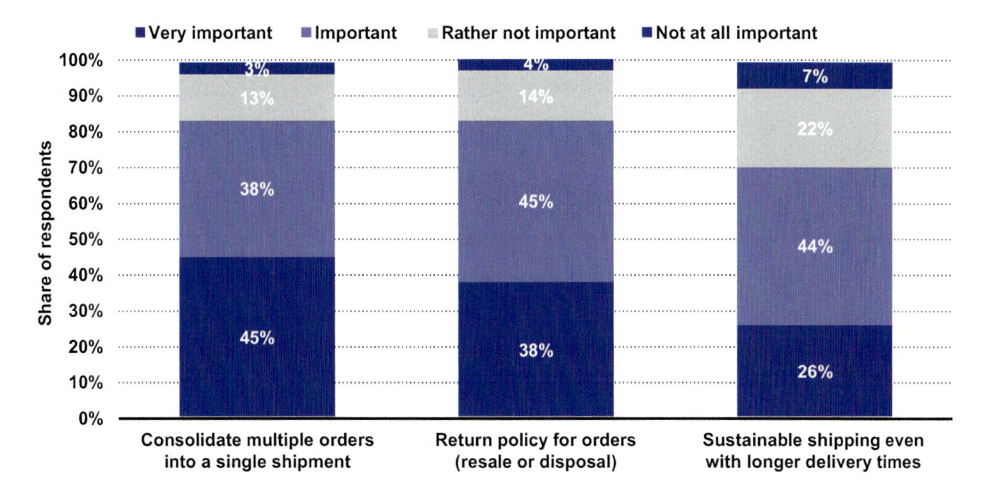

Fig. 1.22 Importance of sustainable order processing among consumers in Germany—2021. (Data source: Statista, 2021h)

- Consumers mostly do not seek information about the return policy—certainly also to avoid cognitive dissonance (see Sect. 1.3.3).

With 66%, significantly fewer consumers agree to a **longer delivery time** if this can improve the sustainability balance (see Fig. 1.22).

The question of why **sustainable shopping for fashion** fails today is also interesting. For this, 829 fashion buyers aged 16 and over in Germany were asked why they do not buy sustainable fashion. Multiple answers were possible (see Fig. 1.23; Statista, 2021i).

It's quite simple: **sustainable shopping for fashion** fails primarily because of the **price.** This is the most important reason for 38% of people. This shows that the communication of the possible superiority of a sustainable fashion item—in terms of environmental compatibility and wear duration—has not yet sufficiently arrived. In addition, higher prices could be offset by a lower shopping frequency. But this also needs to be communicated to the buyers.

28% see the topic "sustainability" as just a **gimmick,** which they themselves do not fall for. 18% prefer **familiar clothes**—regardless of their eco-balance. 15% do not find **sustainable fashion** at their preferred retailer. Here, loyalty to the retailer overlays possible sustainable shopping. 14% criticize the **appearance**—which is easy to understand with many "eco-products". 13% state that they cannot find the **preferred clothes**.

> **Food for Thought** 13% of respondents see no **personal added value** in buying sustainable fashion. This reveals a glaring **communication deficit**. 87% of customers do not realize that they—in the long term—not only do something good for themselves, but also for future generations, when they pay attention to sustain-

Fig. 1.23 Why sustainable fashion shopping fails. (Data source: Statista, 2021i)

ability when buying fashion. This shows that corporate communication has not yet been able to convey these aspects broadly—or did not want to convey them!

What about consumers' willingness to primarily shop at **local businesses** (online or in-store) to reduce their own environmental impact? For this, 10,055 consumers (18 years and older) in eleven countries were surveyed. The results are shown in Fig. 1.24 (see Statista, 2020).

The highest willingness for a **local shopping behavior** is shown by the French. Here, traditionally, more value is placed on good food quality—associated with the acceptance of higher prices. A good third of consumers in Germany (35%) align their behavior locally to act more sustainably. The least willingness to do so is—once again—in the

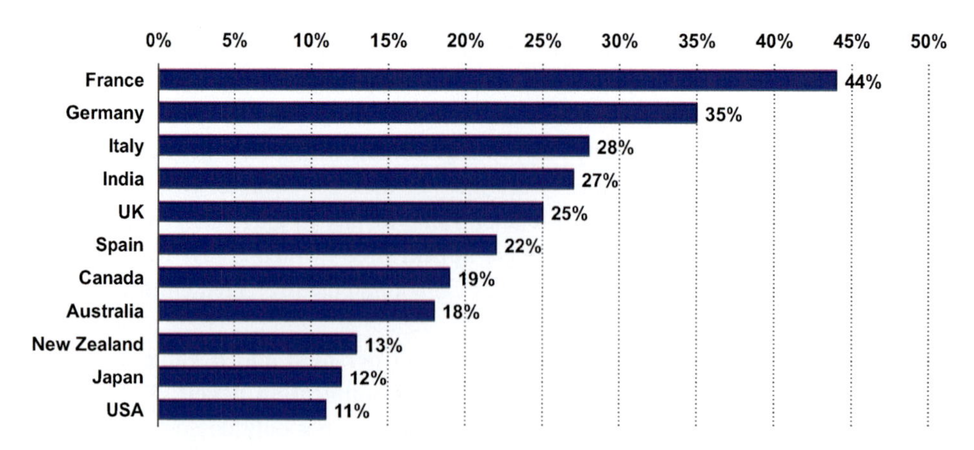

Fig. 1.24 Percentage of consumers who primarily shop at local businesses to reduce environmental impact, by country—2020. (Data source: Statista, 2020)

USA. Here, it is only 11%. Is this because parts of the US population still deny climate change?

An important aspect for purchasing behavior is the **convenience orientation.** This was also shown in a study of the packaging industry. Here, an increasing preference of customers for certain takeaway and delivery offers was identified (see GVM, 2022):

- **Consumers prefer lighter packaging**
  The use of **glass** in packaging decreased by 28% from 1991 to 2020. The use of **ferrous metal** even decreased by 40%. In contrast, the use of **aluminum** (plus 58%) and **paper, cardboard, and carton** (plus 62%) increased. But above all, products are packaged in **plastic** (plus 111%).
  To what extent this development was promoted by consumers or manufacturers cannot be deduced from these figures.
- **Small households demand smaller portion sizes**
  With the increase in smaller household sizes, the contents of packages are getting smaller. However, the necessary packaging itself only becomes slightly smaller.
- **Consumers define further requirements**
  The customers' desire for dosing aids and for resealability of the packaging increases the requirements for contemporary packaging.

A representative survey of more than 1000 people in Germany aged 16 and over shows which **digital and analog habits** they could give up for the sake of climate protection—and which they could not (see Bitkom, 2022b):

- **Internet usage**
  13% of the population could theoretically give up the internet—but hardly anyone actually wants to do that. Only 1% of 16- to 29-year-olds would theoretically give up the internet to protect the climate.
- **Smartphone usage**
  15% could theoretically give up a smartphone for the sake of climate protection. But only 6% of 16- to 29-year-olds and only 32% of those aged 65 and over would actually be willing to do so.
- **Online shopping**
  35% of the population could theoretically give up online shopping. Here, almost the same values are shown across all age groups.
- **Streaming usage**
  36% could theoretically give up streaming clips or movies for the sake of climate protection.
- **Use of airplane and car**
  37% could theoretically give up flying—and 12% could give up the car.
- **Meat consumption**

34% could theoretically completely give up meat consumption. Here, the largest increase was recorded compared to 2021: plus 6%.

This theoretical willingness to act is based on the fact that 84% of the population are concerned about **climate change**. 25% are even "very worried" about climate change—40% are "worried". 19% are at least somewhat concerned about climate change (see Bitkom, 2022b).

Against this backdrop, it is interesting to see what importance the **Internet** has in terms of **CO_2 emissions** today when compared to the emissions of countries (see Fig. 1.25). If the Internet were a country, it would already rank 6th in CO_2 emissions today—even before Germany (see Beuth et al., 2022, p. 58).

A closer look at the **Internet usage** shows that especially with *TikTok* the CO_2 footprint is particularly large (see Fig. 1.26). But other areas of internet usage are also associated with significant emissions, considering the millions of users who often access the internet for many hours a day. The level of emissions also depends on the devices used (see Beuth et al., 2022, p. 58).

> **Note Box** It is difficult to specify exactly how much CO_2 emissions an hour of Ultra-HD video streaming causes, as this depends on various factors. These include the type of device used for streaming and the type of network over which the stream is transmitted. However, an estimate of the CO_2 emissions of an hour of Ultra-HD video streaming could look like this:
>
> - An hour of Ultra-HD video streaming on a smartphone could cause about 0.5 to 1 kg of CO_2.

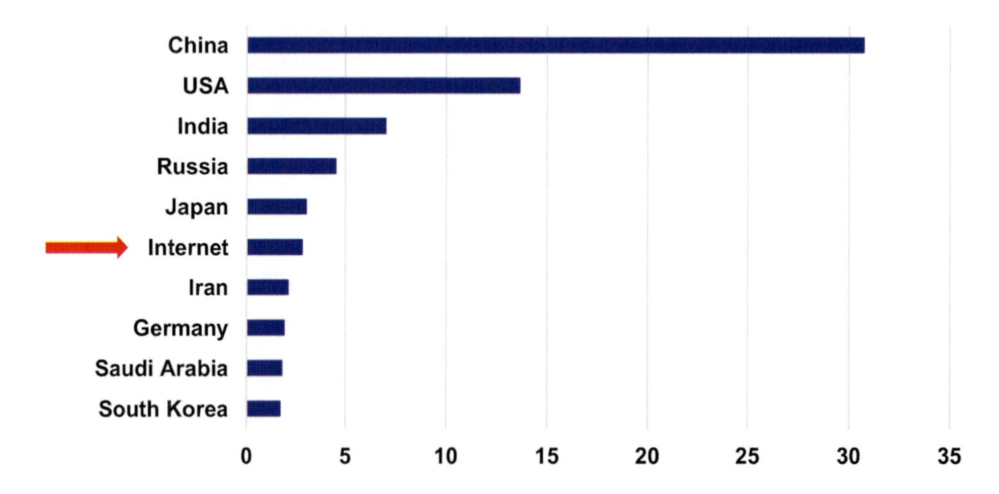

Fig. 1.25 Share of global CO_2 emissions 2020—in percent. (Data source: Beuth et al., 2022, p. 58)

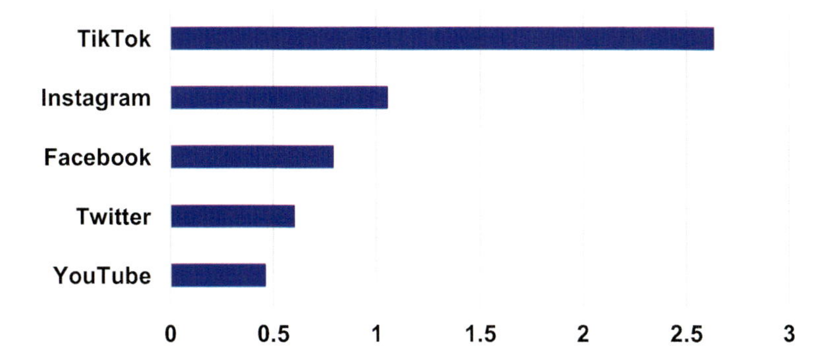

Fig. 1.26 Average CO_2 footprint of different media—in grams per minute. (Data source: Beuth et al., 2022, p. 58)

- An hour of Ultra-HD video streaming on a laptop could cause about 1 to 2 kg of CO_2.
- An hour of Ultra-HD video streaming on a Smart-TV could cause about 2 to 4 kg of CO_2.

An average car causes CO_2 emissions of about 500 g over a distance of two kilometers. So, someone who consumes four hours of *Netflix* on their Smart-TV causes the same amount of emissions as a person who travels between 32 and 64 km by car.

1.3.2 Relevance of the Attitude Behavior Gap

In the discussion about sustainability in purchasing behavior, a look at the **Attitude Behavior Gap** is indispensable. This gap arises when a person's personal values or attitudes (Attitude)—often shown in surveys—do not match their concrete actions (Behavior). Words and actions diverge due to the Attitude Behavior Gap.

Figure 1.27 shows, using the **model of Ajzen and Fishbein** (1973, 2005, p. 194), how behavior is influenced. A person's behavior is influenced by various **background factors**. These include individual and social factors—as well as the level of information. The **individual factors** include personality, emotional intelligence, values, stereotypes, attitudes, and experiences. The **social factors** include education, age, gender, income, religion, and culture. The **level of information** refers to knowledge, media, and options for action.

These background factors influence various beliefs. The **beliefs about behavior** influence the **attitudes towards behavior. Normative beliefs** influence the **subjective norms. Beliefs about control expectations** influence the **perceived behavioral control.**

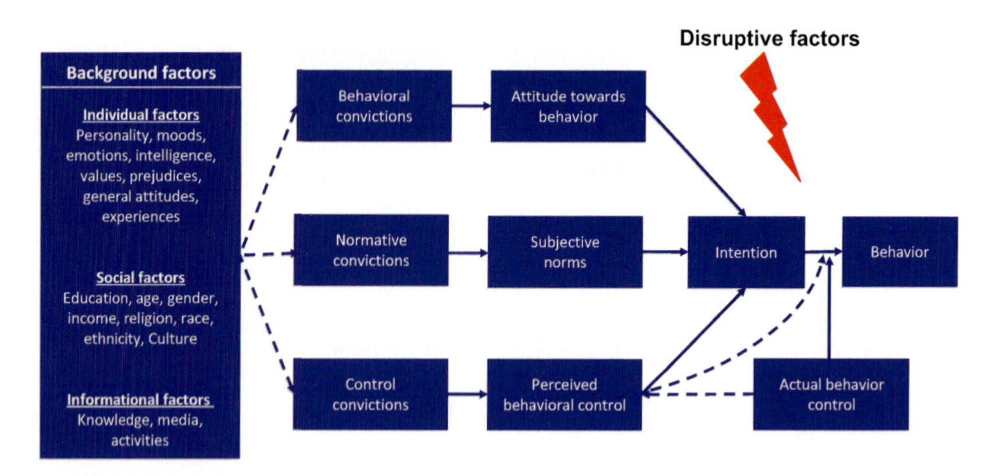

Fig. 1.27 Model of *Ajzen* and *Fishbein* for explaining the Attitude Behavior Gaps. *Source* Based on (Ajzen & Fishbein, 2005, p. 194; 1973)

These three factors influence the **intention to act.** The **behavior** that is actually shown depends not only on the intention but also on the **perceived** and the **actual behavioral control**.

This makes it clear: **Behavior is a complex construct.** Various interfering factors influence the behavior actually shown. If a person has a normative belief that requires protecting the earth, this forms subjective norms. Plastic packaging, disposable dishes, and soy products, for the production of which the Brazilian rainforest was deforested, are then viewed critically. These norms lead to a specific intention to not (anymore) buy such products. If no interfering factors occur, this behavior is consistently implemented in the store. The corresponding intentions and the behavior are further reinforced when people with the same values observe one's own actions (critically). This could be the accompanying person when shopping—or the life partner at home. If behavior deviating from the (common) values were shown here, "sanctions" would be expected. A "norm-based" behavior would be reinforced. Here, the perceived and the actual behavioral control are effective. If, on the other hand, sanctions are absent—or if one's own behavior is perhaps even criticized ("too expensive"), the intentions to act and consequently the specific behavior shown can change. The intention influenced by attitudes, norms, and the perceived intention to act is then not converted into corresponding behavior. The **Attitude Behavior Gap** arises.

There are numerous **interfering factors,** which can prevent norm- and attitude-guided behavior and lead to the Attitude Behavior Gap. First and foremost is the **price.** If—from the buyer's subjective perspective—the price is not appropriate or simply exceeds one's own budget, one's own norms and attitudes are disregarded and other products are bought. This can be referred to as a **price barrier** (cf. these barriers Balderjahn, 2021, p. 209 f.).

Poor **accessibility** to sustainable products can also prevent the intended behavior. Possibly, own behavior patterns have to be overcome to let actions follow the attitude. Here, there is a **habit barrier** because, for example, previously preferred shopping places, providers, and brands have to be changed.

A negative **quality perception** of sustainable products—"don't work", "don't taste", "don't smell good", "don't look good"—can also prevent norm- and attitude-conforming behavior. If one's own needs are prioritized over the actually "appropriate" behavior, there is an **egoism barrier**. If customers question whether the promised sustainability criteria are actually met by the respective offer, this can be referred to as a **trust barrier**.

These are all criteria that directly affect behavior and can represent various **purchase barriers**. This leads to a **discrepancy between intention and behavior.** If in such situations there is also a lack of perceived or actual behavioral control, nothing stands in the way of behavior contrary to one's own intentions. Then fast fashion is bought just like furniture made from unsustainable teak cultivation. Carpets are purchased that may have been knotted in child labor. Then products are also purchased that were made with palm oil from dubious sources. And who can see from which sources (sun, wind power, coal, oil, gas or nuclear) the domestic electricity was obtained?

An important aspect is the question of how **subjective norms** and **attitudes towards certain behaviors** are formed today (see Fig. 1.27). After all, norms and attitudes are significant influencing factors of behavior. The way **information is received and processed** also plays a major role. These norms and attitudes are shaped today not only by parents and school, but to a particularly high degree also by social media and the content disseminated there (see in depth Kreutzer, 2020).

Here, the so-called **FOMO-effect** is to be mentioned—the **Fear of missing out.** This effect describes the fear of people missing out on something important. Which events, which sights, which travel destinations should one not miss? Which products should one have bought, which service should one have used? These impulses no longer come—as in the past—primarily from one's own peer group, but—via social media—from all over the world. Above all, rich, beautiful, carefree and pleasure-seeking influencers worldwide influence norms and attitudes towards certain behaviors. This imprinting occurs especially in the younger generation, whose norms and behavior canon are still in development (see Kilian & Kreutzer, 2022, pp. 239–274). The influence of third parties on norms and behavior is deepened in Sects. 5.2 and 5.3.

> **Note Box** When the **FOMO effect—Fear of missing out**—combines with the **YOLO mantra—You only live once!** –, then consumption and enjoyment know no bounds.
>
> I only live once! I live now! I don't want to wait! I don't want to limit myself! I want everything—but right now! … Cost it what it may!

Against the background of these diverse influencing factors of behavior, we must also be particularly careful with **studies on sustainable behavior**. We must not take everything at face value that was determined there. Especially with prospective (future) buying behavior, there is often a significant discrepancy between **words and actions**. This divergence can often be explained by the phenomenon of social desirability.

A **social desirability** exists when people give the answers they believe will meet with greater social approval than the truthful answers. This behavior leads to **systematic response distortions** in surveys. These effects must always be taken into account when interpreting study results.

How can this systematic divergence of words and actions occur according to Fig. 1.27? Anyone who has so far consumed Argentine beef in large quantities with a preference, roars over the highway with his SUV at 180 km per hour, prefers disposable products, changes his clothing on a monthly basis and/or likes to buy the 53rd pair of shoes, may paint a completely different picture of himself in surveys on sustainable consumption. Since the interviewer often cannot verify the statements of the interviewee, both a perceived and an actual behavioral control are eliminated.

In such a situation, the interviewee can therefore lie through his teeth. This also applies to online surveys. If the respondent still wants to look good, he does not orient his answers to his own norms, but to those that are assumed by the interviewer to be socially opportune. Then what is said is what promises social recognition. After all, most interview partners do not want to out themselves as an "eco-pig"!

Against this background, it becomes understandable why surveys often reveal a willingness to spend more money on organic food. However, the **share of organic food in food sales** in Germany only increased from 3.74% to 6.8% between 2010 and 2021. Although this is a doubling, it is at a low level. In Germany, the per capita sales of organic food per year is €180. The leading countries in per capita sales are Switzerland, Denmark, and Luxembourg. Particularly popular in Germany is the consumption of organic eggs, organic fruit, and organic vegetables (see Statista, 2022). However, driven by high inflation, the German organic market shrank for the first time at the end of 2022. This shows that price can quickly dominate existing sustainability values of customers.

> ▶ **Food for Thought** Human behavior is often driven less by insights than by financial rewards and punishments.
>
> What's the saying?
>
> **As citizens, we often formulate lofty goals. In the supermarket, on the other hand, we are often very stingy!**

Studies repeatedly show that especially younger people desire more sustainable fashion. Thus, 90% of consumers from Generation Z (see Sect. 1.3.1) see companies as responsible for protecting the environment and making a positive societal contribution (see McKinsey, 2019). Members of Generation Z see the need to shop more sustainably in **fashion** and also indicate such behavior. This is partly reflected in a greater interest in **second-**

hand fashion, recycling and the **repair of fashion items.** However, there is often still a large discrepancy between these attitudes and actual behavior. The vast majority still find it very difficult to translate their own values—aimed at sustainability—into corresponding purchase decisions. This is also due to the fact that every second fashion consumer is unsure about what sustainability in fashion actually means (see Zalando, 2021, p. 5).

> **Food for Thought** Consumers often do not know what sustainability means in fashion and other areas. How then can they act sustainably or more sustainably? There is still a large communication gap. And the question arises again whether companies cannot or do not want to fill this gap.

The relevant **study by *Zalando*** already makes clear through the title ***It Takes Two,*** that two parties are needed for sustainable consumption: suppliers and demanders. The study also shows that the study participants most often associate **"sustainable fashion"** with **"guilt feelings"**, but least often with **"fun"**. Such an association must necessarily change if companies want to put sustainability at the center of their business model. With a focus on "fun", fast-fashion providers like *Primark* and *SheIn* are repeatedly setting new sales records (see in depth about SheIn Sect. 4.5).

> **Note Box** The **Generation Z** seems to have no trouble reconciling **Fridays for Future** with **Black Friday**. What's the euphemistic saying?
>
> **Generation Z shows ambivalent behavior!**
>
> Concerns about sustainability can be put aside if a personal advantage is achieved.

> **Food for Thought** Against this background, it is always necessary to check—country by country, target group by target group—how strong the **Attitude Behavior Gap** is in each case. A company can—with a view to the markets it serves—come too late, but also too early with a comprehensive ecological approach. In the latter case, the customers would not yet be ready to actually bear the higher costs or prices associated with more sustainable corporate management.
>
> **He who comes too late is punished by life. But this can also apply if you come too early with a sustainable approach!**

Due to the **discrepancy between words and actions**, sustainable corporate management must comprehensively align itself with the actual **behavior of customers**. In addition, **market research methods** must be used that do not directly, but indirectly, collect relevant behavioral intentions. Here, the distinction between implicit and explicit patterns of effect is important. **Explicit patterns of effect** include the conscious experience and expression of sensations. These can be the subject of rational argumentation. **Implicit**

patterns of effect include subconscious, sometimes also automatic processes of perception and behavior. These elude direct capture.

To avoid the Attitude Behavior Gap, appropriate **implicit measurements** must be carried out to capture implicit patterns of effect. These are procedures for gathering information in which the facts to be investigated are not directly and thus explicitly, but indirectly and consequently implicitly queried. The construct of interest (e.g., the attitude towards sustainability) is determined indirectly. Instead of directly asking about attitudes, an **indirect measurement.** Through appropriate detours, facts can also be determined that are not or only difficultly accessible to a person's consciousness.

Consequently, various **reactions of the test subjects** to selected stimuli must first be distinguished:

- **Explicit reactions** (e.g., statements of the respondents)
- **Implicit reactions** (such as physical reactions that are visible or measurable)

While explicit reactions can be well collected through direct methods of market research, the following methods are used to capture implicit and neurobiological reactions, among others, to capture the effect of various stimuli (words, slogans, images, videos, but also brands) on the test subjects:

- **Measurement of skin resistance** (EDR: electrodermal response or PGR: psychogalvanic response)
  In the measurement of skin resistance, the phenomenon is exploited that a body secretes more sweat with increasing activation. This increases the conductivity for electricity on the skin. These changes can be detected and are also used in the lie detector. Thus, for example, the physical reactions to the visual presentation of the *Dr. Orto Fashionable Health Shoes for Women* compared to *Nike Air Jordan 1 Mid—High Sneakers* can be determined. This captures the physical reactions that are hardly controllable by the test person.
- **Analysis of other bodily reactionsand the emotions triggered**
  How people react to certain stimuli can also be captured by the **change in pupil size.** The pupil initially dilates when high attention is required. Beyond a critical limit, it narrows again, indicating overload. In addition, the **voice frequency** can be recorded. For many people, the voice becomes "higher" with increasing excitement. Such changes can be determined. In addition, it can be determined which **emotional reactions** such as joy, sadness, enthusiasm or also irritations, for example, films like *The Recycling Lie* or *Coca-Cola and the Plastic Problem* evoke. These can also be determined by the shown **facial expression**.
- **Reaction time-based test methods**
  Alternatively, the **reaction time** to certain stimuli can be determined. These reveal the existing association patterns of the test subjects and also established connections. Here, one can also speak of **gut feeling**—after all, our feelings not only influence our

relationships with other people, but also with brands and products. For example, it can be determined whether the word "sustainability" is more associated with "fun" or with "renunciation" or "burden". In addition, it can be recognized which brands are spontaneously associated with the term "sustainability"—and which rather appear with "environmental sinners".

An innovative test concept for this is represented by ***e² Brandreact***. Through this, brands and companies can be positioned in a **five-stage motive space** (cf. Eye Square, 2022):

- **Joy** (triggered by innovations, modernity)
- **Strength** (associated with status, prestige, dominance, power)
- **Security** (conveyed through trust, reliability, order, calm)
- **Sustainability** (transported via environmental awareness, ecology, good conscience)
- **Sensuality** (conveyed through attractiveness, stimulation, sexiness, heart palpitations)

- **Eye tracking** (also **Eye movement analysis,Eyetracking**)
Through **eye tracking**, the viewer's gaze path of mailings, posters, flyers, but also of websites and newsletters is recorded with an eye camera and then evaluated. Through eye tracking, it can also be determined which information is perceived and read . **Fixations** indicate a brief standstill of eye movement. Only at these points can information be absorbed. The occurring **eye jumps** (also called **saccades**) do not allow the eye to perceive information concretely.

Why it is so important to get to the bottom of the norms, motives and attitudes of customers in an indirect way is shown by two examples. In 2019, Coca-Cola announced that it would bring an energy drink to the market. The reasoning for this was convincing (Coca-Cola, 2019): "Coca-Cola Energy was developed by listening to people who told us they wanted an energy drink that tastes more like Coca-Cola than a traditional energy drink." After just one year, distribution was discontinued due to lack of demand. Just listening to the customers is not enough for a company like *Coca-Cola* to develop convincing products.

In fact, *Coca-Cola* had already failed once and thus delivered one of the biggest **marketing and brand flops** of all time: the introduction of ***New Coke*** in 1985. The company itself speaks of: "One of the Most Memorable Marketing Blunders Ever" (Coca-Cola, 2023). On April 23, 1985, the *Coca-Cola Company* took perhaps the greatest risk in the history of the consumer goods industry. The company announced that the recipe for the world's most popular soft drink would be changed to better cater to the preferences of consumers who repeatedly preferred *Pepsi-Cola* in blind tastings. However, only in blind tastings—as soon as the *Coca-Cola* brand was visible, the test subjects liked *Coca-Cola* much better! So much for the **power of the brand!**

The *Coca-Cola Company* introduced the newly formulated *Coca-Cola*, often referred to as "new Coke"—with the first change to the legendary recipe in 99 years. This was the beginning of a mass protest that ended after only 79 days with the return to the original formula. The product was then temporarily called *Coca-Cola classic.*

▶ **Note-Box** We should listen to our customers, but not believe everything they tell us. We need to dig deeper to recognize motives, attitudes, and values!

So far, some considerations on the theoretical readiness for behavioral change. How a **more sustainable lifestyle** can be achieved is shown by a slightly biased comparison of the Gold Standard (2023) in Fig. 1.28. From this comparison, personal **optimization areas for a more sustainable life** can easily be derived.

From a hygienic perspective, the idea of running the dishwasher at only 55°C and the washing machine at only 30°C is critical. The comparison of a train journey from the United Kingdom (UK) to southern France with a flight from the UK to New York is also biased. The correct comparison to the train journey would have been a flight from the UK to southern France. In addition, further information about the specific level of consumption is missing. Nevertheless, this provides a variety of exciting food for thought that we should all personally reflect on!

1.3.3 Relevance of Mental Accounting and Cognitive Dissonance

Mental Accounting refers to the way people categorize different alternatives and their consequences. They can either book these in common or separate mental accounts. Such

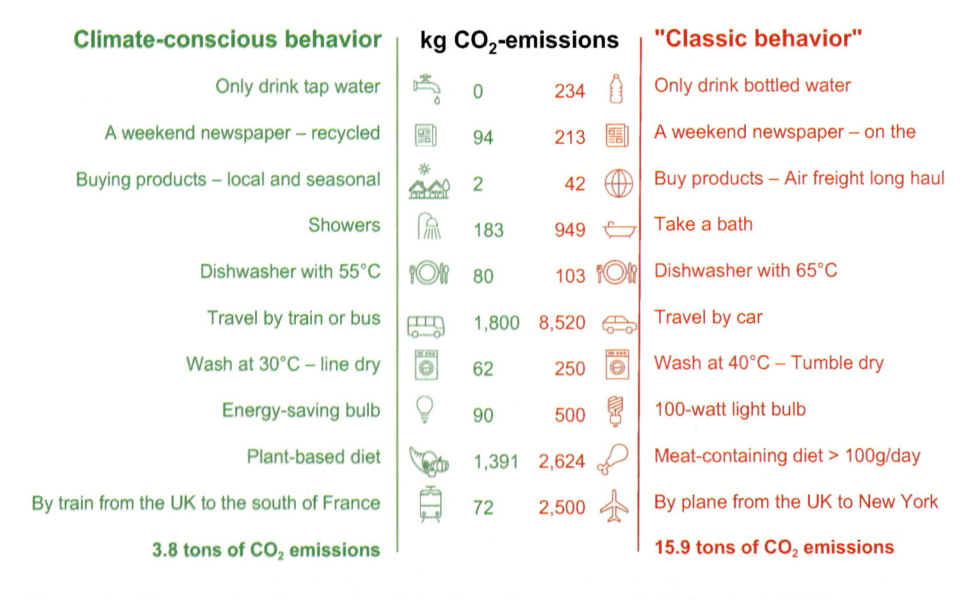

Fig. 1.28 CO_2 emissions from various behaviors. (Data source: Gold Standard, 2023)

mental accounts represent a **valuation framework** that is formed according to individual criteria (cf. fundamentally Tversky & Kahneman, 1981).

Here is an example of **Mental Accounting in finance:** When an employee receives a bonus, he can mentally book this money in different valuation frames. Either on a "special account bonus" to use this money for very specific activities, such as a nice trip or a larger purchase. The money could also mentally be booked on the "account general finances" and used here for everyday expenses.

To implement sustainable corporate management as well as green marketing and green branding successfully, companies must encourage their customers to make green decisions and prefer environmentally friendly products and services. Mental Accounting can help here by contributing to customers viewing green decisions as part of a specific budget that they spend on environmentally friendly products and services. Customers would mentally credit particularly sustainable activities to an **"environmental account"** or an **"Doing-something-good-for-the-environment account"**. Such a "booking" could occur when a consumer pays €3 for six eggs because all sustainability criteria (brother rooster, *Demeter*, organic) are met at the same time. The "credit" earned through this purchasing behavior on the "environmental account" or the "doing-something-good-for-the-environment account" would then be mentally debited when driving a distance by car, even though one could have walked.

The **motto for Mental Accounting** could therefore be:

I have already done something good for the environment. Therefore, I can now confidently drive the car.

The communicative challenge for companies here is to motivate customers to as many credits as possible on the "environmental account"—without this credit being quickly debited again. A high "account balance" on the "environmental account" should represent a value in itself—and give the customer a good feeling!

Cognitive dissonance is also important when focusing on sustainable consumption. The **cognitive dissonance** refers to the mental conflict that arises when people have contradictory thoughts, attitudes, or behaviors. This conflict can lead to people feeling uncomfortable or stressed because they are unable to reconcile their thoughts or behaviors with each other (cf. fundamentally Festinger, 2019).

An example of cognitive dissonance is when someone buys very cheap meat, even though they know it was produced under conditions detrimental to animals and the environment. In this case, there is a cognitive dissonance between the knowledge of the harmful effects of the meat production and one's own purchasing behavior. To resolve this conflict, the person could try to ignore the knowledge of the environmental harm or change their attitudes or behaviors—for example, by buying organic meat.

The emergence of cognitive dissonance can lead people to change their attitudes or behaviors to align them with their thoughts or values. However, it can also lead people to change their thoughts or values to align them with their behavior. The dissonance between the cognition "organic products are better for the environment" and the cognition "I do not buy organic products" could be resolved in the face of inflation-driven high

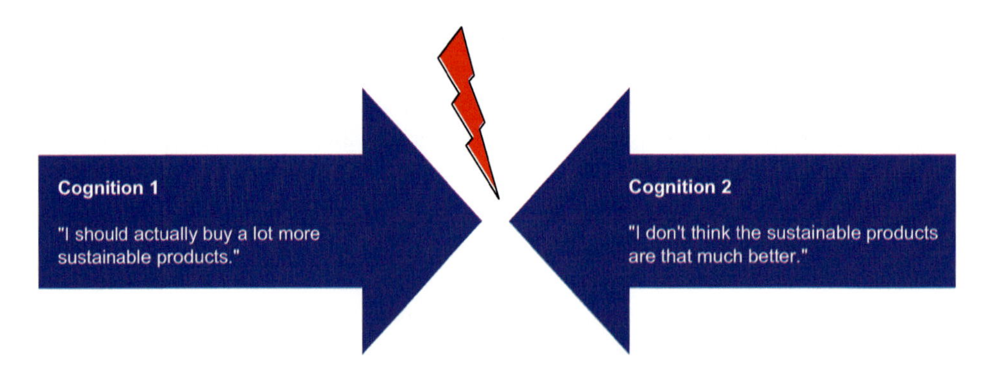

Fig. 1.29 Cognitive dissonance in sustainable consumptionsustainable consumption

food prices as follows: "I would like to buy organic products. However, I can no longer afford them due to the increased prices." The term cognition here refers to a content of consciousness.

To promote sustainable consumption behavior, it is crucial to reduce cognitive dissonances, as can be seen in Fig. 1.29. For this, companies would need to highlight much more the benefits associated with sustainable products for people and the environment. As long as there is still a large information gap, sustainable consumption will remain a niche market. Then customers weigh the second cognition more heavily—and do not change their purchasing behavior.

One way to use cognitive dissonance communicatively to encourage people to consume sustainably is to show them how their current behavior conflicts with their own values and beliefs. This can lead them to feel uncomfortable and be willing to change their behavior to resolve this conflict. More promising, however, are concepts where sustainable consumption is associated with positive associations. How this can be achieved is worked out in the context of green brand management (see Sect. 5.3).

1.3.4 Sustainability for Companies—Backgrounds and Strategies from a Cultural and Consumer Psychological Perspective

Guest contribution by Dirk Ziems, Thomas Ebenfeld, Rochus Winkler, Werner Detering; concept m research + strategies and flying elephant marketing consultancy

The **awareness for sustainability** has reached the mainstream in the last 10 to 15 years. The consequences of global warming, the discussion about the carbon footprint, the issue of plastic avoidance, the problems of palm oil, the relevance of animal welfare and many other topics have become relevant for a majority of consumers. Companies are under pressure to comprehensively realign their activities according to sustainability aspects.

At the same time, consumers need to be made clearly aware of their own responsibility in terms of sustainability.

What are the **psychological backgrounds of this development?** The following reports on the understanding gained from cultural and consumer psychological research on the topic through various studies.

The new sustainability paradigm—part of the cultural-psychological change of era
Actually, it has been known for over 50 years that the prevailing economic system is inevitably heading towards an ecological crisis in the long run. This is because the hitherto dominant **paradigm of the economy** is a **culture of maximization.** This is based on consuming unlimited resources—and that on a planet with limited resources. Nevertheless, the mainstream of the population in the Western-influenced consumer societies has ignored corresponding **warning signs** such as **climate change** and **species extinction** for decades—or treated them as a rather minor issue.

Only in the last one and a half decades has a comprehensive **sensitization of the majority societies** occurred in Europe. In the USA, the situation is still divided. In our psychological consumer and social research, this **change in mood and mentality** has been observed and analyzed over the years in hundreds of multi-hour in-depth interviews. A central finding of this research has been that the **new sustainable orientation** did not simply arise from rational insight. Enlightenment with facts certainly plays an important role. However, a process of change in orientation has been underway in the overall culture for about 15 years, which is even more comprehensive. This can be described as a **psychological change of era from the culture of maximization to the culture of reflection**. This change of era goes beyond economy and sustainability and encompasses the holistic **feeling of life of consumers** (see Fig. 1.30).

In the in-depth interviews, we regularly find that consumers' orientation towards **sustainability** symbolically marks the **departure from the values of the culture of maximization** and the **turning towards the values of the culture of reflection**.

Here is an example: The issue of **plastic waste in the oceans** has become a particularly tangible symbol of the excess of waste, especially in the pre-Corona times. Plastic waste is exemplary for the horrendous all-encompassing contamination of the environment, for the suffering of nature in the form of dying turtles etc. The ubiquitous presence of plastic in everyday consumption illustrates the excess that the culture of maximization of consumption has caused.

In the context of the **circular economy**, the **motive of recycling** represents the glimmer of hope that the culture of consumption can also reflect and reform. With recycling, the excess of resource consumption is tackled and a sustainable economy—corresponding to the natural cycles—is organized.

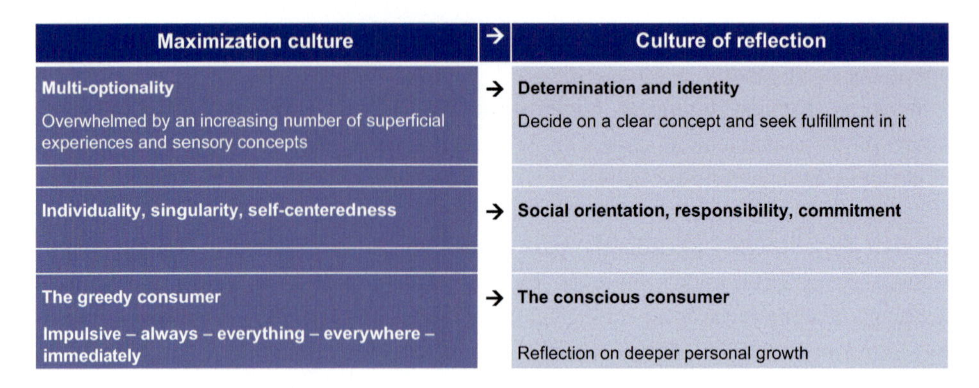

Fig. 1.30 From the culture of maximization to the culture of reflection

The different narratives of sustainability and the demands of consumers on companies

Another finding of our research is that **sustainability** is associated with very different narratives at a deep and often also unconscious level among consumers. The most frequently found **narratives** are presented here:

- **Apocalyptic struggle for existenceand radical renunciation**
  According to this narrative, it is already five past twelve. The train for ecological reversal has already left. Achieving even somewhat tolerable climate goals requires a radical rethink. Therefore, a **radical activism** is necessary (up to soft forms of eco-terrorism) to finally prompt politics, economy, and companies to take decisive action and to shake the mainstream of the consumer population out of their **state of denial**. **Economy and companies** are generally seen in the **role of perpetrators and culprits**. For their greed for profit, they gamble with the basic conditions of life on the planet. Their announcements and measures cannot be trusted. Here, the accusation of greenwashing often arises. Following the apocalyptic narrative, only even more **radical consumer renunciation** and a restriction to the bare essentials, thus **De-Consumption,** can help. However, followers of this thought easily entangle themselves in contradictions, as the media stories about *Greta Thunberg* and her sailing trip to the USA show.
- **Sinand indulgence trade**
  A second narrative is related to the first. In response to the serious warnings and admonitions of science and activists, many consumers feel guilty in their everyday life. They have a **guilty conscience,** that they only inadequately follow the admonitions and sin against the environment and nature. At the same time, a consistent sustainable conversion of consumption is hardly possible. Everyday life and consumption habits are simply resistant to change. Plastic waste can only be avoided to a limited

extent, one remains dependent on the combustion car, one still wants to afford the flight to Mallorca once a year. This leaves them with a constant guilty conscience. However, providers and brands that show progress towards sustainability allow them a **partial relief of conscience.** The purchase of these brands corresponds to a small **indulgence trade.** Against this background, it is also understandable that the sustainability orientation of almost all brands with signals such as packaging conversion, recycling programs, sustainable origin control, regionalism etc. has become so widespread. Due to their latent guilty conscience, the indulgence trade becomes a central component in their everyday consumption for many consumers.

- **Feelgoodand Forgiveness**
  An evolution of the guilt and indulgence trade narrative is the **Feelgood and Forgiveness narrative**, which currently shows the greatest spread and power of all narratives. In the tension field of having sustainable and responsible intentions on the one hand, but not being able to consistently implement them in the change-resistant everyday consumption on the other hand, consumers adopt a **laissez-faire attitude**. They are tired of feelings of guilt and self-accusations and find these also unrealistic and unproductive.
  This **rejection of guilt** we have found particularly often in family interviews. Parents often react here with a cool and indifferent attitude to the accusations of their children from the Fridays-for-Future direction: These demand vegan nutrition at the dinner table, reject the flight to Mallorca and constantly criticize the parents for ecological misconduct.
  Instead of constantly feeling guilty, it is considered ok to be inconsistent in everyday life from time to time. This is the **Forgiveness factor.** After all, one has already changed one's everyday routines in other areas. So one is proud to use only oat milk, to buy the *Conscious Collection* at *H&M*, to aim for an electric car when buying a new one, etc. These are the **Feelgood factors**. For companies and brands, it is important in this context to positively reinforce the Feelgood factors among consumers. Thus, marketing activities such as the establishment of eco-labels and sustainability communities give consumers the positive feeling of being on the side of the right and thus also on the side of progress.
- **Romanticizing Harmony Ideal**
  Other sustainability narratives are less strongly based on the necessities of reorientation and attitudes towards consumption change. They are more guided by **ideals** and **ideal future visions** that are to be achieved with sustainability. In this context, the romanticizing harmony ideal has always been common. Here it is then: back to nature, retreat into the innocent state of nature, find healing in a return to simple preindustrial forms, etc.
  In the consumer world, this narrative is symbolically expressed in various forms of **romanticizing-rural design** or of **craft-like vintage design.** The gastro brand *Marché* stages its gastro sales outlets in airports and motorway service stations—which are symbols of maximization culture—in the form of a quasi-rural ambiance with

wooden carts, oil lamps, and straw bales. The packaging of basic foods of animal origin regularly works with **romanticizing motifs** such as happy cows and chickens, images of farms that could come from children's books, and much more. Also **Slow Food** and **Slow Tourism** are movements that strongly work with **returning and deceleration motives**.

- **Green High-tech Transformationand Personal Growth**
  Another goal-oriented narrative is radically oriented towards the future and technology compared to the backward-looking romanticizing narrative. According to this narrative, the ecological crisis provides the occasion for a comprehensive **technological transformation of the economy and consumption.**
  In the course of the technological transformation, the energy supply is largely converted to renewable sources. Through the **use of eco-high-tech**, all production processes become sustainable. We all drive recycled electric cars with minimal CO_2 footprint on vacation. Consumers gradually change their consumption habits and experience this as personal growth. The enticing thing about this narrative is that thanks to technology as a solution for everything, a **transformation without renunciation and pain** is promised.
  In the automotive sector, *Elon Musk* and *Tesla* are prime examples of this narrative. These brands promise an **exciting departure into the age of sustainability:** Switch to the electric car, but still a driving style as a *Porsche* allows. The **everything-is-possible spirit** of the *Silicon Valley* start-up mentality increases credibility, dynamism, and attractiveness. In the near future, it can be expected that the **sustainability high-tech transformation** will also cover areas such as lab-grown meat, functional food, high-tech textiles, and high-tech housing construction.

The various narratives, the underlying psychological mechanisms, and the derived requirements for companies and brands can be found in Fig. 1.31.

Narrative	→	Psychological mechanisms	→	Requirements for companies and brands
Apocalyptic existence struggle and radical renunciation	→	Radical determination, De-Consumption	→	Counteract radical skepticism with convincing proof points
Sin and indulgences	→	Bad conscience due to own consumption style	→	Enable indulgence trading through virtue signaling
Feelgood and Forgiveness	→	Do not accept feelings of guilt, Emphasize your own progress	→	Strengthen feel-good factors, offer sustainable sublines
Romanticizing harmony ideal	→	Return to an innocent closeness to nature	→	Romanticizing brand staging and packaging; avoid greenwashing
Green high-tech transformation and personal growth	→	Experience yourself as part of an exciting green technological awakening	→	High-tech spirit and demonstrative rule-breaking in favor of sustainability

Fig. 1.31 Narratives, psychological mechanisms, and requirements for companies and brands

The Parallelism of Narratives and the Need for Tailored Strategy Development
In the depth psychological research of *Concept M*, it is repeatedly clear that the narratives just listed are present in consumers simultaneously and in parallel. One and the same consumer can be won over for the vaguely nature-romantic mood of a *Marché* Food Court while he is on his way to a sustainable vacation without a CO_2 footprint in his *Tesla*. At the same time, he feels guilty about buying a drink in a plastic bottle. The reference to 90% recycled raw materials in the packaging gives him a good feeling (feel-good factor). At the motorway petrol station, he then comes across a magazine stand with the latest apocalyptic news about the climate crisis.

The example shows: The everyday life of consumers is far removed from **consistent and consistent orientations** towards sustainability. Companies and brands face the challenge of developing consistent, comprehensive, and tailored sustainability strategies for themselves and according to the specific conditions of their own company and industry sector.

At this point, some **guidelines and principles for the individual development process of a purposeful sustainability strategy** are presented:

- **A sustainability orientation that only includes marketing and communication activities is not sustainable itself.**
  Sooner or later it will turn out that the sustainability claims are only superficial cosmetics and they will be exposed as **greenwashing** if the core of the business processes is not subjected to the **sustainability transformation** (see in depth Chap. 7). If a fashion provider only offers a *Conscious Collection*, while the business model remains anchored in resource-draining fast-fashion cycles, consumers will turn away from the brand in the medium term.
  Short-term strategies must be combined with medium- and long-term strategies to address different sustainability narratives in a targeted and appropriate manner.
  Companies are currently under pressure to immediately address the issue of sustainability with concrete measures that are immediately visible. However, these immediate measures can also quickly appear like pure cosmetics or greenwashing. Therefore, it is necessary to combine the short-term strategy with a medium- and long-term strategy that may also consider the **transformation of the core business** (see corresponding measures in Chap. 4).
  A **gas station brand** can offer a specific gasoline product in the short term that increases the proportion of sustainably grown vegetable oils. Here, the narratives of indulgence and feel-good factor are addressed. In the medium term, the gas station chain can plan a transformation to a network with charging stations for electric vehicles. The underlying narrative is the green transformation. In the long term, the oil company that operates the gas station network can enter into hydrogen production and distribution. Here too, the narrative would be the green transformation. All measures

and plans should be communicated comprehensively today to bind consumers to the **sustainability profile of the brand** (see Sect. 5.3).

- **Substantial proof pointsinstead of empty promises**
  A big problem in the past was often that the **sustainability communication** was limited to announcements that were not followed by proof points as a "reason to believe". Thus, corporate communication was full of claims like "CO_2-neutral by 2045". But this was just the minimum targets prescribed by the EU. How the companies would shape the path to CO_2 neutrality, what they would have to achieve concretely on the way there, remained unclear.
  In the consulting projects of *Concept M*, it becomes visible again and again that model or pilot projects, which are set up as a future workshop for sustainability, have a great radiance on consumers. In these model and pilot projects, both nature-romantic and green-high-tech-related narratives can be addressed. This is the case, for example, with model projects for new forms of animal husbandry in "small families" (nature-romantic idyll) or with pilot projects for paper production with hydrogen.
- **Pace-setting role insustainability competition**
  Companies can strive for sustainability in many fields. But only a few activities have a **defining pace-setting role** (see Sect. 5.4). Many other efforts are either difficult for consumers to understand, less relevant, or less credible. It can be embarrassing when sustainability measures are communicated that are seen as a **hygiene factor** and thus taken for granted. *Tesla* once advertised that only 17% of child labor was involved in their batteries, which was seen as sheer cynicism by consumers.
  What needs to be done? In the **development of the sustainability strategy**, it is always advisable to check with consumers and other stakeholders the individual potential fields of action for their relevance, credibility, and potential suitability as a pace-setter for the brand or company (see Sect. 5.1).
- **Current crisis times intensify and modify the role of sustainability**
  Given the short-term timing of the news cycles, it may seem that sustainability issues are being pushed into the background from the consumer's perspective by current crisis topics such as Corona, Ukraine war, and inflation. The interviews make it clear again and again: This is not the case!
  Consumers understand or at least suspect that the current crises are part of the **holistic system crisis of the maximization culture**. This maximization culture has brought about a high degree of dependency and uncontrolled interweaving. This includes, for example, dependence on cheap Russian gas, dependence on China, and global supply chains. These **dependencies** are now taking their toll and a reorientation towards more sustainable and considered economic processes is all the more urgent. This includes, for example, the replacement of the fossil economy with a hydrogen economy, even if it is still in development. Also, a change in diet to less meat consumption is indicated, even if this means partial renunciation.
  Nevertheless, the mentioned crises naturally have an influence on sustainable consumption. Thus, sustainably produced agricultural products (such as organic meat)

are currently under strong pressure because they are becoming unaffordable for many people. A challenge that currently arises is therefore the **democratization of sustainability** and the **cost reduction of sustainable products.** This can also lead to innovative ways such as **car sharing** of electric cars and online commerce of **ecological (start-up) products**.

Food for Thought Sustainability has gained importance in recent years and is more than just a trend. Especially for young target groups—as consumers and employees alike—it is indispensable that companies and brands take on and demonstrate **environmental responsibility**. However, the topic offers both great opportunities and stumbling blocks. In order to be able to profile themselves sustainably, companies need a strategy that is credible and comprehensible and thus leads to relevant results. The depth psychological research and implementation-oriented consulting of *Concept M* can make an important contribution to this.

1.4 Status Quo of Corporate Alignment Towards Sustainability

What is the **status of a Green Economy** in German companies? How many companies already have a "green" strategy? In 2021, business owners and managers in 363 companies were asked: "Is there a strategy in your company for the path to green change?" The results are shown in Fig. 1.32 (see Statista, 2021a). A good quarter (26%) of the companies are pursuing a "green strategy" in a targeted manner. 17% say that they have a "green strategy", but do not implement it consistently. 32% are currently developing such a strategy. Another quarter indicates that they do not have such a strategy.

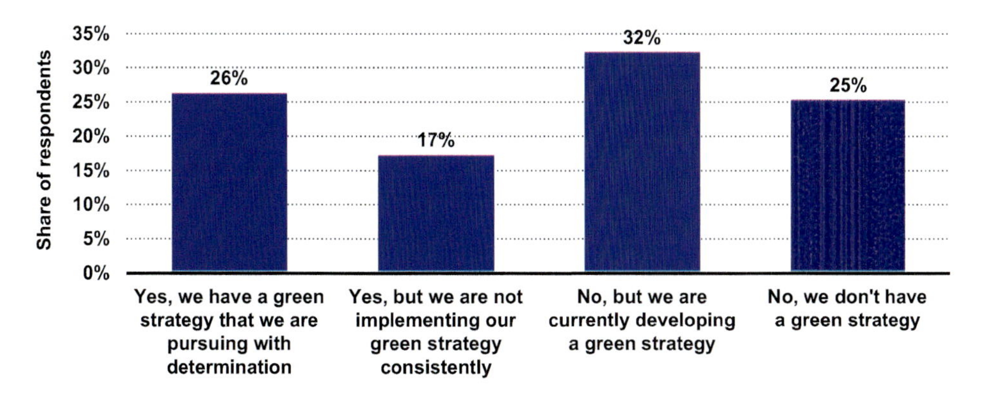

Fig. 1.32 Green Economy: Share of German companies with a green strategy—2021. (Data source: Statista, 2021a)

This result shows that only 26% are pursuing their "green strategy" in a targeted manner. This offers the opportunity to differentiate oneself in long-term competition. However, due to inconsistencies in buyer behavior, this does not guarantee competitive advantages. The other values are surprising to alarming: 17% have a "green strategy" without implementing it consistently. There is a great risk of greenwashing here (see Chap. 7). 32% are currently developing such a strategy—thus they have set out on the path. 25% have no strategy and are not developing one. There is a risk that they will disappear from the market in the future—either due to legal requirements or due to a lack of capital and/or customers.

> **Note Box** No decision is also a decision and looking away is not a strategy!

The **Federal Government** is aiming for a **greenhouse gas neutrality** in Germany from the year 2045. These greenhouse gases include, among others, carbon dioxide, methane, nitrous oxide, sulfur hexafluoride, perfluorinated hydrocarbons and partially halogenated fluorocarbons (see OECD, 2022). From the year 2045, no more greenhouse gases should be emitted than are bound elsewhere. Already by 2030, these emissions should decrease by 65% compared to 1990 (see Federal Government, 2022). What has already been achieved and what still needs to be done on this path is shown in Fig. 1.33 (see Federal Environment Agency, 2022). The sectors are delimited there according to the categories of the ***Federal Climate Protection Act.***

What is the companies' stance on this ambitious **goal setting in terms of sustainability?** Answers to the questions "Does your company want to be completely climate-

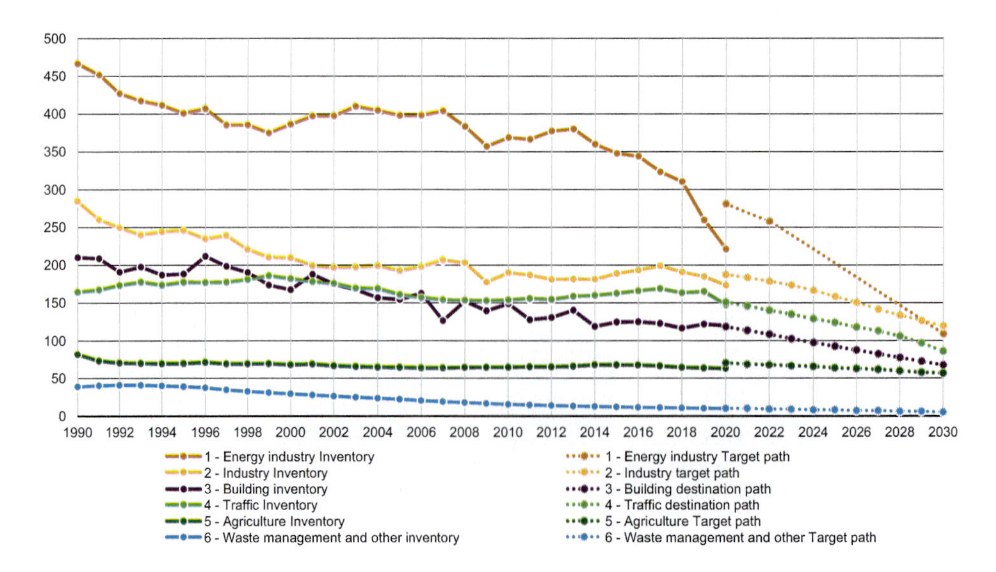

Fig. 1.33 Development and achievement of greenhouse gas emissions in Germany—in million tons of CO_2 equivalents. (Data source: Federal Environment Agency, 2022)

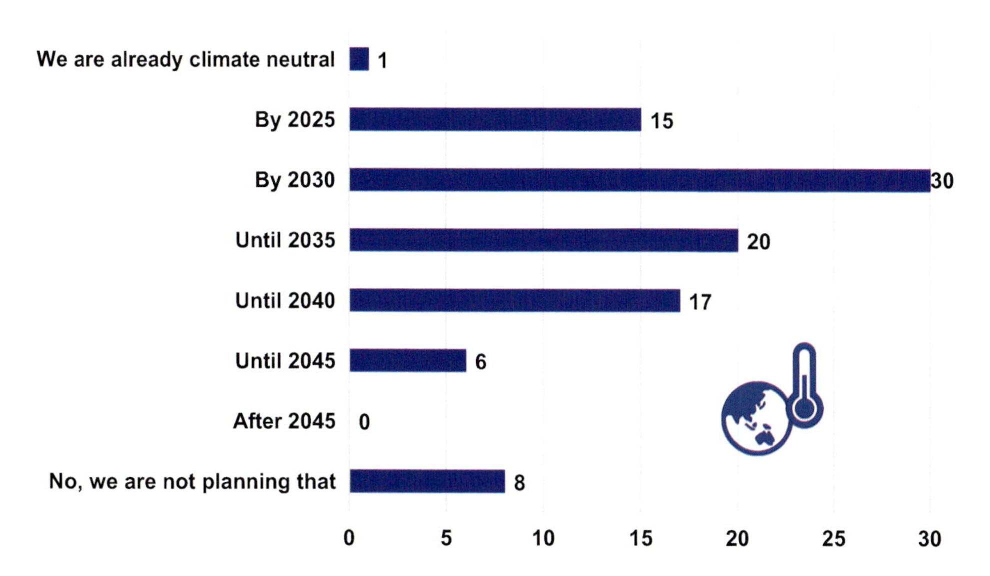

Fig. 1.34 Time of the targeted climate neutrality of German companies in %—2022. (Data source: Bitkom, 2022a)

neutral in the future? If yes: by when?" are provided by a representative survey of 506 companies with 20 or more employees in Germany (see Bitkom, 2022a). It becomes clear that the German economy is strongly committed to climate protection and wants to exceed the political goals to a large extent. Figure 1.34 shows that 45% of the companies want to be climate neutral by 2030. Another 37% aim to achieve this by 2040. The majority of companies thus support the political goal. Only 8% of companies in Germany cannot or do not want to achieve climate neutrality by 2045. Exactly 1% of companies already consider themselves to be climate-neutral.

Given these goals, it is exciting to determine to what extent companies are already pursuing a **strategy for sustainability and climate protection** today. The picture in Germany regarding the question "Does your company pursue a strategy for sustainability and climate protection?" is shown in Fig. 1.35 (see Bitkom, 2022a). Only 21% have a strategy for sustainability and climate protection for the entire company. Another 31% have such a strategy at least for individual areas. 37% lack a sustainability strategy, but its development is planned. 9% lack this strategy—and development is not planned!

What is exciting is the **importance of digitalization in the transformation process** (see Bitkom, 2022a). The question was asked: "What importance do digital technologies and applications have in the context of this strategy?" Every company that already pursues a concrete sustainability strategy (52%) or plans a sustainability strategy (37%) relies on **digital technologies.** For 24%, digital technologies are even crucial for the implementation of sustainability goals. For another 27%, they are of "very great importance", for 42% of "rather great importance". Only for 4% do digital technologies have a

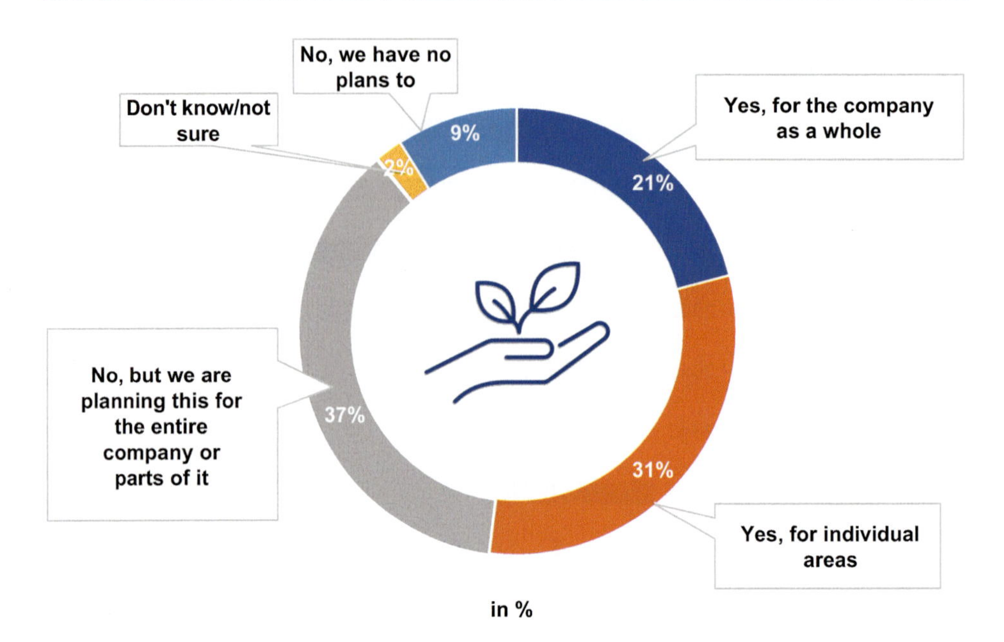

Fig. 1.35 Implementation of a strategy for sustainability and climate protection in Germany—2022. (Data source: Bitkom, 2022a)

"rather low importance" for the implementation of their own sustainability strategy (see Fig. 1.36).

> **Note Box** In total, 93% of companies in Germany rely on digital technologies to achieve climate neutrality. This makes it clear: An ecological transformation cannot succeed without digital transformation.

Many companies can already recognize **positive climate effects from digitalization measures**. In 77% of companies, the **CO_2emissions** have decreased overall due to the use of technologies and applications. The following picture emerges in detail (cf. Bitkom, 2022a):

- For 71% of respondents, **cloud computing** contributes to the reduction of CO_2 emissions. This is achieved because the operation of servers, storage, and applications in a large data center can usually be carried out more efficiently than operation in each individual company.
- 52% see great potential for more climate protection in the **Internet of Things** (IoT). By networking devices and machines via the internet, energy efficiency can be increased (cf. on the Internet of Things Fig. 4.7).

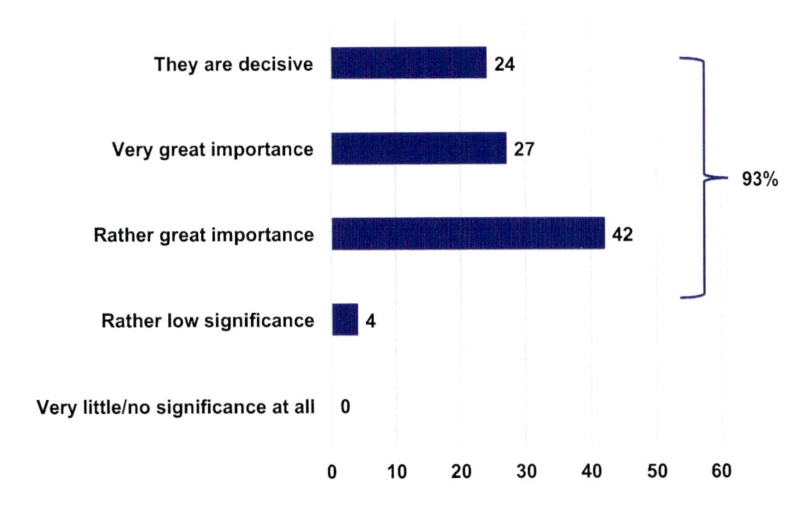

Fig. 1.36 Importance of digitalization for the climate strategy in Germany in %—2022. (Data source: Bitkom, 2022a)

- Further savings potential can be exploited through **Big Data and Analytics** (51%) and through the **automation of business processes** (51%).
- For 47%, **video conferences** contribute to the reduction because the (daily) commute to work and business trips are eliminated.
- 36% can further leverage climate protection potential in their own company through the **use of Artificial Intelligence**. This is achieved through intelligent building management and self-optimizing control of production processes. An important AI application is predictive maintenance—the anticipatory maintenance (cf. in depth Kreutzer, 2021, pp. 255–258).
- Overall, 75% of the companies surveyed see **digitalization** primarily as an **opportunity for sustainability and climate protection.** 21% see more risks here.
- 91% demand the **training of IT professionals,** to supplement **climate and sustainability aspects**.

Which **measures for more climate protection and sustainability** are companies in Germany already implementing internally (cf. Bitkom, 2022a)?

- 49% largely refrain from **printouts,** to save paper and other resources (keyword "paperless office").
- 47% of companies use **energy-efficient hardware** (e.g. monitors and printers).
- 39% rely entirely or partially on **home office,** to reduce commuting to work.
- 28% allow the **private use of company devices** (such as smartphones and laptops) to conserve resources.

The following question is also interesting: To what extent is the **refurbishing in the IT** already being used, which is part of the circular economy, to reintroduce used, professionally refurbished IT devices back into the usage cycle? The following figures show that there is still great growth potential in refurbishing (cf. Bitkom, 2022a):

- Only 4% of companies use refurbished products—and even then only in individual cases.
- Another 13% are considering their use at least for the future.
- 25% have dealt with the use of refurbished IT products, but then decided against it.
- 53% completely reject the use of Refurbished IT.
- However, 68% of companies see Refurbished IT as an important contribution to saving resources.
- 56% argue that as many companies as possible should at least consider the use of refurbished IT products.
- 51% see Refurbished IT as an important future topic.

> **Note Box** A central component of the circular economy—the **refurbishing**—has not yet arrived in companies. However, every (IT) device that is used longer in private households or in companies helps to reduce the **ecological footprint** of people and companies.

Interesting are the **reasons why companies want to become more sustainable**. The following results were determined for this (cf. Bitkom, 2022a):

- 63% of companies that already pursue a sustainability strategy or plan to do so, list **climate protection** as their primary goal.
- 60% want to lead by example and also expect positive **effects on their own reputation.**
- 52% specifically want to improve their **image.**
- 32% wish to become more attractive as an **employer.**
- 39% aim to maintain their own **competitiveness,** by better meeting the expectations of their customers with a planned or existing sustainability strategy.
- 28% expect a **potential for savings** through sustainable action.
- 25% point to the **desire of business partners** for sustainable action.
- 33% simply become more sustainable in order to comply with corresponding government regulations.

What **accompanying political measures** do companies want to support a **sustainable corporate management** (cf. Bitkom, 2022a)?

- 96% of companies wish that politics would promote the **expansion of renewable energies.**

- 79% wish for more **consulting services** from politics to become climate-neutral through digitalization.
- 58% demand that the **state leads by example** and pays attention to sustainability in the public sector when purchasing IT services and digital devices.
- 52% demand financial **incentives for investment in digital technologies,** that contribute to greater sustainability.

Further **demands on politics** are formulated by Bitkom itself (2022a). These include:

- Further **funding programs,** to specifically initiate digitalization measures with a positive sustainability effect (e.g., through the use of digital twins; cf. in depth Kreutzer, 2021, p. 290 f.)
- **Super-depreciations** can promote the use of new technologies.
- The **provision of Green Data** should be accelerated to provide the "fuel" "data" for the ecological transformation. This includes, for example, publicly available data on the state of the environment, energy consumption, and mobility data. Such data facilitate the development of sustainability innovations and sustainable business models.

> **Note Box** "Consistent digitalization is the key to successful climate policy" (Rohleder, 2022).

1.5 The Phenomenon of External Effects and External Costs

The **activities of companies** (such as the production of beverages, cosmetics, washing machines, and trucks) as well as the **use of products** (e.g., cars, game consoles, printing machines) and the **use of services** (including air and sea travel, streaming, house construction) are associated with harmful emissions into the environment, without being taken into account in the companies' or customers' calculations. These negative entries are referred to as **external effects**—specifically as **negative external effects.** These are effects (economic) decisions have on the environment, for which the polluters do not pay and for which the affected parties receive no compensation.

> **Note Box** In the case of **external effects,** there is no relationship regulated by market mechanisms between the perpetrators of these effects and those affected or suffering from them.

Such external effects occur when the **production** and **marketing** of products consume (scarce) resources (water, soil, etc.) and/or pollute (air, rivers, sea, landscape, etc.). The general public bears the cost of eliminating the corresponding pollution. The corresponding costs are externalized, i.e., shifted to external partners. Therefore, we speak of **external costs**—often also of **ecological follow-up costs** or **social costs.** Since these costs are borne by third parties, such external effects and the associated costs do not factor into the decision-making process of the perpetrators—unless companies are legally obliged to do so.

> **Note Box** **External costs** arise when the costs of an action are not borne by the perpetrator, but by the general public. These costs are thus outsourced from the perpetrator's area of responsibility (companies, customers) and thereby externalized.

Negative external effects and the associated external costs occurred when the company ***Coca-Cola*** switched the sale of various brands of glass bottles to PET bottles. The company itself saved high costs for the production, retrieval, cleaning, and reuse of the glass bottles. The disposal of billions of plastic bottles was largely imposed on the general public, overwhelming many countries. The costs for uncollected and unrecycled bottles are borne by all of us—through polluted seas, beaches, cities, green spaces, and other littered habitats.

Such **negative external effects** can also occur when **using an offer**. Since users do not have to bear the costs for the external effects (such as when driving or on flight or ship journeys for the emissions caused), these are also not taken into account in the purchasing process. Here too, the customer is offered a product or service at a price that is not ecologically appropriate.

The example of *Coca-Cola* can also be used here. While buyers of glass bottles were motivated by a deposit to return the bottles to collection points, this incentive does not apply to PET bottles in markets that do not use a deposit system or where such a system is not widely used. The plastic bottles are disposed of after a single use—wherever. Often, the waste systems in the affected countries were not prepared for the influx of PET bottles because both the collection possibilities and recycling concepts were lacking and are still lacking worldwide.

> **Note Box** When external effects occur in the production and/or use of products or services, their price does not reflect the ecologically appropriate price.
> This leads to a veritable market failure!

The example of *Coca-Cola* underscores why the presence of external effects is referred to as a **market failure**. Through the **externalization of costs**, the costs end up with (uninvolved) third parties. As a result, resources are directed into industries that incur more costs than are reflected in the internal company calculations. After all, the ecological and social costs caused by the company are not taken into account in their calculations. These costs often have to be borne by third parties. The market fails because there is a **cost-induced misallocation of resource allocation**. Companies that do not care about external costs are rewarded with profits. If you're thinking of *Coca-Cola* now, you're probably not wrong!

> **Food for Thought** How credible, trust-building, and solution-relevant can the note on *Coca-Cola* bottles and many other packages be, stating that they are recyclable? The reference to a **recycling potential** is not identical to the **actual extent of recycling!**
>
> Here lies a massive **thinking and understanding error**—which is deliberately accepted by the advertising companies. And for consumers, the recycling symbol contributes to the **feel-good factor!**
>
> **A win-win situation for companies and customers—but at the expense of the environment.**

There are similar examples from many other companies and brands. On the packaging of **cookies from*Bahlsen*** it says:

"This packaging is 100% recyclable. Help out and dispose of the cardboard box in the waste paper and the foil and the cookie tray in the yellow bag or yellow bin if possible."

This is intended to give the customer a good feeling of being able to act sustainably with this product. However, even if everything is disposed of as recommended, this does not mean that the plastic parts, even in a sorted state, are not still "thermally recycled"—in other words, burned or even deposited.

> **Note Box** Here again, the rightly criticized term **"recyclable"** is pointed out. Recyclable only denotes a possibility for recycling, but says nothing about the actual handling of waste materials!

The negative external effects lead to a **(partial) destruction of the environment and biodiversity.** This results in floods and water shortages, rising sea levels, crop failures, poisoned air and soils, and a species extinction of unimaginable proportions. This is marked as **Phase 1** in Fig. 1.37. These developments are increasingly causing a partial **internalization of costs (Phase 2).** If power plants cannot be cooled sufficiently due to a water shortage and therefore have to be shut down, this leads to higher costs. If ships on the Rhine can only sail with half the load due to low water levels, this causes higher costs. If production areas and companies sink in water or snow, high costs are incurred. The Corona pandemic is partly also explained by the loss of biodiversity due to humans increasingly encroaching on biological ecosystems. The costs of the lockdown have burdened not only consumers but also millions of companies to a great extent. And these **costs of environmental destruction** will dramatically increase in the future. One thing is becoming increasingly clear:

> **Nature is now fighting back!**

However, it should be taken into account that the **causers of the external effects** and those affected by an **internalization of costs** are usually not identical. For example, the rise in sea levels caused by humans leads to islands in the South Seas sinking into the

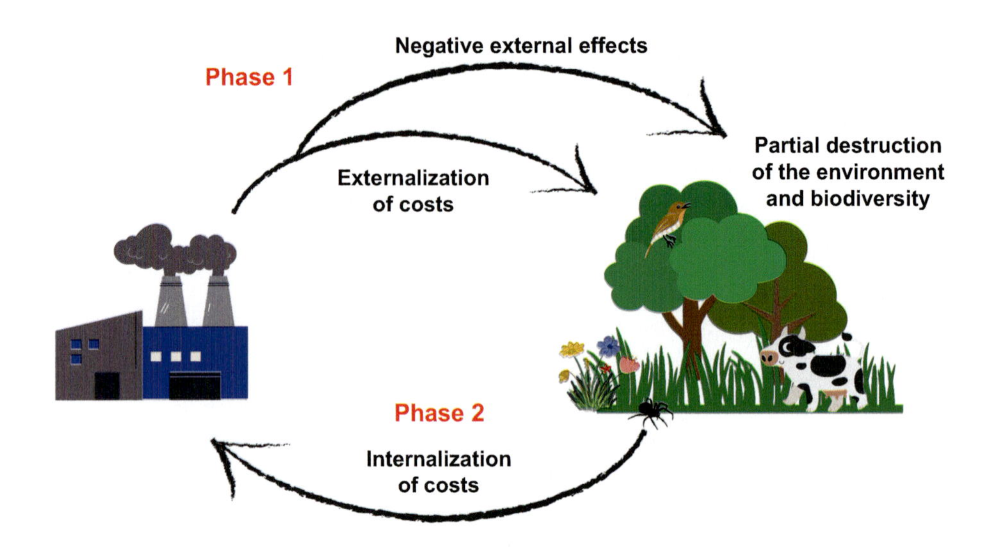

Fig. 1.37 Negative external effects lead to an internalization of costs

sea. The people living there lose their habitat without having caused this themselves. The **globalization of human intervention** in ecological systems leads to a **globalization of the internalization of costs**—without regard to national borders and causers!

Against this background, the legislator is increasingly taking action to avoid or at least reduce the externalization of costs. Approaches to this will be presented in Chap. 2.

Questions you should ask yourself

- What is the importance of sustainability at all levels of our value chain in our company?
- What is the importance of sustainability in the minds of our executives and employees at various hierarchical levels?
- What do we actually understand by sustainability?
- How well known are the starting points of the circular economy in our company?
- How can we increase knowledge about the circular economy in our company?
- Is there a training agenda on the topic of sustainability? Who is responsible for it?
- Do we have a sustainability strategy?
- Whose area of responsibility is the development of a sustainability strategy?
- Is the topic of "sustainability" also anchored at the highest corporate level? And if so, how? And if not, what would need to happen?
- Are budgets and personnel provided for the development and implementation of a sustainability strategy? And if so, how comprehensive? And if not, what would need to happen to secure the provision of resources?

- Are specific sustainability goals formulated for the company?
- Do we know how our different customer segments feel about the topic of "sustainability"?
- What do customers, partners, financiers, and shareholders expect from us in terms of "sustainability"?
- To what extent do external effects occur in the production of our products and services?
- What external effects are associated with the use of our products and services?
- How do we want to deal with these external effects in the future?
- What measures have our competitors already taken?
- What do our customers expect in terms of the external effects caused by us?
- Where in the company do all threads on "sustainability" come together?

References

Ajzen, I., & Fishbein, M. (1973). Attitudinal and normative variables as predictors of specific behavior. *Journal of Personality and Social Psychology, 27*(1), 41–57.

Ajzen, I., & Fishbein, M. (2005). The influence of attitudes on behavior. In D. Albarracín, B. T. Johnson, & M. P. Zanna (eds.), *The handbook of attitudes* (pp. 173–221). Lawrence Erlbaum Associates.

Balderjahn, I. (2021). *Nachhaltiges Management und Konsumverhalten* (2. edn.). UVK.

Beuth, P., Hoppenstedt, M., & Rosenbach, M. (2022). Wenn Rechner heizen. *Der Spiegel, 31*(2022), 56–59.

Bitkom. (2022a). 9 von 10 Unternehmen setzen ihre Klimaziele mit digitalen Technologien um. https://www.bitkom.org/Presse/Presseinformation/Digitalisierung-und-Klimaschutz-in-Wirtschaft-2022. Accessed 27 July 2022.

Bitkom. (2022b). Internet, Auto, Online-Shopping: Worauf die Deutschen fürs Klima verzichten könnten – Und worauf nicht. https://www.bitkom.org/Presse/Presseinformation/Worauf-Deutsche-fuers-Klima-verzichten-koennten. Accessed 11 Nov 2022.

Bundesregierung. (2022). Generationenvertrag für das Klima. https://www.bundesregierung.de/breg-de/themen/klimaschutz/klimaschutzgesetz-2021-1913672. Accessed 27 July 2022.

Coca-Cola. (2019). First coke-branded energy drink to launch in the U.S. in 2020. https://www.coca-colacompany.com/news/coke-energy-drink-launches-in-2020. Accessed 4 Aug 2022.

Coca-Cola. (2023). New coke: The most memorable marketing blunder ever? https://www.coca-colacompany.com/company/history/the-story-of-one-of-the-most-memorable-marketing-blunders-ever. Accessed 2 Jan 2023.

Earth Overshoot Day. (2022). This year, Earth Overshoot Day lands on July 28. https://www.overshootday.org/. Accessed 21 July 2022.

EPEA. (2023). Gemeinsam die Welt von morgen gestalten. https://epea.com/ueber-uns/cradle-to-cradle. Accessed 3 Jan 2023.

European Parliament. (2021). Circular economy: Definition, importance and benefits. https://www.europarl.europa.eu/news/en/headlines/economy/20151201STO05603/circular-economy-definition-importance-and-benefits. Accessed 3 Jan 2022.

Eye Square. (2022). Implizite Methoden. https://www.eye-square.com/de/implizite-methoden/. Accessed 4 Aug 2022.

Festinger, L. (2019). *Theorie der Kognitiven Dissonanz* (3. edn.). Hogrefe.

Global Footprint Network. (2022). Country trends. https://data.footprintnetwork.org/#/countryTrends?cn=5001&type=earth. Accessed 21 July 2022.

Global Standard. (2023). Global organic textile standard. https://global-standard.org/de. Accessed 2 Jan 2023.

Goebel, J. (2021). So will Amazon die massenhafte Vernichtung von Retouren beenden. In: Wirtschaftswoche. https://www.wiwo.de/unternehmen/dienstleister/outlet-statt-vernichtung-so-will-amazon-die-massenhafte-vernichtung-von-retouren-beenden/27480108.html. Accessed 3 Aug 2022.

GVM. (2022). *Entwicklung von Konsumverhalten, Aufkommen und Materialeffizienz von Verpackungen.* GVM.

Kilian, K., & Kreutzer, R. T. (2022). *Digitale Markenführung.* Springer Gabler.

Kreutzer, R. T. (2020). *Die digitale Verführung. Selbstbestimmt leben trotz Smartphone, Social Media & Co.* Springer Gabler.

Kreutzer, R. T. (2021). *Toolbox für Digital Business. Leadership, Geschäftsmodelle, Technologien und Change-Management für das digitale Zeitalter.* Springer Gabler.

Kunststoffe.de. (2023). Begriffsdefinitionen für das werkstoffliche Recycling. https://www.kunststoffe.de/a/grundlagenartikel/begriffsdefinitionen-fuer-das-werkstoffl-285262. Accessed 3 Jan 2023.

Land, K.-H. (2018). *Erde 5.0. Die Zukunft provozieren.* FutureVisionPress

McKinsey. (2019). The state of fashion 2019. https://www.mckinsey.com/industries/retail/our-insights/the-influence-of-woke-consumers-on-fashion. Accessed 24 Mar 2022.

McKinsey. (2021). *Consumers' sustainability sentiment and behavior before, during and after the COVID-19 crisis.* Consumer Research Germany.

OECD. (2022). OECD sustainable manufacturing indicators. https://www.oecd.org/innovation/green/toolkit/oecdsustainablemanufacturingindicators.htm. Accessed 30 Aug 2022.

Raue, T. (2022). Die Leute wollen Hühnchen essen – aber halt kein Fleisch. *Gault & Millau, 1,* 16–18.

Rohleder, B. (2022). 9 von 10 Unternehmen setzen ihre Klimaziele mit digitalen Technologien um. https://www.bitkom.org/Presse/Presseinformation/Digitalisierung-und-Klimaschutz-in-Wirtschaft-2022. Accessed 27 July 2022.

Simon-Kucher & Partner. (2022). Akteure mit einem wichtigen Einfluss auf die Nachhaltigkeit. *Nachhaltiges Deutschland,* 3.

Stahel, W. R. (2019). *The circular economy: A user's guide.* Routledge.

Statista. (2019). Welche aktuellen Sachverhalte zum Thema Nachhaltigkeit halten Sie für relevant? https://de.statista.com/statistik/daten/studie/1040564/umfrage/relevante-themen-und-nachhaltigkeit-aus-sicht-des-lebensmittelhandels-in-deutschland/. Accessed 30 Mai 2022.

Statista. (2020). Share of consumers that primarily shop at local businesses (online or in-store) to reduce their environmental impact in 2020. https://www.statista.com/statistics/1192365/consumers-that-shop-locally-to-reduce-environmental-impact/. Accessed 30 Mai 2022.

Statista. (2021a). Gibt es eine Strategie in Ihrem Unternehmen für den Weg zum grünen Wandel. https://de.statista.com/statistik/daten/studie/1283459/umfrage/deutsche-unternehmen-mit-gruener-strategie/. Accessed 30 Mai 2022.

Statista. (2021b). Relevance of sustainability as a purchasing criterion in Germany in 2021. https://www.statista.com/statistics/1288668/relevance-of-sustainability-as-a-purchase-criterion-germany/. Accessed 1 June 2022.

Statista. (2021c). How important is sustainability to you when making purchasing decisions for apparel, fashion, and footwear. https://www.statista.com/statistics/1303946/sustainability-importance-apparel-purchase/. Accessed 1 June 2022.

Statista. (2021d). Generationen in Deutschland nach Zustimmung zu Aussagen über Einkaufen, Konsum und Markenartikel im Jahr 2021. https://de.statista.com/statistik/daten/studie/1133562/umfrage/umfrage-zum-konsumverhalten-und-markenbedeutung-nach-generationen/. Accessed 30 Mai 2022.

Statista. (2021e). Anzahl der Personen in Deutschland, die beim Einkaufen darauf achten, dass die Produkte aus fairem Handel (Fair Trade) stammen. https://de-statista-com.ezproxy.hwr-berlin.de/themen/4331/nachhaltiger-konsum/. Accessed 1 June 2022.

Statista. (2021f). Share of consumers willing to untertake sustainable online shopping habits in German-speaking countries in 2021. https://www.statista.com/statistics/1270577/sustainable-online-shopping-in-german-speaking-countries/. Accessed 30 Mai 2022.

Statista. (2021g). Elektroschrott. https://de-statista-com.ezproxy.hwr-berlin.de/statistik/studie/id/101889/dokument/elektroschrott/. Accessed 27 July 2022.

Statista. (2021h). Importance of sustainable order fulfillment among shoppers in Germany in 2021. https://www.statista.com/statistics/1288886/importance-of-sustainable-order-fulfillment-germany/#:~:Text=Importance%20of%20sustainable%20order%20fulfillment%20in%20Germany%202021,combine%20multiple%20online%20orders%20into%20a%20single%20shipment. Accessed 30 Mai 2022.

Statista. (2021i). Nachhaltiges Modebewusstsein scheitert am Preisschild. https://de.statista.com/infografik/26399/gruende-fuer-den-verzicht-auf-nachhaltige-mode-in-deutschland/. Accessed 1 June 2022

Statista. (2022). Anteil von Bio-Lebensmitteln am Lebensmittelumsatz in Deutschland in den Jahren 2010 bis 2021. https://de-statista-com.ezproxy.hwr-berlin.de/statistik/daten/studie/360581/umfrage/marktanteil-von-biolebensmitteln-in-deutschland/. Accessed 4 Aug 2022.

Statista (2022b). Share of consumers more likely to buy from a brand with a clear commitment to sustainability in 2021. https://www.statista.com/statistics/1305896/share-of-consumers-more-likely-to-buy-from-sustainable-brands/. Accessed 1 June 2022.

Tversky, A., & Kahneman, D. (1981). The framing of decisions and the psychology of choice. *Science, 211,* 452–458.

Umweltbundesamt. (2022). Treibhausgasminderungsziele Deutschlands. https://www.umweltbundesamt.de/daten/klima/treibhausgasminderungsziele-deutschlands#internationale-vereinbarungen-weisen-den-weg. Accessed 4 Aug 2022.

United Nations. (1987). *Our common future.* United Nations.

United Nations. (2020). Sustainable development goals. https://www.un.org/sustainabledevelopment/wp-content/uploads/2019/01/SDG_Guidelines_AUG_2019_Final.pdf. Accessed 14 Febr 2023.

United Nations. (2022). *The sustainable development goals report 2022.* United Nations.

Zalando. (2021). *It takes two.* Zalando.

Institutional and Legal Framework Conditions and Requirements for Sustainable Action

2

When the facts change, I change my mind.

John Maynard Keynes

Abstract

The focus on sustainable corporate governance is no longer an autonomous decision of the responsible managers. More and more legal requirements are being defined to align companies towards more sustainable action. The most important legal framework conditions for sustainable corporate governance are presented in this chapter. Especially in this field, a high level of dynamism can be observed, so it is always necessary to pay attention to updates of the regulations outlined here.

2.1 European Commission's Green Deal

A project of the ***European Commission*** that is relevant not only for companies based in Europe, is called the name ***Green Deal***. This initiative aims to ensure the **transition to a modern, resource-efficient and competitive economy**. The following goals are pursued (see European Commission, 2022):

- By 2050, the EU should emit **no net greenhouse gases**.
- The **growth** of the EU should be **decoupled from resource use**.

At the same time, it should be ensured that no one, neither human nor region, is left behind in this process. To this end, a third of the investments from the recovery package ***NextGenerationEU*** and the seven-year budget of the European Union (EU) with a total

© The Author(s), under exclusive license to Springer Fachmedien Wiesbaden GmbH, part of Springer Nature 2024
R. Kreutzer, *The Path to Sustainable Corporate Management*,
https://doi.org/10.1007/978-3-658-43974-3_2

volume of 1.8 trillion € should flow into the *Green Deal*. The European *Green Deal* aims to achieve the following **sustainability goals**:

- Provision of clean air and clean water, ensuring healthy soil and preserving biodiversity
- Construction of renovated, energy-efficient buildings
- Provision of healthy and affordable food
- Ensuring further public transport
- Availability of cleaner energy and state-of-the-art clean technologies
- Offering durable products that can be repaired, recycled and reused
- Creating future-proof jobs and imparting the skills necessary for the transition
- Building a globally competitive and crisis-resistant industry

All institutions involved in economic processes can certainly agree on these goals. Whether the defined paths to these goals can always convince and what "pains" may be associated with them will be discussed below.

2.2 ESG Criteria

The so-called **ESG criteria** are increasingly becoming the **benchmark for sustainable corporate governance** in the twenty-first century. The letters "**E**" for **Environment**, "**S**" for **Social** and "**G**" for **Governance** define—derived from the dimensions of sustainability in Chap. 1—specific **areas of action for companies** in the fields of Planet, People and Profit. Today, it is no longer sufficient to run a company "only" profitably in the long term. However, a financially healthy company is still a prerequisite for long-term survival in the market. In addition, however, further requirements must be met, which are increasingly being raised by **investors**—and increasingly also by **customers** and other **stakeholders**. Companies that have their funding cut off, that cannot find qualified employees, or whose customers are leaving, cannot survive. In addition, the legislator defines further requirements for companies.

2.2.1 What are the ESG Criteria?

The contents of the **ESG criteria** are described as follows:

- **"E"** stands for **Environment** in the sense of environmentally compatible or environmentally friendly action (keyword "ecological sustainability").
- **"S"** stands for **Social** in the sense of behavior that not only corresponds to the aspects of occupational safety and health step, but also includes social commitment (keyword "social sustainability").

- **"G"** stands for **Governance** in the sense of sustainable corporate governance (keyword "economic sustainability").

The detailed contents of the ESG criteria are shown in Fig. 2.1.

This description of the ESG criteria shows that two of the requirements highlighted here have already been discussed in the context of ecological and social sustainability. The "Governance" criterion brings an additional requirement to the fore, which is to be understood as part of economic sustainability.

The **ESG criteria** are **verifiable requirements,** which are increasingly being incorporated into investment decisions, will be incorporated, or should be incorporated. The aspect of "verifiability" is important here because **greenwashing** should be avoided. Greenwashing occurs when companies communicatively cloak themselves in a "green mantle"—but their actions are still not environmentally or socially compatible (see Chap. 7).

2.2.2 Scope of the ESG Criteria

Which companies are already obliged to deal with the ESG criteria today? In 2018, the EU Commission presented an **action plan for sustainable finance**—the **Sustainable Finance Action Plan.** This obliged institutional investors to consider sustainability criteria in investment decisions and customers' sustainability preferences.

ESG criteria

Environment	Social	Governance
▪ Reducing the impact of business activities on climate change ▪ Protection of natural resources ▪ Increasing the efficiency of resource use ▪ Implementation of a circular economy ▪ Use of renewable energies ▪ Production of sustainable products ▪ Use of sustainable technologies and processes ▪ Sustainable building management ▪ Sustainable water management ▪ Sustainable mobility and logistics concepts	▪ Respect for human dignity and compliance with human and employee rights ▪ Safe and ergonomic workplace design ▪ Non-discrimination ▪ Diversity ▪ "Fair" treatment and payment of employees - throughout the entire supply chain ▪ Comprehensive training and development opportunities for employees ▪ Renouncing cooperation with authoritarian governments ▪ Assuming social responsibility - beyond the company's core performance ▪ Fair dealings with customers	▪ Publication of the company's relevant values and guidelines ▪ Compliance with the relevant laws and regulations ▪ Legally compliant payment of taxes ▪ Transparent documentation of the processes for managing and controlling the company ▪ Existence of clearly comprehensible remuneration and promotion guidelines ▪ Implementation of communication geared towards transparency - internally and externally ▪ Fairness in competition ▪ Independent control bodies

Fig. 2.1 ESG criteria

In 2019, the European legislator, with the **Disclosure Regulation** (DisclosureVO) and in 2020 with the partially supplementary **Taxonomy Regulation** (TaxonomyVO) various **transparency obligations with regard to sustainability aspects** were decided, which came into force on 01.03.2021 and 01.01.2022 respectively. Investors are thus to receive a sound information basis in order to be able to make their investment decisions also taking into account sustainability aspects. In this way, financial resources are to be directed towards sectors that are driving a shift towards greater sustainability. This is intended to contribute to the implementation of the **Sustainable Finance Action Plan**. The goals of the Paris Climate Agreement are to be achieved in this way.

The **Disclosure Regulation** (Regulation EU 2019/2088) is a **regulation on sustainability-related disclosure obligations in the financial services sector** (see European Union, 2019). Sustainability risks and sustainability factors distinguish between company-related and product-related transparency obligations. **Sustainability risks** are defined here as events or conditions in the areas of environment, social or corporate governance **(ESG)** that can adversely affect the value of an investment. The **sustainability factors** include environmental, social and employee concerns. Also, respect for human rights and the fight against corruption and bribery are included.

The **company-related transparency obligations** are solely directed at participants in the financial market. These include fund managers, providers of securities and credit institutions that offer portfolio management. This also includes investment or insurance advisors who offer advisory services on financial products. These so-called "relevant companies" must publish certain information on sustainability risks and factors on their website according to the Disclosure Regulation. These cover the following areas:

- Information on the inclusion of sustainability risks in advice and investment decisions
- Information on the impact of sustainability on the company and its activities
- Information on the consideration of sustainability risks in remuneration policy

*The **product-related transparency obligations*** refer to the financial products offered. The defined "relevant companies" must cover the following areas in pre-contractual information (such as in a sales prospectus):

- Information on how sustainability risks are taken into account in advice and investment decisions
- Assessments of how sustainability risks affect the return on financial products
- Statements on how, for example, ecological or social characteristics are met when a financial product is advertised with such characteristics. In these so-called **light green financial products**, ecological-social aspects are taken into account in the investment process—for example, as exclusion criteria. If necessary, information on an index used as a benchmark can also be mentioned here.
- Statements on how the goals of a sustainable investment to be achieved by financial products are to be achieved. The so-called **dark green financial products** pursue spe-

cific sustainability goals (such as reducing greenhouse gases). If necessary, it can also be explained here to what extent an index used as a benchmark is aligned with the achievement of the goals.

- For offers that are neither light nor dark green financial products, only information on how to deal with potential sustainability risks must be provided. These can be general explanations on the website and in the sales documents.

The Taxonomy Regulation (Regulation EU 2020/852) is a regulation on the establishment of a framework to facilitate sustainable investments and to amend the Disclosure Regulation (see European Union, 2020). The Taxonomy Regulation defines criteria to determine the degree of ecological sustainability of an investment. In addition, product-related transparency obligations under the Disclosure Regulation are supplemented and expanded. For example, it must be stated to what extent the financial product invests in ecologically sustainable activities. The pursuit of environmental goals must also be specified. Environmental goals include, for example, adaptation to climate change, sustainable use and protection of water and marine resources. The transition to a circular economy is also part of this.

Since August 2, 2022, financial service providers must inquire about their customers' respective sustainability preferences when providing investment advice and managing financial portfolios. These details must be taken into account when selecting possible investment alternatives (e.g., investment funds, insurance investment products, pension products). This is defined in the Delegated Regulation of the European Commission (see European Union, 2021).

Through this, the legislator wants to give the customer the opportunity to determine the extent of sustainable investments in their financial assets. Such a sustainable investment must contribute to the realization of the ESG sustainability factors. The customer can also stipulate that at least the significant negative impacts on ESG factors are taken into account.

Food for Thought

If it is mandatory to inform about the sustainability of investments in financial assets, then all companies must strive for the sustainability of their own actions. Only then can access to the financial resources necessary for further development be secured in the future.

In the future, many **medium-sized companies** in Germany could also be obliged to report on their sustainability risks. The EU Commission is already working on a **revision of the CSR Directive,** to expand its scope. As early as 2023, all listed and also non-capital market-oriented large companies could be obliged to report on non-financial information. This is provided that the EU's proposal is implemented. This could lower the size criterion "number of employees" to 250 employees. This would significantly increase the number of companies required to report (see Kögler, 2021; see Sect. 2.2).

> **Food for Thought**
> The challenge is: **Increasing transparency along the entire value and usage chain.**
> The motto is: **Transparency of everything!**

2.2.3 Definition of Individual ESG Criteria

The **"E"** in the ESG criteria stands for **Environment** and consequently for the environment. With regard to this criterion, it is examined to what extent a company's activities affect the environment. This includes, for example, the overall **resource consumption** and the **efficiency of this consumption.** In addition, the **emission of greenhouse gases and other pollutants** is examined.

These criteria have become increasingly important in recent years. This is due to the decreasing **reserves of important raw materials.** On the other hand, more and more responsible politicians, corporate representatives, and consumers now realize that **climate change** is man-made. Therefore, it can only be slowed down by humans themselves. However, climate change can no longer be reversed. The drought and heat phases of recent years in Europe, but at the latest the disaster in the Ahr valley in 2021, have shown many that climate change is not happening somewhere—far away—but right on our doorstep. There, but also worldwide, we have to realize that climate change is not only associated with environmental problems, but also with humanitarian and economic threats to entire countries and continents.

However, not only companies contribute to resource consumption and emissions. **Private households** also consume scarce resources and emit pollutants—when living, heating, eating, surfing, streaming—simply when living.

> **Note Box** No one should shift the responsibility for curbing climate change solely to companies! And no one should only demand changes from companies or politics. Everyone should start with themselves.

To check to what extent a company is currently dealing with the topic of **"Environment"**, the following questions can be asked, which partially overlap:

- Are concepts for reducing the impact of corporate actions on **climate change** being developed and/or implemented (including reduction of CO_2 or aiming for CO_2 neutrality)?
- Are measures for the **protection of natural resources** being used within the entire value chain or are such measures being planned?

- Are strategies for increasing the **efficiency of resource use** being developed and/or implemented (including protection of biodiversity)?
- Is investment being made in concepts for **circular economy** to promote reuse or further use as well as refurbishing, refabrication/remanufacturing (keyword **"Circular Economy"**)?
- Is the **use of renewable energy** being planned and/or implemented?
- To what extent are **sustainable products** being produced (sustainability of the materials used, of the product itself or of the product use, to overcome a linear economy)?
- Are **sustainable technologies** and **sustainable processes** being used or is their use being planned?
- Is a **sustainable building management** being used or is it being planned?
- Is a **sustainable water management** being used or is such a system being planned (including water saving, reuse/treatment, environmentally friendly wastewater treatment)?
- Are **sustainable mobility and logistics concepts** being used or are such concepts being planned?

The extent to which a company is already dealing with such requirements can be determined using various standards. To introduce an **environmental management system** according to **ISO 14001**, the responsibilities and processes of operational environmental protection must be defined (see Federal Environment Agency, 2020). This includes the processes for planning, implementation, and control.

In addition, responsibilities as well as behaviors and procedures must be documented in writing. Even more comprehensive is the orientation towards the **environmental management system EMAS**. EMAS stands for **E**co-**M**anagement and **A**udit **S**cheme, the "Community system for environmental management and environmental audit" (see EMAS, 2023). Companies that participate in this must not only meet the requirements of ISO 14001. In addition, core indicators of environmental protection must be recorded. Also, an environmental statement must be published every year. In addition, the environmental management system must be validated by an approved environmental auditor.

> **Note Box** The "E" in the ESG criteria stands for the consideration of environmental criteria in corporate management.

The **"S"** in ESG criteria stands for **Social.** Here, it is necessary to examine to what extent a company is currently dealing with the social and societal aspects of its own activities. To check this, the following questions can be asked, which also partially overlap:

- Are **human dignity, human rights** and **workers' rights** respected (including prohibition of forced and child labor)?
- Is there a safe and ergonomic design of workplaces to ensure the **health protection of employees**?

- Is the principle of **non-discrimination** implemented in all areas of the company?
- Does the company promote **diversity?**
- Is there a **"fair" treatment and payment of employees**—within the entire supply chain?
- Are comprehensive **offers for further education and training** provided to employees?
- Does the company refrain from **cooperating with authoritarian governments**?
- Does the company take on **social responsibility**—beyond the core performance of the company (keyword **"Corporate Social Responsibility"**), which can be reflected in donation and sponsorship activities?
- Does the company stand for **fair treatment of customers?**

These standards have found their expression, for example, in the **OECD Guidelines for Multinational Enterprises** (see OECD, 2022a). Also, the **ILO core labor standards** and the ten **principles of the UN Global Compact** formulate corresponding requirements (see ILO, 2022; United Nations, 2022). These are also laid down in the **ISO 26000**, a guide to the social responsibility of organizations (see ISO, 2022a).

> **Note Box** The "S" in ESG criteria stands for the consideration of social/societal criteria in corporate governance.

The **"G"** in ESG criteria stands for **Governance** i.e., responsible corporate governance. The following—partially overlapping—questions can be asked to clarify the status quo:

- Does the company publish the relevant **values** and **guidelines,** that underlie corporate governance (e.g., equal opportunities for promotions)?
- Does the company comply with the relevant laws and regulations **(compliance)** and does it implement specific measures to prevent corruption, bribery, fraud, money laundering, etc.?
- Is a **lawful payment of taxes** ensured?
- Are the **processes for control and monitoring** transparently and comprehensibly documented (including **risk and reputation management**)?
- Are there comprehensible **compensation and promotion guidelines**?
- Is there a communication strategy aimed at **transparency** both internally and externally?
- Does the company strive for **fairness in competition**?
- Are the independent **control bodies** (e.g., the advisory board, board of directors, or supervisory board) composed in a balanced (diverse) manner?

Recognized standards have already been defined for this area as well. This includes the **German Corporate Governance Code** (see DCGK, 2023). This set of rules primarily contains recommendations and suggestions for good corporate governance for listed companies. A particular focus of the new version is on sustainability. In this context, the opportunities and risks associated with social and environmental factors should be particularly highlighted.

The previously mentioned **UN Global Compact** also contains relevant regulations for governance, here with a focus on corruption prevention (opposition to all forms of corruption, including extortion and bribery). The **ISO 37000** defines international standards for "Good Governance" (see ISO, 2022b). The **G20/OECD Principles of Corporate Governance** support policy makers in evaluating and improving the legal, regulatory and institutional framework of corporate governance (see OECD, 2022b). Further guidance is provided by the **German Sustainability Code,** which defines a cross-sector transparency standard for reporting on corporate sustainability activities (see DNK, 2022). Guidelines for the preparation of sustainability reports are also defined by the **Global Reporting Initiative** (see GRI, 2022).

▶ **Note Box** The "G" in the ESG criteria stands for the consideration of specific criteria in corporate governance itself.

The company *Cubemos* (2023) offers an online **ESG quick check,** to check how your own company is positioned in this regard. This facilitates entry. The entry page can be seen in Fig. 2.2.

2.2.4 Adherence to ESG Criteria—Essential for Long-Term Survival

The extent of the necessary engagement with the various criteria underscores their relevance—for the long-term survival of both companies and humanity alike. As strategic investors are interested in the long-term survival of companies, ESG criteria are increasingly becoming the focus of **investment decisions**. To facilitate investment decisions, sustainability rating agencies are used. Unlike rating agencies such as *Fitch, Moody's* and *Standard & Poor's*, a sustainability rating is not created on behalf of the issuers, but on behalf of the investors.

In summary, it can be stated that **ESG investments** are moving more into the center. ESG funds and ESG ETFs are being launched and ESG ETF savings plans are being designed. The key term here is: **sustainable wealth creation.**

But also **customers**—albeit still at a low level—are demanding "fair and sustainable offers" (see Sect. 1.3). However, the desired "good conscience" is already much further along than the actual purchase. Otherwise, providers like ***Primark*** and ***SheIn,*** who

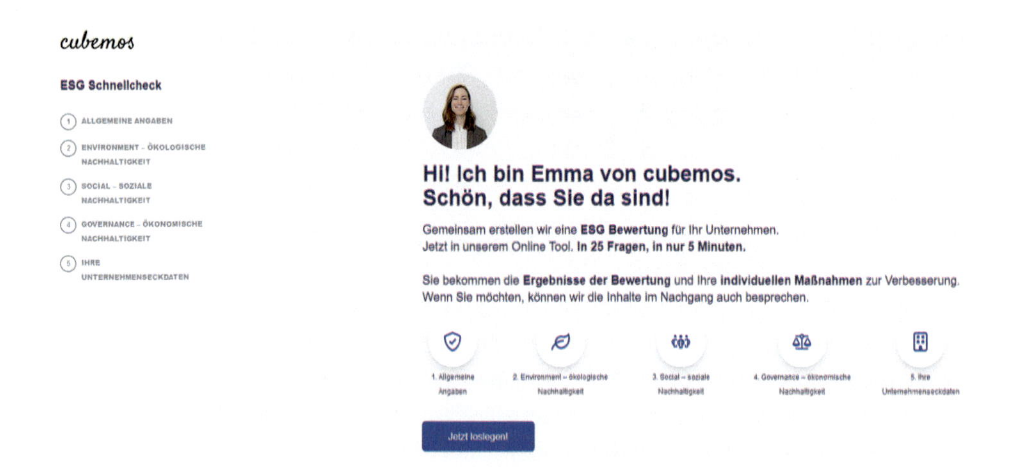

Fig. 2.2 ESG Quick Check. (© with kind permission from cubemos GmbH)

stand for Fast and Ultra Fast Fashion, would not continue to achieve great success (see Sect. 4.5). But a rethinking process has begun, albeit very slowly …

> **Note Box** Adherence to ESG criteria is no longer just a nice-to-have element. It has become a "must-have" in many areas!

2.3 Corporate Sustainability Reporting Directive (CSRD)

The Corporate Sustainability Reporting Directive **(CSRD)**, published in draft form in April 2021, will significantly expand ESG reporting for companies based in the EU. This is a **directive on corporate sustainability reporting.** It regulates the annual non-financial reporting for certain companies on sustainability issues (see European Commission, 2021; Federal Association for Sustainable Economy, 2022; Kreher & Gnändiger, 2022).

The Corporate Sustainability Reporting Directive replaces the currently applicable **Non-Financial Reporting Directive (NFRD).** EU-based companies must account for Environmental, Social, and Governance (ESG) issues in a **"Sustainability Statement",** which is to be integrated into the management report. With this, the European Commission is establishing a **uniform framework for the reporting of non-financial data** for the first time.

The Corporate Sustainability Reporting Directive establishes the concept of the **double materiality ("Double Materiality")**. As a result, companies must report on the following aspects (see Fig. 2.3):

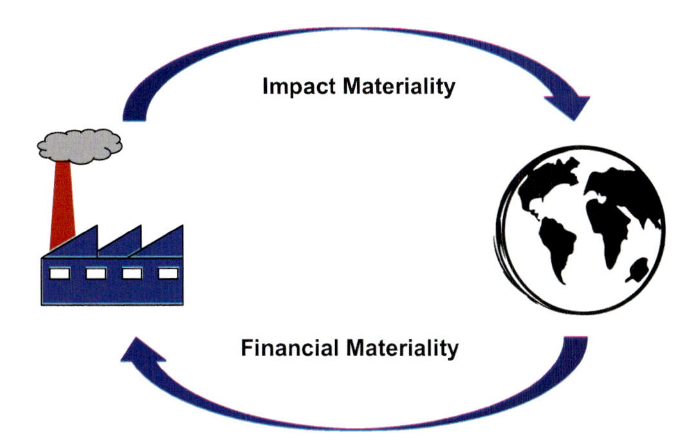

Fig. 2.3 Principle of double materiality ("Double Materiality")

- **Impacts of the company's own economic activities on the environment and people ("Impact Materiality")**
- This refers to the influence on and the risks to people and the environment caused by the company's activities.
- **Impacts of environmental changes on the company's own economic activities ("Financial Materiality")**
- Here, the risks of climate change itself, which affect the company's activities, must be assessed.

"Impact Materiality" and **"Financial Materiality"** are equally important for reporting. This includes considering current and potential impacts. When assessing a company's impacts, the **entire supply chain** and **various time horizons** must be covered. Also, the **impacts on all stakeholders** of a company must be analyzed. The stakeholder onion model can be used to identify the relevant stakeholders and their expectations (see Fig. 8.4).

In addition, detailed **information on sustainability goals** and corresponding **key figures** are required. The Corporate Sustainability Reporting Directive is based on the **Sustainable Finance Disclosure Regulation (SFDR)** and the **EU Taxonomy.** However, it often turns out that the companies' information on this is not sufficient. Sometimes, reports omit information that is important for investors and other interest groups. At the same time, the information provided by different companies is often difficult to compare. This undermines trust in the information provided.

Quality problems in sustainability reporting mean that sustainability risks to which companies are exposed are not always clearly visible. This is unsatisfactory for both investors and customers. Investors need reliable and meaningful information to fulfill their own disclosure obligations under the regulation on the disclosure of sustainable finances.

The rule is: If the **market for green investments** is to be credible, investors need comprehensive information about the sustainability of the companies in which they invest or want to invest. If such information is lacking, investments cannot flow into environmentally friendly activities and companies. High-quality and reliable public reporting by companies should help provide comparable and robust **data on companies' sustainability**.

So far, the **rules on non-financial reporting** in Europe have applied to "public interest entities". These include listed companies as well as banks and insurance companies. The **Non-Financial Reporting Directive** was therefore primarily relevant for large companies. To create greater transparency regarding sustainability, sustainability reporting is being extended to more companies. As a result, the number of **companies required to report in Europe** will increase from 11,700 to over 50,000 (including insurance companies and credit institutions). In Germany alone, 15,000 companies will then have to consider the **rules on non-financial reporting**. Non-EU companies that have a subsidiary or a branch in an EU state are also expected to fall under the Corporate Sustainability Reporting Directive.

Companies (regardless of their capital market orientation) are obliged to report according to the Corporate Sustainability Reporting Directive if they meet two of the following three criteria:

- more than 250 employees and/or
- net sales of more than €40 million and/or
- balance sheet total of more than €20 million

Capital market-oriented/listed SMEs are to implement the CSRD with an extended deadline ("simplified standards"). All **other SMEs** are **exempt from the reporting obligation** and can report voluntarily ("voluntary and simplified SME reporting standards").

The European Parliament calls on the Commission in its draft amendment to oblige SMEs from **high-risk sectors** (clothing, shoes, agriculture and mining) to **report on sustainability**. Whether SMEs in such high-risk sectors (regardless of their capital market orientation) will be subject to a reporting obligation depends on ongoing political negotiations.

According to the current status, the **reporting obligation** is to come into force as follows (cf. CSR reporting obligation, 2022):

- On January 1, 2024 for companies already subject to the Non-Financial Reporting Directive. The first report is due in 2025.
- On January 1, 2025 for large companies currently not subject to the Non-Financial Reporting Directive. Here, a first report must be prepared in 2026.
- On January 1, 2026 for listed SMEs as well as for small and non-complex credit institutions and proprietary insurance companies. A first report must be submitted here in 2027. There is an opt-out option until 2028.

The affected companies will have to provide information on more than 130 aspects of their environmental and social behavior in the future. In extreme cases, more than 800 different indicators are needed for this (cf. Wildemann, 2022, p. 16).

What is the common saying regarding the number of criteria to be measured?

> Twenty is plenty!

> **Food for Thought**
> It is to be hoped that the **competitive advantages** promised by the proponents of these comprehensive regulations through sustainable corporate governance and "green offers" will actually materialize. Because it cannot be ruled out that less sustainably thinking customers or customers with less purchasing power will opt for products and services that meet lower standards—and are therefore cheaper to purchase. Such offers can penetrate the EU market from other countries and displace law-abiding EU companies from the market.
>
> Which argumentation the market will follow will only be visible in several years. By then, certain regulations are hardly reversible. And (new) market structures in one direction or another may have already established themselves.

2.4 Supply Chain Act—Supply Chain Due Diligence Act

The **Supply Chain Act** (**LkSG**) or the **Supply Chain Due Diligence Act** defines corporate due diligence obligations to prevent human rights violations in supply chains. Specifically, the set of rules is called **Act on Corporate Due Diligence in Supply Chains**. It was announced on 22.07.2021. The aim of this law is to improve the international human rights situation. To this end, specific requirements are set for the management of supply chains. The aim of the legislator is to define for the companies "... a clear, proportionate and reasonable legal framework for fulfilling human rights due diligence obligations" (BMAS, 2022).

> **Note Box** Through the Supply Chain Act, the legislator forces companies to give **social and ecological sustainability** a special importance. In addition, companies are encouraged to systematically consider the **ESG criteria**.

Opinions on whether a clear, proportionate, and reasonable legal framework has been achieved to fulfill human rights due diligence obligations vary widely. The requirements of the Supply Chain Act are intended to be internationally compatible and to align with the due diligence standard ("Due Diligence Standard") of the UN Guiding Principles.

The Supply Chain Act establishes an **obligation to make efforts**. However, this does not imply **an obligation to succeed** or **a guarantee liability**.

The law adopted for Germany comes into effect on 01.01.2023. From 2023, it initially applies to companies with at least 3000 employees domestically. This includes approximately 900 companies. From 2024, it also applies to companies with at least 1000 employees domestically. By then, the law will cover approximately 4800 companies in Germany. However, all other companies should also familiarize themselves with the legal requirements—at least if they themselves are suppliers to large companies. The German Supply Chain Act is to be adapted to a future European regulation to prevent competitive disadvantages for German companies. This will result in even more extensive obligations to monitor supply chains for violations of human rights and environmental protection in the future, as the EU wants to design the obligations even more comprehensively.

This law obliges companies based in Germany to take on more responsibility in their supply chains and to comply more comprehensively with their due diligence obligations. The due diligence obligations not only apply to the **own business area.** The **actions of direct suppliers** and the **actions of indirect suppliers** must also be examined. The responsibility of companies thus covers the entire supply chain. The special **due diligence obligations of the companies** include the following measures (cf. BMAS, 2022):

- An **risk management** must be organizationally anchored to identify, avoid or minimize the risks of human rights violations and environmental damage. For risk management, **preventive and remedial measures** are prescribed. **Risk analyses** must be carried out regularly.
- A basic declaration on corporate human rights strategy must be adopted.
- **Preventive measures** must be developed for the own business area as well as for the direct suppliers.
- A **quick implementation of remedial measures** is required when legal violations are detected.
- There is an **obligation to establish complaint procedurescomplaint procedures,** to record and process legal violations.
- An **obligation to document and report** on the fulfillment of due diligence obligations is defined.

The Supply Chain Act includes a **catalog of internationally recognized human rights conventions.** From this, guidelines and prohibitions for the actions of companies are derived. This is intended to prevent a violation of protected legal positions. This includes, among other things, prohibitions of forced labor, child labor, and slavery. Also prohibited are, among other things, the withholding of an appropriate wage, disregard for occupational health and safety, and disregard for the right to form trade unions or employee representations. It also prohibits denying access to food and water or unlawfully depriving land and livelihoods. With the introduction of the law, these UN Guiding Principles are legally anchored.

If companies do not comply with these legal obligations, substantial fines can be imposed. Natural persons who have intentionally or negligently violated due diligence obligations face fines of up to €800,000. For legal entities or associations of persons with an average annual turnover of more than €400 million, the fine framework increases to up to 2% of the average annual turnover. The reference value here is the worldwide turnover of all natural and legal persons, if they act as an economic unit. With a turnover of €400 million, a fine of up to eight million € would thus be possible. Companies that have had to pay a fine can be excluded from the award of public contracts. Compliance with the law is checked by the *Federal Office for Economic Affairs and Export Control (BAFA)*. This also provides further information on implementation (see BAFA, 2022).

The supply chain law derives strict **duty of care rules** not only for the directly affected company, but also for the immediate and indirect suppliers. With regard to the **immediate suppliers**, the following obligations arise:

- Contractual assurance to respect human rights
- Contractual agreement on relevant control mechanisms
- Offer of training and development programs
- Training of the affected business areas
- Conducting an annual risk analysis

With regard to the **indirect suppliers, duty of care obligations arise in the event of a specific occasion.** Then the following steps are necessary:

- Conducting an event-related risk analysis
- Development of concepts for minimizing or avoiding the risk
- Initiation and anchoring of appropriate preventive measures against the perpetrator

The aim of the legislator is to strengthen the rights of affected people in supply chains. The financial and organizational expenses are to be borne by the companies. To what extent these regulations create internationally "fair" and above all uniform competitive conditions will only be shown over time. In addition, the ministry's page states so "precisely" (BMAS, 2022):

> "The appropriate manner of an action that meets the duty of care is determined by company-specific criteria."

What **consequences** the Supply Chain Due Diligence Act can have is made clear by *Jörg Wellmeyer*, board member of the Africa Association and managing director of the construction company *Strabag International*, (see Wellmeyer, 2022, p. 19):

> "Strabag in Africa is like a traveling circus. When we start somewhere, we organize the drinking water treatment for our construction site, we build accommodations. Then I go to the local carpenter and order twenty desks for the offices. I don't buy them in Germany, I order them from the local carpenter. With the supply chain law, I have to ask him: Where does the wood come from? Who cut down the tree? Did he also have protective equipment? Where and how were the screws produced? In the future, we will have to check how the suppliers of our canteen produce vegetables or poultry. This is extreme."

The consequence?

> "We only finish the projects that have already started. We no longer apply for new tenders in Africa. Strabag is ending the classic construction business in Africa."

The result?

Above all, Chinese companies will be pleased to have one less competitor in the fight for projects.

> **Food for Thought**
> The supply chain law forces German and European companies to adhere to very specific standards in their activities. However, it does not oblige companies from other countries. Now European companies will encounter competitors from other countries who do not have to adhere to such standards and can therefore produce more cost-effectively. This results in systematic distortions of competition to the detriment of European companies.
>
> When European and non-European competitors bid for projects, for example from the World Bank, the European companies may no longer be able to compete with the offers from Chinese companies, because the latter do not have to meet the same (bureaucratic) requirements. Is this what the legislators intended—or is it just accepted?

How do German consumers feel about these developments? It is interesting to look at which **measures within their supply chains German textile companies abroad** should be obliged to take—from the perspective of German consumers. For this purpose, 1004 people over 14 years old were surveyed. The specific question was: "German companies often have textiles, which they sell in the European Union, produced abroad and use long supply chains for this purpose, i.e., numerous production steps with different suppliers and producers abroad. Should German textile companies, in your opinion, be obliged by politics to…? Do you fully agree, rather agree, rather disagree or not agree at all?" The results of the answers "fully agree" and "rather agree" are shown in Fig. 2.4 (see Statista, 2021).

The vast majority of consumers believe that the obligations asked about here should be introduced. Again, the question of social desirability arises (see Sect. 1.3.2). It also remains open whether consumers are aware of the implications of such obligations—as desirable as they may be—on the competitiveness of companies and especially on the prices they themselves have to pay.

How the requirements of the supply chain law can be met is discussed in more detail in Sect. 4.4.1.3.

> **Food for Thought** A common criticism of the supply chain law is that it would shift the state's duty to protect human rights onto companies.

2.5 Circular Economy Act and Packaging Act

A central basis for sustainable corporate management is also the **Circular Economy Act** (KrWG). In § 1, the goal is defined as follows: "The purpose of this Act is to promote the circular economy to conserve natural resources and to ensure the protection of humans and the environment in the generation and management of waste" (Federal Ministry of Justice, 2021a). At the same time, this Act is intended to promote the achievement of the European legal target specifications of Directive 2008/98/EC of the *European Parliament* and the *Council* (see Federal Ministry of Justice, 2021a).

From the text of the law, the following **goal hierarchy for waste management** can be derived:

1. Avoidance of waste
2. Preparation of materials for reuse
3. Material recovery through recycling
4. Other recovery, e.g., energy recovery ("burning") or landfilling

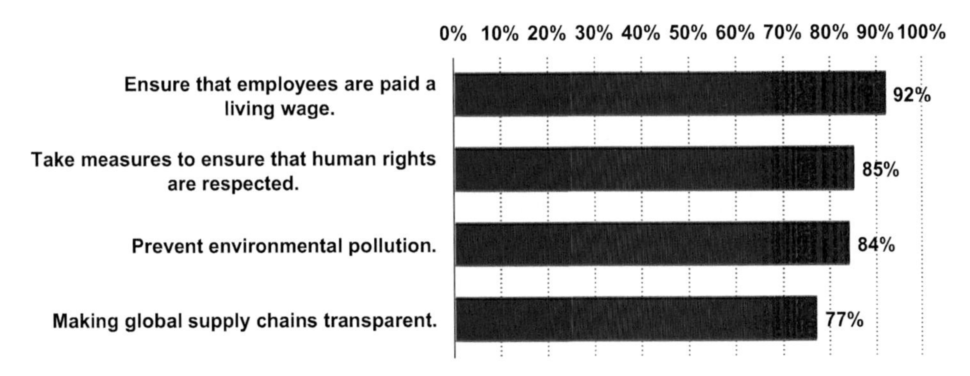

Fig. 2.4 Requirements for the supply chains of German textile companies abroad by consumers in Germany—2021. (Data source: Statista, 2021)

The primary goal of the Circular Economy Act is therefore the **reduction of waste quantities.** Contrary to the term "Circular Economy Act", the focus is not on the further utilization of unavoidable waste, but really on avoidance.

The **Packaging Act**—specifically the law on the placing on the market, the return and the high-quality recycling of packaging (VerpackG)—extensively interferes with the entrepreneurial design possibilities for packaging. In § 1 VerpackG, the waste management objectives of this law are defined in the form of requirements for product responsibility according to § 23 of the Circular Economy Act for packaging. This is intended to avoid or reduce the **impact of packaging waste** on the environment. The following objectives are defined for this purpose (cf. Federal Ministry of Justice, 2021b):

- Companies are encouraged to primarily avoid packaging waste. In addition, reuse and recycling should be enabled.
- A joint collection of packaging waste and other similar household waste close to households is aimed at in order to gain additional valuable materials for high-quality recycling.
- The proportion of beverages filled in reusable beverage packaging should be strengthened to avoid waste and the recycling of beverage packaging in closed cycles should be promoted. Reusable beverage packaging should reach a share of 70%.
- In addition, market participants should be protected from unfair competition.
- This law is also intended to achieve the European legal target specifications of Directive 94/62/EC on packaging and packaging waste, which define concrete requirements.

Strict requirements apply to the **use of recycled plastics for food**. It is defined which plastics may be used as so-called **food contact materials**. The Union's guide to Regulation (EU) No. 10/2011 on plastic materials and articles intended to come into contact with food stipulates that no processed old plastic (recyclate) from packaging waste may be used in direct contact with food (cf. BVL, 2022). This is intended to prevent pollutants from entering or adhering to the food via the packaging plastics.

Directive (EU) 2019/904 of the *European Parliament* and the *Council* of 5 June 2019 on the reduction of the impact of certain plastic products on the environment regulates, for example, the following in § 6 under product requirements:

> "For beverage bottles according to Part F of the Annex, each Member State ensures that
>
> a) from 2025, beverage bottles mainly made of polyethylene terephthalate ("PET bottles") consist of at least 25% recycled plastic, calculated as an average of all PET bottles placed on the market in the territory of the respective Member State;
>
> b) from 2030, these beverage bottles consist of at least 30% recycled plastic, calculated as an average of all beverage bottles placed on the market in the territory of the respective Member State."

This defines binding requirements at the packaging level.

The goals of the EU Commission go much further. It wants to ensure that by 2030 **all packaging in the EU is recyclable**. In addition, many more packages should be reused. Today, an average of around 177 kg of packaging waste per person per year is generated in the EU. With almost 226 kg per person per year, Germany is even at the top. To prevent further growth of packaging waste, certain packaging is to be completely banned. Which ones these will be will become apparent in the course of the ongoing legislative process (cf. o. V., 2022, p. 19).

In recent years, it has become apparent that the legislator is increasingly shaping the legal framework conditions for companies towards sustainability. Every company is well advised to install a **legal monitoring** system. Only in this way will it be possible to assess the relevance of the constantly changing and often also tightening legal framework conditions for one's own company—and to act accordingly.

Questions you should be asking

- Who in our company deals with the ESG criteria?
- Are any existing activities consolidated in one place?
- How regularly is it determined which requirements our company must meet and from when?
- How thoroughly is it checked which of our customers must meet corresponding requirements and from when?
- Are corresponding information systematically collected and merged?
- How extensively do the requirements of the supply chain law affect us?
- Is it already regularly determined what impact the supply chain law has on our reporting obligations?
- What requirements for our company result from the circular economy law and the packaging law?
- Are these requirements consolidated in one place in our company for evaluation (legal monitoring) and prepared for the responsible managers for near-term implementation?
- Is a concerted approach achieved for the entire company here?

References

BAFA. (2022). Lieferketten. https://www.bafa.de/DE/Lieferketten/Ueberblick/ueberblick_node.html. Accessed 21 July 2022.

BMAS. (2022). Sorgfaltspflichtengesetz. https://www.bmas.de/DE/Service/Gesetze-und-Gesetzesvorhaben/gesetz-unternehmerische-sorgfaltspflichten-lieferketten.html. Accessed 21 July 2022.

Bundesministerium der Justiz. (2021a). Gesetz zur Förderung der Kreislaufwirtschaft und Sicherung der umweltverträglichen Bewirtschaftung von Abfällen. https://www.gesetze-im-internet.de/krwg/BJNR021210012.html. Accessed 8 Sept 2022.

Bundesministerium der Justiz. (2021b). Gesetz über das Inverkehrbringen, die Rücknahme und die hochwertige Verwertung von Verpackungen. https://www.gesetze-im-internet.de/verpackg/. Accessed 8 Sept 2022.

Bundesverband Nachhaltige Wirtschaft. (2022). Corporate Sustainability Reporting Directive (CSRD). https://www.bnw-bundesverband.de/corporate-sustainability-reporting-directive-csrd/#1651070161554-3e93d19b-251d. Accessed 6 June 2022.

BVL. (2022). Leitfaden der Union zur Verordnung (EU) Nr. 10/2011 über Materialien und Gegenstände aus Kunststoff, die dazu bestimmt sind, mit Lebensmitteln in Berührung zu kommen. https://www.bvl.bund.de/SharedDocs/Downloads/03_Verbraucherprodukte/lebensmittelkontaktmaterialien/leitfaden_eu_10_2011_Kunststoff.html. Accessed 8 Sept 2022.

CSR-Berichtspflicht. (2022). Vorgaben für das Nachhaltigkeitsreporting von morgen. https://www.csr-berichtspflicht.de/csrd. Accessed 29 Sept 2022.

Cubemos. (2023). ESG-Schnellcheck. https://esg-check.cubemos.com/. Accessed 3 Jan 2023.

DCGK. (2023). Deutscher Corporate Governance Kodex. https://www.dcgk.de/de/. Accessed 2 Jan 2023

DNK. (2022). Deutscher Nachhaltigkeitskodex. https://www.deutscher-nachhaltigkeitskodex.de/. Accessed 26 Jan 2022.

EMAS. (2023). Ressourcensparendes Umweltmanagement mit EMAS. https://www.emas.de/was-ist-emas. Accessed 1 Jan 2023.

Europäische Kommission. (2022). Europäischer Grüner Deal. https://ec.europa.eu/info/strategy/priorities-2019-2024/european-green-deal_de. Accessed 6 Dec 2022.

Europäische Union. (2019). Verordnung (EU) 2019/2088 des Europäischen Parlaments und des Rates. https://eur-lex.europa.eu/legal-content/DE/TXT/HTML/?uri=CELEX:32019R2088&from=de. Accessed 26 July 2022

Europäische Union. (2020). Verordnung (EU) 2020/852 des Europäischen Parlaments und des Rates. https://eur-lex.europa.eu/legal-content/DE/TXT/HTML/?uri=CELEX:32020R0852&from=DE. Accessed 26 July 2022.

Europäische Union. (2021). Delegierte Verordnung (EU) 2021/1253 der Kommission. https://eur-lex.europa.eu/legal-content/DE/TXT/?uri=CELEX%3A32021R1253. Accessed 26 July 2022.

European Commission. (2021). Questions and Answers: Corporate Sustainability Reporting Directive Proposal. Brussels. 21.4.2021.

GRI. (2022). Global Sustainability Standards Board. https://www.globalreporting.org/. Accessed 25 Jan 2022.

ILO. (2022). ILO Kernarbeitsnormen. https://www.ilo.org/berlin/arbeits-und-standards/kernarbeitsnormen/lang--de/index.htm. Accessed 31 Jan 2022.

ISO. (2022a). ISO 26000. Social responsibility. https://www.iso.org/iso-26000-social-responsibility.html. Accessed 30 Jan 2022.

ISO. (2022b). ISO 37000:2021. Governance of organizations – Guidance. https://www.iso.org/standard/65036.html. Accessed 27 Jan 2022.

Kögler, A. (2021). ESG-Berichtspflicht dürfte bald auch Mittelständler treffen. https://www.dertreasurer.de/news/finanzierung-corporate-finance/esg-berichtspflicht-duerfte-bald-auch-mittelstaendler-treffen-2019251/. Accessed 5 Aug 2022.

Kreher, M., & Gnändiger, J.-H. (2022). Erhöhte Anforderungen an ESG-Berichterstattung, Börsen-Zeitung. https://www.boersen-zeitung.de/unternehmen-branchen/erhoehte-anforderungen-an-esg-berichterstattung-341cfc12-a134-11ec-8ac6-554b894b6cb7. Accessed 6 June 2022.

o. V. (1. January 2022). Mindestens Recycling, besser Wiederverwertung. *Frankfurter Allgemeine Zeitung*, p. 19.

OECD. (2022a). OECD-Leitsätze für multinationale Unternehmen. https://www.oecd.org/berlin/publikationen/oecd-leitsaetze-fuer-multinationale-unternehmen.htm. Accessed 2 Febr 2022.

OECD. (2022b). G20/OECD-Grundsätze der Corporate Governance. https://www.oecd.org/daf/g20-oecd-grundsatze-der-corporate-governance-9789264250130-de.htm. Accessed 28 Jan 2022.

Statista. (2021). Zu welchen Maßnahmen sollten deutsche Textilunternehmen bezüglich ihrer Lieferketten im Ausland verpflichtet werden? https://de.statista.com/statistik/daten/studie/1238484/umfrage/umfrage-zu-lieferketten-von-deutschen-textilunternehmen-im-ausland/. Accessed 1 June 2022.

Umweltbundesamt. (2020). ISO 14001 – Umweltmanagementsystemnorm. https://www.umweltbundesamt.de/themen/wirtschaft-konsum/wirtschaft-umwelt/umwelt-energiemanagement/iso-14001-umweltmanagementsystemnorm. Accessed 1 Febr 2022.

United Nations. (2022). United Nations Global Compact. https://www.globalcompact.de/ueber-uns/united-nations-global-compact. Accessed 30 Jan 2022.

Wellmeyer, J. (29. November 2022). „Wir beenden das klassische Baugeschäft in Afrika". *Frankfurter Allgemeine Zeitung*, p. 19

Wildemann, H. (31. October 2022). Die Pflicht zur Tugend machen. *Frankfurter Allgemeine Zeitung*, p. 16.

Integration of Sustainability Concepts into the Purpose Definition of Companies

3

Nothing in the world is as powerful as an idea whose time has come.

Victor Hugo

Abstract

To establish sustainable corporate management in the long term, the central aspects of a sustainability strategy must be anchored in the company's purpose definition. To enable this, the central aspects of purpose-oriented corporate management are explored in depth here. In addition, it is shown how concrete behavioral rules and leadership styles can be derived from the purpose definition.

3.1 The Definition of an Entrepreneurial Purpose

To ensure a comprehensive alignment of the company towards sustainability, appropriate guiding principles and visions for the company must first be defined. This may also involve further developing the existing **company values**. Overarching is a **Purpose** for the entire company to be developed, which carries aspects of sustainability within it.

▶ **Reminder Box** The central challenge is: **Start with the why!**

Every company should provide convincing answers to the question of the "Why?" of entrepreneurial activity—an answer to the question of the Purpose. **Purpose** or **company purpose** refers to the fundamental goal of a company. The purpose of a company should

© The Author(s), under exclusive license to Springer Fachmedien Wiesbaden GmbH, part of Springer Nature 2024
R. Kreutzer, *The Path to Sustainable Corporate Management*,
https://doi.org/10.1007/978-3-658-43974-3_3

differ from that of other companies and have a **positive societal impact** (see Fig. 3.1; see Sinek, 2011; further Illner, 2021; Kreutzer, 2021, pp. 136–138).

At its core, the **Purpose** is about the purpose of the company and consequently the question of the "Why": Why do we exist as a company? The answer to this should not be: "To make a profit." Profit can only ever be the result of company activity—not its actual core. The purpose of a company is to think bigger. The core of a **Purpose-oriented corporate management** lies in the belief that companies are more than mere economic subjects aimed at achieving the highest possible profits. Instead, companies should also serve a higher purpose and make a positive contribution to society and the environment. Purpose-oriented corporate management sees the company as part of a larger community and aims to align the interests of all stakeholders—customers, employees, suppliers, investors, and the broader society. The purpose serves as a guideline for all decisions and actions.

Such a purpose could, for example, be to solve a specific societal problem or to bring about a positive change in the world. Purpose-oriented corporate management thus pursues the goal of leading the company in such a way that it not only achieves economic success but also has a **positive societal impact**. In the course of a strategic alignment towards sustainability, a reference to sustainability should already be established in the **Purpose definition**. To **develop a purpose oriented towards sustainability**, the following questions can be answered:

- What is at the core of the company's sustainability or how can sustainability become the core of the company's activities?
- What contribution does the company make to a "better world"?

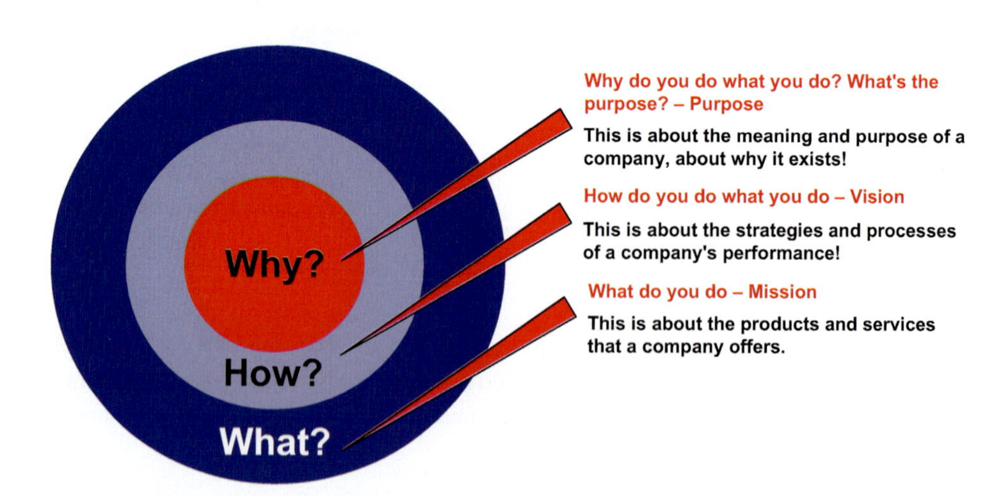

Fig. 3.1 Development of the entrepreneurial Purpose: Start with the why!

Ideally, this **Purpose** offers a high **identification potential**. On the one hand, for the employees and executives who are already on board. The purpose should also motivate them to go the "extra mile" because they fully identify with the cause of "their" company. On the other hand, a convincing purpose can be interesting for potential employees and executives, but also for investors and cooperation partners. Working for a company that has a convincing direction can be a value in itself because the company's contribution to "the big picture" becomes visible and tangible. If the purpose is also visibly and above all comprehensibly lived outwards, this increases the attractiveness of the company and its offers also for the customers.

From this corporate purpose, the **vision** is derived. This vision refers to what the company wants to achieve in the future. It is a long-term perspective that describes the goals and dreams of the company. Therefore, it is about the strategies and business processes, and thus the core activities of a company (question of "how?"). To **develop a sustainable vision**, the following questions can be answered:

- How can the entire value chain be designed more sustainably?
- How are the company's own processes aligned with sustainability—in the areas of procurement, R&D, production, logistics, sales, disposal?
- What contribution does the company itself make to sustainability?
- What contribution does the company make to contribute to the sustainability of others?
- Can the company's own business model be supplemented with sustainability offers?
- Can a new business model have sustainability at its core?
- ...

With the **mission**, the results to be achieved can be described (question of "what?"). The mission of a company describes how the company wants to achieve its vision. It indicates the steps the company will take to achieve its goals and establishes the fundamental values and principles on which the company's actions are based. The mission indicates what the company stands for and what it does. To **develop a sustainable mission**, the following questions can be answered:

- How can the company's own products and services be designed more sustainably?
- How can a more sustainable use or a more sustainable use of own products and services be ensured?
- Through which measures can the company contribute to the circular economy?
- ...

It is crucial that purpose, vision, and mission are intensively linked and form a **consistent unit**. This unit of effect is an indispensable prerequisite for the purpose to support the development of a corporate identity, offers a high potential for identification for many stakeholders, and mobilizes many people to support the achievement of the purpose.

A convincing example of a sustainability-oriented corporate policy is provided by ***BASF***. This company has defined the following statement as its **background for action**. This brings together purpose, vision, and mission in a powerful form (BASF, 2023a):

> "We want to contribute to a world that provides a viable future with improved quality of life for everyone. Therefore, we support our customers and society with chemistry that makes the best possible use of existing resources. To improve our contribution to a sustainable future, we measure the overall impacts of the economic, environmental, and social aspects of our business activities with our Value-to-Society method. We support the United Nations in implementing the UN Sustainable Development Goals (SDGs) and commit to the Paris Climate Agreement.
>
> Climate change is the greatest challenge of our time, therefore we aim for net-zero CO_2 emissions by 2050. To achieve this, we will become more efficient in our production and energy use, we will increase our use of renewable energy, and we will accelerate the development and use of new CO_2-free processes for the production of chemical products.
>
> We create value for our customers with Accelerator solutions that contribute to sustainability in the value chain. With our Circular Economy Program, we want to decouple growth from resource consumption. Circular solutions reduce waste, save fossil resources, and help reduce CO_2 emissions.
>
> We take responsibility in the supply chain and in our production. We work with our suppliers to assess and improve their sustainability performance, and we involve our suppliers in our Supplier CO_2 Management Program. In our production, we use resources efficiently based on our integrated production network—our Verbund. We act responsibly and protect the environment by reducing emissions and waste. We support the protection of functioning ecosystems such as forests and oceans, for example through our commitment in the *Alliance to End Plastic Waste*. We implement sustainable water management at our production sites in water stress areas and at our Verbund sites.
>
> Respecting human rights in our own operations and business relationships is the basis of our social responsibility. We ensure a safe working environment. We welcome diversity, promote an inclusive workplace, and encourage all employees to engage genuinely."

This statement addresses all central **areas of action of a sustainable corporate management**. From this, very concrete goals can initially be derived. *BASF* is striving for the **transition to a more circular economy**. For this, *BASF* focuses on three areas of action (see BASF, 2023a):

- Use of circular, i.e., recycled and renewable raw materials
- Establishment of new material cycles
- Implementation of new business models

From the year 2025 onwards, 250,000 tons of recycled and waste-based raw materials are to be processed each year—instead of fossil raw materials. By the year 2030, *BASF* wants to double its sales with solutions for the circular economy to 17 billion euros.

Through these formulations and the defined and externally communicated goals, *BASF* takes responsibility for itself. The company will be measured by its words. At the same time, the value-based statements serve as a guideline for the actions of executives and employees.

Zalando (2022) has defined its sustainability-oriented purpose as follows:

> "Reimagining fashion for the good of all."

This is a big promise, considering the problems associated with fast fashion along the entire value chain (see Sect. 4.4.1.2). The sustainability-oriented strategy here is: *do. MORE*. By implementing this strategy, *Zalando* wants to become a **sustainable fashion platform**—with a net positive impact on people and the earth. *Zalando* aims to give back more to society and the environment in the long term than it takes.

For this, Zalando (2022) has set comprehensive goals. By 2025, the **emissions from its own business activities** are to be reduced by 80% compared to 2017. By 2023, waste in **packaging** should be minimized. In addition, more materials should be reused. The use of **single-use plastic** should be completely avoided. In addition, by 2023, the **principles of the circular economy** should be applied and the lifespan of at least 50 million fashion products should be extended. For this, *Zalando* launched the so-called ***Zalando Zircle*** in 2019 (see Sect. 4.4.3.2).

These examples make it clear that the topic of "sustainability" has already found its way into companies' purpose definitions.

3.2 Derivation of Behavioral Rules and Leadership Styles

Purpose, Vision, and Mission must not be merely **wall decorations in the executive suites**, but must become the **guiding principle for all managers and employees** of the entire company. For this, it is essential that the top management works towards the implementation of the Purpose—and in a way that is perceptible to all. In addition, the entire incentive structure of the company must be derived from the defined Purpose, Vision, and Mission. This means that the monetary and non-monetary incentives of the company are tied to results that are helpful and necessary on the path to sustainability. In order for Purpose, Vision, and Mission to become binding and guiding, they must be translated into specific **behavioral rules** and **leadership styles** (see Figs. 3.2 and 3.3).

These guidelines for corporate action are particularly about clarifying legality and legitimacy (see Wiesner, 2016, p. 6). With **legality**, it is about the **legal permissibility of an action**, a tolerance, or an omission. What is "legal" is defined by laws, regulations, and rules (see Chap. 2). **Legitimacy** on the other hand refers to the **belief in or trust in the lawfulness of the action**. What is "legitimate" is much harder to grasp. After all, legitimacy is shaped by values, norms, and habits. These differ even within a country like Germany: Is beer a food and can therefore be consumed during working hours, or is it

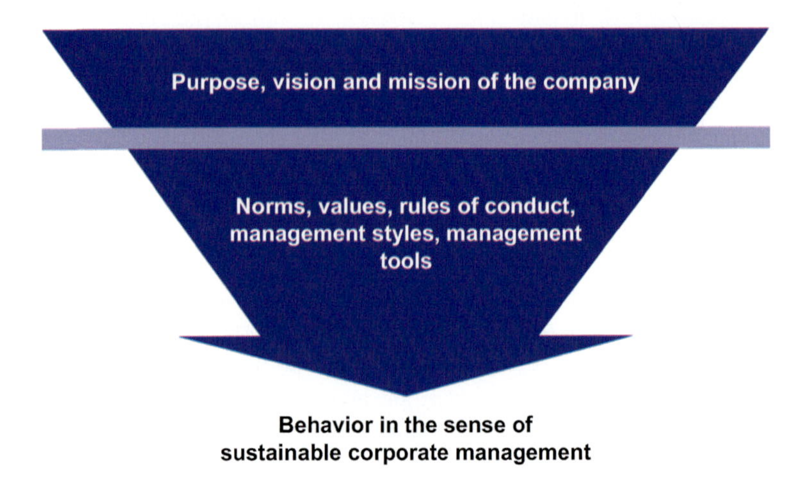

Fig. 3.2 Supporting instruments for aligning a company towards sustainability

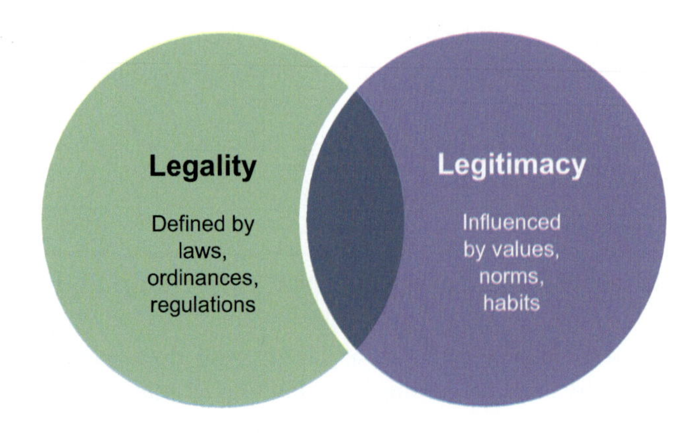

Fig. 3.3 Relationship between legality and legitimacy

simply forbidden as an alcoholic beverage? Values, norms, and habits differ significantly between different cultural circles. Here, one is reminded of the status of women in society—or the acceptance of “favors” or bribery as part of normal business interaction in many countries (see Fig. 3.2).

To make clear statements here, so-called **Codes of Conduct** are to be derived from the defined goals and decisions. This is intended to ensure that central points of orientation are incorporated into daily actions. Here, the example of *BASF* is to be used again to illustrate how such a **concretization into guidelines for action** can take place. Under Code of Conduct it says there (BASF, 2023b):

"We treat people fairly and with respect."

We treat people with fairness, consideration and respect. Our aim is to ensure that each individual feels valued, and fully supported in achieving their personal best. Always, and without exception.

Human rights, labor and social standards

We value people and respect human rights. As a company aiming for profitable growth, we have made the decision to pursue sustainable value creation, which means striving to positively contribute to the protection and promotion of human rights and people's well-being.

We commit to internationally agreed-upon standards, such as the United Nations' Universal Declaration on Human Rights, the UN Guiding Principles on Business and Human Rights, the OECD Guidelines for Multinational Enterprises and the Tripartite Declaration of Principles Concerning Multinational Enterprises and Social Policy (MNE Declaration) of the International Labor Organization (ILO).

In our own business activities, we avoid causing or contributing to adverse human rights impacts. As a participant in numerous global value chains, we are dependent on partners and demand that they likewise respect human rights and the associated international labor and social standards (ILSS). We offer to help our partners in their efforts to meet their human rights responsibilities.

Respect in the workplace

Everyone at BASF should always feel valued and respected. So, we expect every one of us avoid saying or doing anything that is humiliating, condescending, offensive or otherwise disrespectful to our colleagues. This is fundamental for our motivation and dedication at work.

We built our team on talent and appreciate the differences in our team because they make us stronger and are essential to our success. We promote an inclusive environment that embraces diversity of all kinds, including a wide variety of backgrounds, thoughts, perspectives, demographics, ethnicities, and origin.

We do not tolerate discrimination or harassment against anyone based on grounds such as age, race, color, sex, sexual orientation, gender identity/expression, national origin, religion, disability, genetic information, or any other personality traits or preferences. This governs all our employment decisions such as recruiting, hiring, promotions, benefits, disciplinary actions or terminations.

We value the health and safety of people above all else.

Wherever we do business, we act responsibly—not just complying with all relevant regulations but going the extra mile to reduce risks and minimize our environmental impact.

Environmental protection, health and safety

We are committed to energy efficiency and climate protection, and are constantly working to develop sustainable solutions for our business operations and for our customers. Across all our operations and in close collaboration with our suppliers, we work to achieve the highest standards of health and safety, and to maintain the trust of our employees, customers, business partners and other stakeholders. In the event of any kind of incident or emergency, we are well prepared to take whatever action is necessary.

We support, among others, the UN Sustainable Development Goals Climate Action (SDG 13), Responsible Consumption and Production (SDG 12) and Zero Hunger (SDG 2). And to this end, we work continuously to reduce the greenhouse gas emissions from our business activities and reduce emissions to water from our production processes.

We aim to use products and resources in the best way possible across the entire value chain and act responsibly when we operate our plants and when our goods are transported around the world.

A business that demands we take extra care

In terms of environmental protection, health and safety, we are aware that the nature of our business demands that we take exceptionally good care to reduce risks and prevent accidents. The protection of people and the environment is our top priority. Our core business—the development, production, processing and transportation of chemicals—demands a responsible approach. We systematically address risks with a comprehensive ***Responsible Care Management System***. We expect our employees and contractors to know the risks of working with our products, substances and plants and handle these responsibly.

We review the safety of our products from research and development through production and all the way to our customers' application. We continuously work to ensure that our products pose no risk to people or the environment when they are used responsibly and in the intended manner.

A commitment shared with our partners and suppliers

We do not just set ourselves ambitious goals for safety and security, health and environmental protection; we expect our business partners to aim equally high. In particular, we count on our suppliers to be fully engaged with these goals, and work with them to improve their sustainability performance."

Such **Codes of Conduct** provide all managers and employees with more concrete points of orientation for daily actions. If these guidelines are communicated aggressively and their compliance is demanded day by day, they can become the central **framework for action** for everyone.

> **Note Box** Codes of Conduct make a crucial contribution to ensuring that the values and norms of a company are also taken into account in daily actions.

In the context of sustainable corporate management, it is also necessary to examine what type of leadership should be used. Here, two **leadership styles** are particularly relevant (cf. Kreutzer, 2021, pp. 452–456; Scholz, 2014, pp. 1077–1199):

- **transactional** (i.e., exchange-oriented) **leadership style**
- **transformational** (i.e., changing) **leadership style**

A **leadership style** encompasses all actions and behaviors with which a supervisor confronts his employees and which he uses to achieve certain results. In the case of the **transactional leadership style**, leadership takes place in the sense of an exchange process or a trade between managers and employees (see Fig. 3.4; see Morhart et al., 2012, p. 398). Here, individual **transactions are at the center.**

The underlying principle of the **transactional leadership style** is: **do ut des** ("I give so that you give."). The supervisor defines the expectations and goals, while the employees receive a reward in return for achieving them. It relies on goal agreements, against which the performance of the employees is measured at regular intervals. This **Manage-**

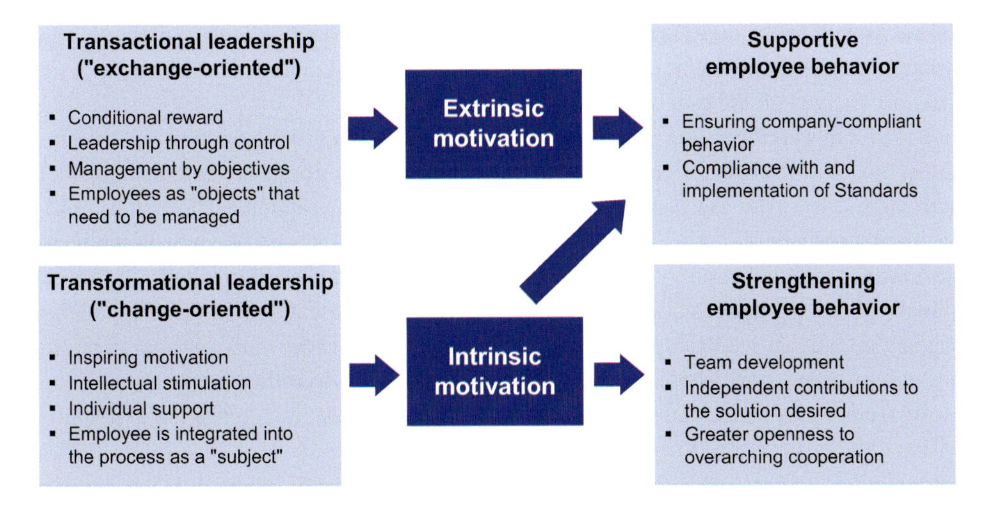

Fig. 3.4 Modes of action of transactional and transformational leadership

ment by Objectives style of leadership is often supplemented by a **Management by Exception** (i.e., "leadership in exceptional cases"); here the supervisor only intervenes in case of significant deviations from specifications.

Companies that rely on a **transactional leadership style** set standards for how employees should behave. This includes the mentioned codes of conduct. This makes it unmistakably clear to the employees what tasks and behaviors are expected of them. Depending on performance, appropriate positive or negative consequences can be expected (see the upper course in Fig. 3.4). Companies that rely on a transactional leadership style set behavioral standards for sustainable corporate management, such as how employees should behave. This includes regulations for home office obligations and "recommendations" for commuting by public transport as well as for dealing with resources within the company itself. This makes it unmistakably clear to the employees what tasks and behaviors are expected of them (see Morhart et al., 2012, p. 392).

In the case of the **transformational leadership style**, the transformations of performance delivery are at the center (see the lower course in Fig. 3.4). These are **change and transformation processes.** The transformational leadership style focuses on the "soft" factors and uses the insight that employees can also be motivated by the prospect of self-realization. The transformational leadership style aims to transform the needs and goals of the employees so that they put their own interests behind the company's goals. Consequently, managers and employees do not face each other as opponents, but as **supporters** in pursuing the common goals derived from purpose, vision, and mission.

This is achieved primarily by leaders who convey a **meaningful purpose** and a **compelling vision**, act as role models themselves, and actively support the intellectual and personal development of the employees. If employees are put at the center of the ecological as well as the digital transformation process of companies, a higher affective, i.e., emotion-based, commitment to the company can be achieved than with a leadership style that only focuses on the execution of narrowly defined tasks.

For a high-performance corporate leadership—not only with regard to sustainability goals—a **hybrid of the models** is recommended to combine the advantages of both approaches (see Fig. 3.4). Through components of transactional leadership, **goal-oriented behavior** can be enforced to a certain extent. Thus, the desired **behavioral standards** can be ensured in all areas (e.g., in the form of codes of conduct). However, guidelines and regulations should only be emphasized to the extent that the components of transformational leadership can also have an effect. The leader must succeed in building **commitment and identification (also) with the sustainability goals** through the "soft" factors of transformational leadership. At the same time, a high level of **self-motivation** must be promoted to achieve these goals.

> **Note-Box** Through **transactional leadership**, a **goal-oriented behavior** and the consideration of **behavioral standards** can be ensured. The **transformational leadership** aims to promote **commitment** and **identification** (also) with the **sustainability goals** and increase **self-motivation**.

Questions you should ask yourself

- Does our company have a purpose, vision, and mission—in written form?
- Are these communicated comprehensively internally to become action-guiding?
- Do purpose, vision, and mission aim at sustainability in an ecological, social, and economic sense?
- What needs to be done to give sustainability a greater importance in the company?
- How well known are the starting points of the circular economy in our company?
- Are codes of conduct formulated to support the transfer of purpose, vision, and mission into daily action?
- Has it been worked out in workshops how purpose, vision, and mission can be taken into account in everyday work?
- Are aspects of transactional leadership and transformational leadership used in parallel to achieve (also) sustainability goals?
- Is there a training agenda to work towards a more sustainable design of purpose, vision, and mission and their implementation in daily business?
- Where is the responsibility for these questions anchored in the company?

References

BASF. (2023a). Wir schaffen Chemie für eine nachhaltige Zukunft. https://www.basf.com/global/de/who-we-are/sustainability.html. Accessed 2 Jan 2023.

BASF. (2023b). Our code of conduct. https://www.basf.com/global/en/who-we-are/organization/management/code-of-conduct.html. Accessed 2 Jan 2023.

Illner, K. (2021). *Purpose, Sinn und Werte*. Haufe.

Kreutzer, R. T. (2021). *Toolbox für Digital Business. Leadership, Geschäftsmodelle, Technologien und Change-Management für das digitale Zeitalter*. Springer Gabler.

Morhart, F., Jenewein, W., & Tomczak, T. (2012). Mit transformationaler Führung das Brand Behavior stärken. In T. Tomczak, F.-R. Esch, J. Kernstock, & A. Herrmann (eds.), *Behavioral Branding. Wie Mitarbeiterverhalten die Marke stärkt* (pp. 389–406). Springer Gabler.

Scholz, C. (2014). *Personalmanagement. Informationsorientierte und verhaltenstheoretische Grundlagen,* (6. edn.). Vahlen.

Sinek, S. (2011). *Start with why: How great leaders inspire everyone to take action*. Pinguin Group.

Wiesner, K. (2016). *Faires management und marketing*. De Gruyter.

Zalando. (2022). Nachhaltigkeit. https://corporate.zalando.com/de/nachhaltigkeit. Accessed 24 Mar 2022.

4 Identification and Exploitation of Sustainability Potentials in One's Own Value Chain and in the System of Value Chains

We should always remember that strategies ultimately lead to the desired results!

Manager Wisdom

Abstract

To achieve sustainability in a company, the entire value chain must be aligned with the circular economy. This can be facilitated by anchoring it in a system of value chains. For sustainable companies, a multitude of action options are available. This starts with the sustainable design of the offerings and extends through supplier management, production processes, to the design of logistics and distribution. Where emissions cannot be avoided, compensation solutions are available. In addition, the company's own business model can be rounded off with sustainability solutions—or entirely new business models focused on sustainability can be developed. Concurrently, involvement in relevant organizations can complement sustainable corporate management.

In recent years, the pressure on all companies to act more sustainably overall has intensified. This is due to the **increasing environmental awareness of customers,** the **higher environmental sensitivity of investors** as well as the diverse **legal requirements regarding sustainability and transparency of business operations.** The increase in environmental disasters and their (financial) impacts have also increased the pressure on companies to act (see Chaps. 1 and 2).

In responding to these challenges, companies can pursue very different strategies. Modifying a concept of crisis management by Sapriel (2019), **sustainable corporate management can be understood as a cultural issue** (see Fig. 4.1).

© The Author(s), under exclusive license to Springer Fachmedien Wiesbaden GmbH, part of Springer Nature 2024

R. Kreutzer, *The Path to Sustainable Corporate Management*,
https://doi.org/10.1007/978-3-658-43974-3_4

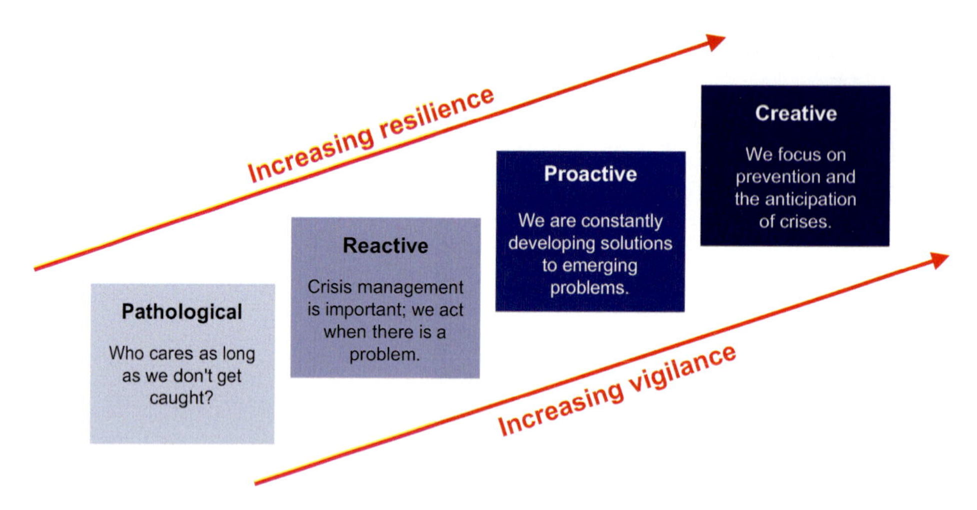

Fig. 4.1 Sustainable corporate management as a cultural issue

Every company is called upon to examine how it wants to act in sustainable corporate management. This examination can be based on various guiding principles:

- **Pathological approach**
  The company acts without regard for sustainability until it gets caught. Then it laments or complains and, if necessary, pays or works off imposed penalties. Point improvements are then implemented—but only as far as absolutely necessary. And until it gets caught again ...
- **Reactivestrategy**
  In such a corporate culture, a reaction is only made when a problem occurs. Here, crisis management is relied upon. Companies implement legal obligations as late as possible—and only these. In the reactive strategy, nothing is implemented in advance due to a lack of awareness for considering sustainable aspects. Only when there is pressure to act from within or outside is a reaction made.
- **Proactiveapproach**
  With a proactive approach, solutions for emerging problems are developed early on. This can relate, for example, to the availability of resources (oil, gas, water, electricity, employees). In addition, the often multi-year legislative processes are actively followed in order to prepare early for new legal regulations. At the same time, work is already being done on more sustainable processes as well as products and services long before a company is obliged to do so.
- **Creativeapproach**
  Companies already take hints ("Weak Signals") of possible crises as a stimulus to develop creative solutions to their management early on. Here, no" one waits for new

answers to be called for or for corresponding obligations to be defined by legal frameworks. The company becomes creatively active and may even cannibalize its own performance areas in order to find sustainable answers for the future (see in depth Kreutzer, 2021).

> **Note Box** The greater the **vigilance** in the company regarding the issue of sustainability, the more proactive and creative a company can act. This also increases the **resilience,** i.e., the resistance of the company.

But how can a company become "green" (sustainable) and still make a profit if it wants to act not only reactively, but above all proactively or even creatively? For this, the **(global) value chains** must be moved to the center of the sustainability discussion.

4.1 Value Chain and System of Value Chains

An indispensable starting point for identifying and exploiting sustainability potentials is the own **value chain**. Its basic concept was developed by *Michael Porter* (see Porter, 2004, pp. 59–92). Here, we also speak of **value chain** and **Value Chain**. By using a value chain analysis, the following goals can initially be achieved:

- **Identification of the causes of competitive advantages**—both in one's own and in other companies
- **Recognition of potentials for achieving competitive advantages for one's own company** (focus: sources of additional customer benefit, starting points for improving one's own cost situation and/or for developing new business models)

In addition, the following goal is now to be defined:

- Analysis and exploitation of sustainability potentials within one's own value chain

In the first step of value chain analysis, the aim is to identify **starting points for achieving competitive advantages** for one's own company. The value chain analysis focuses on the core of the business model. The central elements of a business model's value chain include areas that particularly contribute to the differentiation of the offer in the eyes of the customers. In addition, attention should be paid to the fields of the value chain that are associated with particularly high costs. These activities can come from various company areas. Each of these activities is initially assigned to one of the following classes:

- **Core processes** (also called **directactivities**)
  The activities to be assigned to the core processes are directly involved in value creation for the customer.
- **Supporting processes** (also called **indirect activities**)
  These activities only indirectly contribute to performance creation. A classic example of this is the company infrastructure, which can include the HR department, controlling or the R&D department.
- **Quality assurance**
  This category includes the activities that contribute to ensuring high quality in various company areas. These include, for example, quality tests as well as the ongoing monitoring of production and the continuous analysis of marketing and sales activities. The identification and verification of the exploitation of sustainability potentials is also located here. The steps for exploitation itself are to be directly integrated into the core processes and the supporting processes.

When assigning activities to one of these categories, the specific company situation and the industry being analyzed must always be taken into account. Therefore, the assignment shown in Fig. 4.2 is only an exemplary implementation (see Porter, 2004, p. 62).

> **Note Box** The assignment to supporting processes and core processes must be made in each company according to the respective priorities.

The **value chain** is based on Porter's core idea (2004, p. 63) that every company can be described as a **collection of activities** through which products or services are developed, produced, communicated, distributed and delivered. The company-specific design of the

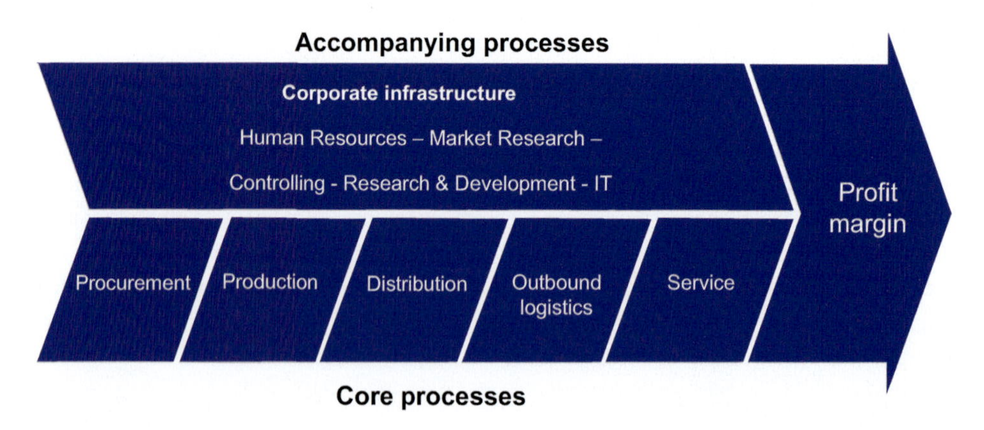

Fig. 4.2 Basic concept of a value chain

value chain has a direct impact on the achievable **profit margin** of the company (see Fig. 4.2).

Any change in the design of the accompanying processes, but especially the core processes themselves, influences this profit margin. This results from the difference between the revenue generated and the costs used to create value. The more efficient or effective the use of resources, the greater the **value creation** achieved for the company in terms of the achievable profit. This makes it clear that any type of activities along the value chain can be a source of competitive advantages. The leverage is particularly large in the core processes.

Most often, a company's value chain is not (any longer) used in isolation. The value chains of companies are often linked in various ways with the upstream and downstream value chains of suppliers and customers. Together they form a **system of value chains** (also **value creation network;** see Fig. 4.3; Porter, 2004, pp. 59–61).

How intensively such networks have already been established or will be established in the future varies greatly from industry to industry. Every company is called upon to identify **optimization potentials**—also with regard to sustainability—that can be achieved by linking with upstream and downstream value chains. This aspect of linking different value chains through a **computerization of manufacturing technologies** is at the core of the developments of **Industry 4.0**. The goal here is often to develop an "intelligent factory" **(Smart Factory).** The **informational integration of suppliers and customers** is intended to enable or facilitate the company's own ability to respond more quickly to changes in supply and demand. This also applies with regard to violations of the requirements of the supply chain law and supports the achievement of circular economy goals (see in depth Kreutzer, 2021).

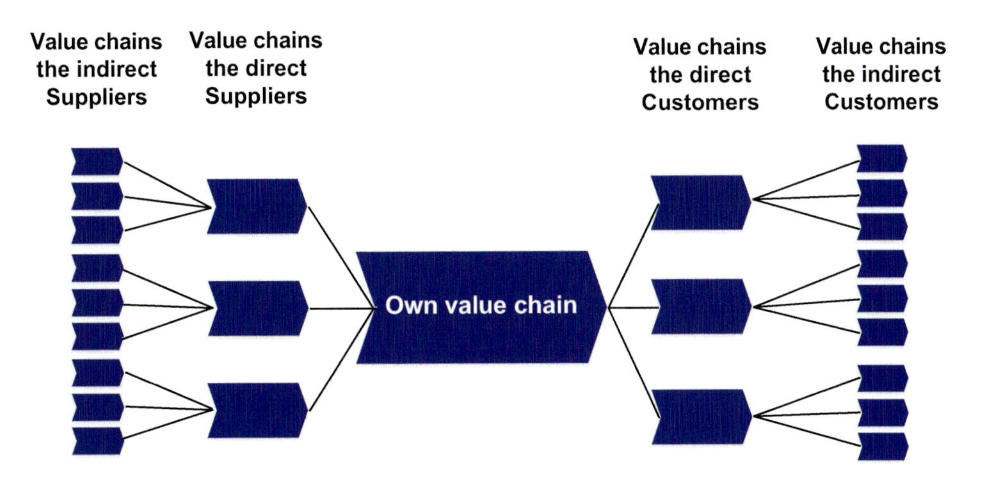

Fig. 4.3 System of value chains

4.2 Exploring Sustainability Potentials in One's Own Value Chain

The **value chain analysis** can focus on a product group, a service area, or individual strategic business units in larger companies. Only such a focus allows specific starting points for improving one's own competitive position to be identified.

In the **first step of the analysis**, the following areas of questions are addressed to capture the **status quo of the company**:

- Which activities are to be distinguished in the context of value creation?
- Which of these activities represent core processes, accompanying processes, or quality assurance processes?
- What costs are associated with the different activities?
- To what extent do these activities contribute to improving the competitive position?
  - What customer benefit is generated?
  - What cost advantages are achieved?
- Are the activities common in the industry? If not, do they generate a visible customer benefit and/or cost advantages?
- Are the activities of the own value chain optimally coordinated and linked with each other?
- Are there overlaps or avoidable dependencies?
- Are potential synergies unused?
- Is the own value chain aligned with the value chain of the own suppliers?
- Does the value chain integrate those of the own customers?
- What sustainability potentials can be identified in the own value chain?
- How comprehensive and early are compliance and violation of sustainability requirements recognized and recorded?
- To what extent are concepts of the circular economy already implemented in the own company?
- How does the own company act in terms of "sustainability"—pathologically, reactively, proactively and/or creatively?
- By which KPIs can compliance with sustainability guidelines be checked?

The answers to these questions provide clues to the **optimization of the cost structure** or to the **exploitation of differentiation potential** for achieving competitive advantages. In addition, **sustainability potentials** can be systematically identified. The value chain analysis can—by comparing one's own value chain with that of competitors—provide further important clues for further development. Benchmarking can also contribute to this. This attempts to identify best-of-breed solutions in companies outside of one's own industry in order to develop creative solution approaches.

In this context, in the **second step of the analysis**, in direct **comparison with relevant competitors or with benchmarks from other industries**, answers to the following questions can contribute:

- What options are there for designing the value chain within or outside of one's own industry?
- How is the same value creation process designed by competitors?
- What connections have competitors already made with upstream and/or downstream value chains?
- What costs of a value creation stage are compared to which competitive advantages in one's own company—and how is this to be assessed in competition?
- What costs of a value creation stage are compared to which customer advantages in one's own company—and how is this to be assessed in competition?
- Which value creation stages must be performed by the company itself and which can be outsourced (to suppliers, to outsourcing partners or to customers)?
- How do competitors and other companies exploit sustainability potentials?
- What (new) technologies are involved in the exploitation of sustainability potentials?
- To what extent are concepts of the circular economy already being implemented in other companies?
- How do competitors act and how do particularly successful companies act in terms of "sustainability"—pathologically, reactively, proactively and/or creatively?

Based on the differences identified here in the design of value chains or value creation chains, in the **third step of the analysis**, concrete fields of action for improving one's own competitive position can be identified. Here, clues for **reducing costs** are to be identified, for example by identifying product features and/or service fields that generate costs but no relevant customer benefit. This can save resources. There may also be clues as to how costs can be reduced by reducing the variety of variants, by reducing the depth of value creation, by modular production or by merging production volumes. This can also save resources.

Secondly, **fields of action for differentiating one's own services** in competition can be identified. Finally, **clues for optimizing the interfaces of value chains**—within and outside one's own company—can be identified. This can reveal activities that should appropriately be delegated to suppliers or customers outside of one's own area of responsibility. In addition, activities that have been outsourced so far may be identified that should be reintegrated in a value-adding and/or cost-reducing manner. In all these steps, the effects on resource consumption must be recorded.

Thirdly, very concrete **approaches to exploiting the sustainability potentials** are to be identified. The following questions need to be answered:

- Which **concepts of the circular economy** can be used (additionally)?
- Is attention paid to a **sustainable design** in the conception of products and services (including packaging)?
- To what extent can **renewable energies** be used?
- Can the **reuse and recycling** of valuable materials or products be expanded?
- Is **refurbishing** used to extend the life cycle of products or entire systems through **quality-assured overhaul and repair**?
- Can used devices and systems be brought up to the quality standard of a new device through **refabrication** or **remanufacturing**?
- Can a **decomposition** be carried out to feed modules of a device or system into another use cycle?
- Can own products be subjected to **upcycling**?
- To what extent can **recycling** be promoted for the reprocessing of valuable materials?

> **Note-Box** The value chain analysis is an important tool for identifying cost differences and differentiation points between various companies and indicators for sustainable corporate management. In this process, one's own business model is compared in depth with that of competitors. In addition, creative impulses are gained by looking at the value chains of other industries.

These fields of action illustrate that value chain analysis is closely linked to the design of the business model. An important supplementary contribution to the design of the action options determined here can be provided by the insights gained through the use of a SWOT analysis. The basics of the **SWOT analysis** can be found in Fig. 4.4 (see Kreutzer, 2021, pp. 86–116).

Which questions need to be answered in the synthesis of the external and internal perspective in the SWOT analysis is shown in Fig. 4.5.

An exemplary result of a **SWOT analysis for exploring the sustainability potential** of a company can be found in Fig. 4.6. Here it becomes visible that not all opportunities in the market can be exploited if these opportunities meet weaknesses. At the same time, risks can be well managed if they can be mastered by the strengths of the company.

> **Note-Box** The SWOT analysis leads to further exciting insights to identify sustainability potentials. At the same time, it can be determined to what extent a company is capable of exploiting them.

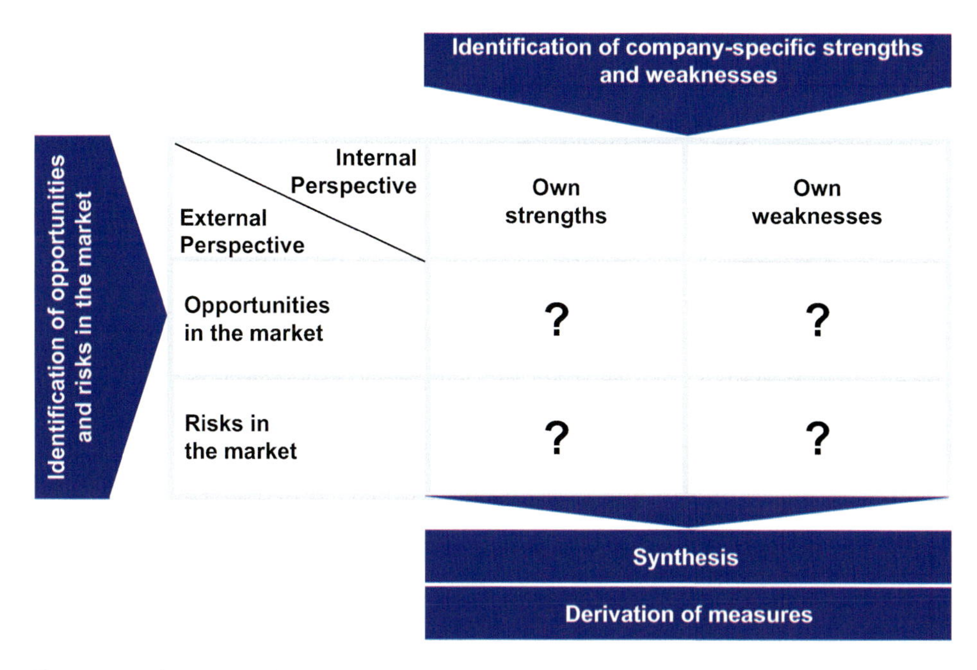

Fig. 4.4 Basic concept of the SWOT analysis

External perspective / Internal perspective	Own strengths	Own weaknesses
Opportunities in the market	What opportunities in the market can we make the most of based on our strengths?	Which opportunities in the market are we unable to benefit from due to our weaknesses?
Risks in the market	Which risks in the market can we benefit from due to our strengths?	Which market risks hit us particularly hard due to our weaknesses?

Fig. 4.5 Synthesis of the external and internal perspective in the SWOT analysis

4.3 Installation of a Digital (Informational) Value Chain

The **development** of the **Internet of Things** or the **Internet of Everything** has a significant influence on the design of the value chain of many companies. The components of the Internet of Everything can be found in Fig. 4.7.

What developments are emerging in the Internet of Everything?

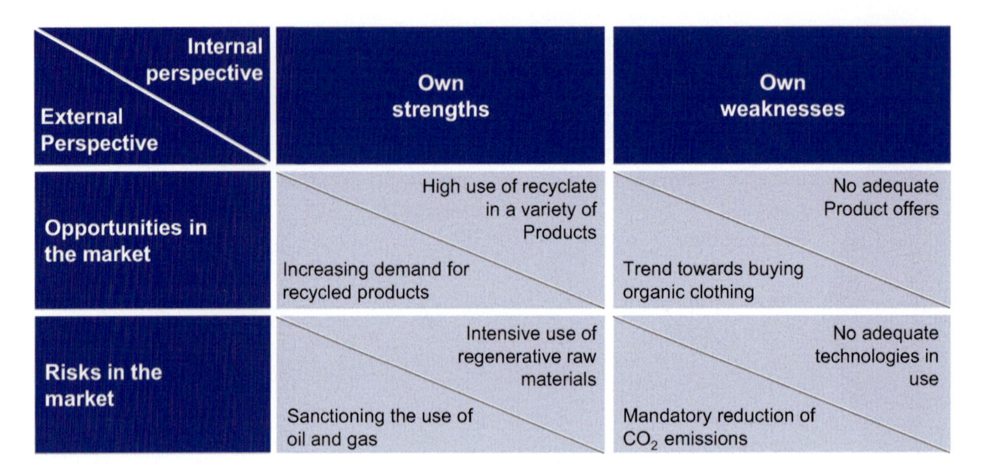

Fig. 4.6 Synthesis of opportunities/risks and strengths/weaknesses

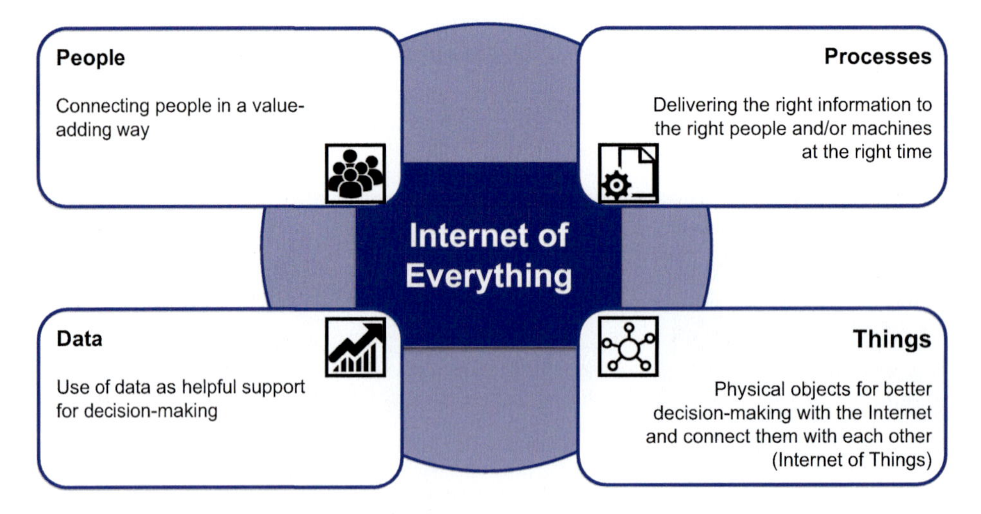

Fig. 4.7 Components of the Internet of Everything

- **Things**
  The starting point of the Internet of Everything was the **Internet of Things** (IoT; cf. Firouzi et al., 2020; Chui et al., 2010). The entry into this development was created by more and more objects becoming "smart" **(Smart Devices** or **Connected Devices).** "Smart" means that these objects are connected to the internet. Today, dolls, toothbrushes, watches, refrigerators, washing and coffee machines, speakers, radios, televisions, as well as shutters, heaters, air conditioners, escalators, elevators, and cars have an internet connection. In addition, there are many millions of machines and produc-

tion facilities that are also connected to the internet. The latter is referred to as the **Industrial Internet of Things** (IIoT).
The number of **Internet-of-Things devices** worldwide is expected to almost triple from 9.7 billion in 2020 to more than 29 billion IoT devices in 2030. In 2030, the highest number of IoT devices will be found in China, with about five billion devices in the hands of consumers. IoT devices are used in all industries, with the consumer segment accounting for about 60% of all IoT-connected devices in 2020 (cf. Statista, 2022b).
The most important **industries** with currently more than 100 million connected IoT devices are the electricity industry, the gas, steam and air conditioning supply, the water supply and waste management, the retail and wholesale trade, transport and storage, and public administration. Other use cases with more than one billion IoT devices by 2030 include connected (autonomous) vehicles, IT infrastructure, plant monitoring and control, and smart grids (cf. Statista, 2022b). The biggest **driver of networking intensity** are the **industrial applications** of the Industrial Internet of Things.
- **Processes and Services**
The internet not only offers the possibility to create a **connection of one's own value chain with various supplier and production stages**. The internet also enables a comprehensive **connection to end customers,** to network them with the production process. Today, companies are much more encouraged than before to think and act upstream and downstream in the design of their processes!
Upstream describes the procedural and informational penetration of upstream production stages. This possibility is of particular importance in the implementation of the requirements of the supply chain law (cf. Sect. 2.4). **Downstream** refers to this process towards the end customer. Up- and downstream concepts provide interesting starting points for the establishment of a circular economy (cf. Sect. 1.2). In this way, ecosystems can be created in procurement, research & development, production, logistics, and marketing that provide value-added offers for the respective partners involved. Comprehensive networking often achieves higher speed, greater accuracy, and thus increased relevance of offers. At the same time, the associated **closed information cycles** provide important prerequisites for sustainable corporate management.
- **Data**
Through the use of **sensors** and other **measuring instruments** in all areas of human life and across all stages of value creation, not only the quantity of data increases, but also their quality and the speed of provision. After all, many of these data will be available in real time. This allows important data to be obtained in the course of extraction, manufacturing, further processing, marketing, and usage processes. These are indispensable for many questions of sustainable corporate management.
To this end, the companies involved must use **Big Data Analytics** or **Business Analytics** to avoid drowning in the information tsunami. Only then can it succeed in gaining the insights indispensable for decisions (cf. Kreutzer, 2021, pp. 294–337).

- **People**
  Today, people have the ability to be permanently connected to the internet via laptops, smartphones, tablet computers, smart glasses, smart watches, and other so-called wearables (portable devices with internet connection). Many people are online 24/7 in this way—and often also addressable 24/7. This not only changes information, communication, purchasing, and learning behavior, but also the accessibility during the customer journeys. This allows further data to be collected about the acquisition and use of products and services, which are relevant for expanding the sustainability of one's own services. If the metaverse—the so-called walk-in internet—prevails, even more people will be online for longer. The data stream available for business applications will then increase significantly again (cf. in depth Kreutzer & Klose, 2023).

This comprehensive digitalization makes it equally possible—as in many cases also necessary—to supplement the classic value chain with a **digital (informational) value chain**. This is especially indispensable with regard to achieving sustainability goals. Figure 4.8 shows how the physical value chain is permeated and enriched by a digital value chain.

The digital value chain promotes the **informational networking** between the internal company areas and contributes to overcoming the often still existing silo mentality in internal and external relations. This silo mentality results in each area working only for itself and no thinking and action taking place in networks. The digital value chain creates

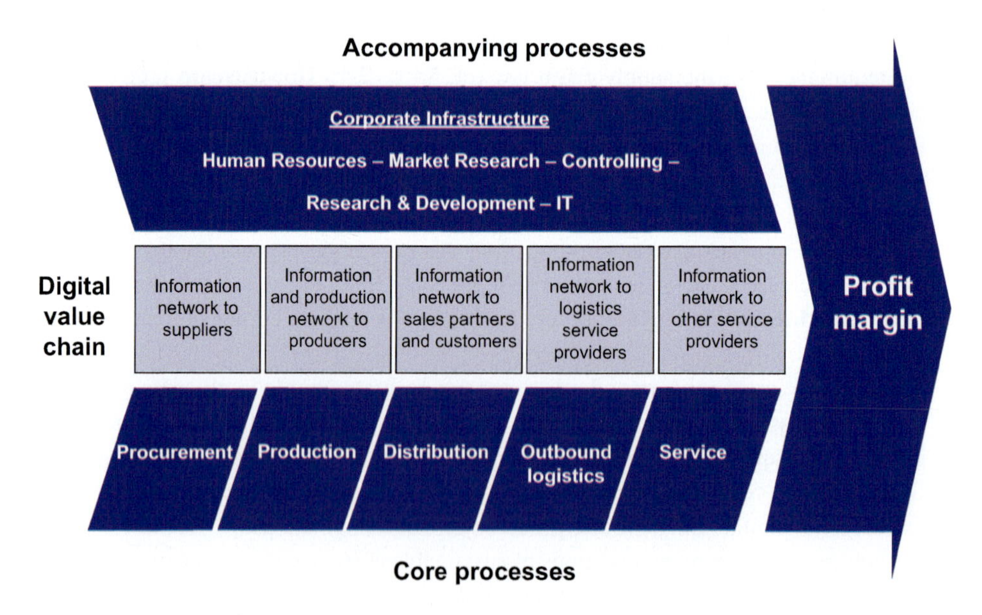

Fig. 4.8 Merging of physical and digital value chain

the (informational) prerequisites for enabling the already described **development of systems of value chains**.

A **digital value chain** is often the prerequisite for being able to identify and handle potentials for profit increase, cost reduction, and sustainability in the system of value chains at all. Often, essential sustainability goals can only be achieved through networking beyond one's own company.

> **Note Box** The **physical value chain** is increasingly permeated by a **digital value chain**. This can realize various efficiency and effectiveness reserves in value creation. In addition, sustainability potentials can be identified and exploited. Moreover, this analysis provides clues for further development or supplementation of the existing business model.

4.4 Action Options for Sustainable Corporate Management

Companies can choose between various **action options for a sustainable corporate management** (see Fig. 4.9):

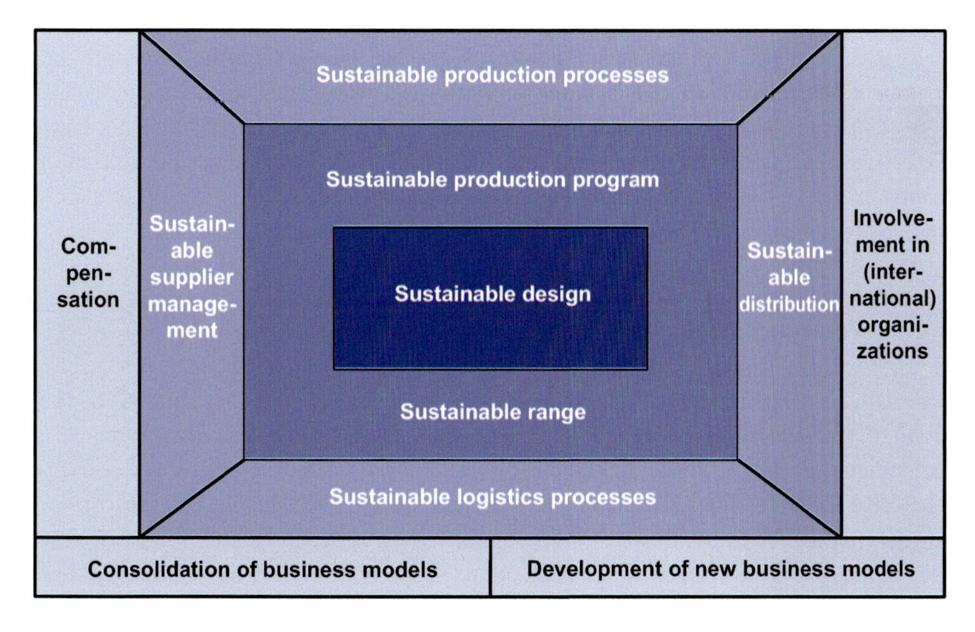

Fig. 4.9 Action options for sustainable corporate management

- Focusing on individual areas of the circular economy (including sustainability in design, production program/assortment, production, suppliers, logistics, sales)
- Rounding off the existing business model with concepts of the circular economy
- Development of new business models for the circular economy
- Provision of compensation services
- Engagement in (international) organizations to promote the circular economy

When selecting the appropriate action options, it should be taken into account to what extent a company is already striving to implement concepts of the circular economy or sustainability in general. The **focus on individual areas of the circular economy** is advisable when companies first approach this topic. In a later phase, it can be checked whether a **rounding off of the existing business model with concepts of the circular economy** can be done credibly.

Established companies and start-ups are—in parallel—called upon to examine the possibility of **developing new business models for the circular economy**. There is no longer any doubt about the urgency of getting the circular economy moving. The more politics and investors focus on this issue, the more comprehensive the financial resources available for corresponding company foundations will be.

Companies usually cannot avoid all negative emissions at once. In addition, there are production processes where such emissions cannot be avoided in the long term or not comprehensively. In both cases, companies can strive to balance harmful emissions through **compensation services**.

Regardless of the aforementioned steps or in parallel, **engagement in (international) organizations promoting the circular economy** can be advantageous for companies. This contributes—ideally—positively to the corporate image. On the other hand, larger projects can be successfully mastered through cooperation. In addition, valuable information exchange can take place to identify and utilize further fields of action in the circular economy early on.

4.4.1 Focus on Individual Areas of the Circular Economy

Especially when entering sustainable corporate management, the **focus on individual areas of the circular economy** is a sensible approach. Here, it is examined which of the options sustainable design, reduction of waste in production and distribution, use of renewable resources, optimization of the life cycle through sustainable use, reuse, refurbishing, remanufacturing, decomposition, recycling or upcycling should be used. Also, possibilities for optimizing waste collection and waste recovery are to be examined here (see Fig. 1.8).

In addition, the compensation of harmful emissions plays a role. However, **compensation** is not part of the circular economy. After all, nothing is returned to the system in compensation. Compensation can only be an intermediate stage on the way to **Net Zero**.

> **Note Box**
> The order of priority in sustainable corporate management is—based on the **5-R rule:**
>
> - **Refuse**
> - **Reduce**
> - **Reuse**
> - **Repurpose**
> - **Recycle**
>
> In addition, possibilities of the **compensation** should be examined. However, compensation is not part of the circular economy because no cycles are closed through compensation.

Starting points for the sustainable design of marketing and sales are discussed separately in Chap. 5 due to their great relevance and not here as part of the value chain.

4.4.1.1 Sustainable Design

In **sustainable design**, already in the creative phase, it is taken into account which resources are needed for development, production, packaging, logistics, sales, consumption or use, and disposal. Here, a view of the entire life cycle from the perspective of sustainability is already taken at the birth of a product or service. Such action is or would be a sign of truly comprehensive corporate responsibility.

> **Note Box**
> The **goal of Sustainable Design,Eco Design,Eco-Design** or **Green Design** is to comprehensively capture the economic, ecological, and social effects of a product or service and improve them for the benefit of all stakeholders (including the environment). The aim is to provide customer benefits with as little resource use and as few emissions as possible.
>
> A sustainable design ensures that **sustainability** is already embedded in the **core of the offer**. The materials used also contribute to this. On the one hand, they can increase the lifespan of the products. On the other hand, recycling can be enabled or at least facilitated by decomposability and the use of mono-fibers, mono-materials, and non-toxic surfaces.

In order to encourage companies towards a more sustainable design, an **Ecodesign Directive** has already been drafted. This is intended to help reduce the negative impacts of **energy-related** products on the environment throughout their entire lifecycle. Since the products on offer often differ considerably in terms of their environmental impacts, even when their functions and performances are similar, there are significant **optimization potentials** in favor of the environment.

In order to exploit these potentials, **minimum requirements for product design** have been established. These requirements must be met when products are offered on the European internal market. Uniform EU-wide regulations are intended to prevent diverging national legal provisions from becoming trade barriers. The **Energy-Related Products Act (EVPG)** implements the EU directive into German law (cf. Federal Environment Agency, 2022a).

Which products fall under this law? **Energy-related products,** that meet the following criteria, are—with the exception of the mobility sector—generally affected by the directive (cf. Federal Environment Agency, 2022a):

- Annual sales volume in the EU of at least 200,000 units
- Significant environmental impacts of the respective product according to the strategic priorities of the community set out in Decision No. 1600/2002/EC
- Significant potential for improving environmental compatibility without excessively high costs

All other companies that do not fall under this Ecodesign Directive should also strive for a **sustainable design** of their products and services. Such an **ecodesign** can be described with the following criteria:

- **Resource-saving**
  Resources should be saved in development, production, logistics, sales, use, and disposal. Resources can be saved, for example, by using renewable energies and resources in the mentioned process stages.
- **Durability**
  The design of products should be aimed at a long usage phase. Therefore, all companies are called upon to refrain from **built-in or planned obsolescence of products**. The keyword here is "Built-in Obsolescence". In the case of **planned obsolescence**, the aging of a product—the obsolescence—is deliberately brought about by the manufacturer through the design. For this purpose, so-called **predetermined breaking points** are built into a product to artificially shorten the product's lifespan. This can be seen, for example, in the batteries of smartphones, whose performance—often after the warranty period—decreases massively. Accusations are also made against smartphone manufacturers that the battery life of the devices is artificially shortened through operating system updates in order to stimulate a new purchase.
  Planned obsolescence of products or services is also achieved by not using innovations at the time of their market readiness, but at a staggered time. This motivates customers, for example, to buy a new smartphone, a new digital camera, or a new smartwatch every six or twelve months because each contains a technological superiority. The already sold products are systematically and deliberately devalued through **technological obsolescence**—to motivate customers to buy new ones.

By the way: **Fashion** is the prime example of planned obsolescence. Fashion trends change the cuts, the colors, etc. For jeans, sometimes low-rise is in, then high waist is the trend, then flared jeans are the trend, then ripped jeans are a must-have, before bootcut, flare, and skinny jeans become mandatory. Even those who want to keep up with fashion to some extent "can't" wear certain clothes without being considered mega-out. This makes "actually" still wearable products "unwearable"! Fashion is a gigantic destruction machine for clothing items that are far from the end of their lifespan.

- **Repairability**
  How little many manufacturers are still willing to consider **possibilities for repair** in the design can easily be seen in glued-in batteries or otherwise welded products. These often cannot be opened and thus repaired without simultaneously destroying them. This has nothing to do with a **sustainable product design**.
  However, it also depends on the customers whether they already pay attention to the fact that a **repairability** is given when purchasing. After all, it is the active customers who decide which products can be successfully marketed in the long term. Perhaps soon a **"right to repair"** will be established—considerations for this already exist for the following reasons (see European Parliament, 2022; Droste & Höwelkröger, 2022):
  - Based on a *Eurobarometer* survey, 77% of consumers in the EU would prefer to repair their products rather than buy new ones. However, this is often not possible due to the often **high repair costs** and the **lack of service**, so that devices have to be replaced or disposed of.
  - **Legal termination of planned obsolescence.**
  - **Electronics** are today the **fastest growing source of waste** in the EU. Of the 3.5 million tons of old devices collected in 2017, only 40% were recycled.
  - In Germany alone, almost **400,000 tons of electronic waste** are produced each year. Every 58 minutes, electronic waste weighing as much as a medium-haul aircraft—about 42 tons—is produced in Germany. This results in electronic waste as heavy as 23 aircraft in a single day. And: Only just over a fifth of the defective devices are repaired.
  - The repair of electronic devices significantly **reduces resource consumption and greenhouse gas emissions**.
- Fair and socially provided services
  Already in the design phase, it can be taken into account which countries, which suppliers and which raw materials stand for a fair and social interaction with the involved persons. Already in this early phase, the course can be set in such a way that one becomes independent of unscrupulous partners by designing products and services differently. This also applies to the integrated logistics partners, who implement different social standards (see in depth Wiesner, 2016).

- **Possibility for Recycling—Upcycling or Downcycling**
  The different recycling concepts with their different sustainability have already been discussed (see Sect. 1.2). Already in the design phase, these possibilities should be considered in order to support a circular economy as comprehensively as possible.
  An important impulse for this is to give articles a **digital product passport**. This could record which plastics and other components are included that are suitable for recycling or further use. Such a product or material passport is already being used in buildings, for example (see Sect. 4.4.1.2).

> **Food for Thought** A sustainable design can also cause customers to use a product less frequently in order to avoid emissions. That would be an exciting idea to reduce the intensive use of smartphones!

In the context of ecological design, a look must also be taken at the so-called **rebound effect**. A resource-saving design, efficient production and the anchoring of products and services in the circular economy can lead to an undesired result: Customers change their behavior and demand such an offer more strongly or use it more intensively. As a result, the advantages of such an offer aimed at sustainability can be partially or completely compensated. In this sense, rebound stands for rebound or recoil.

Selected examples are intended to illustrate this rebound effect. In recent decades, the **consumption of fuel** per 100 kilometers driven in cars has been continuously reduced due to technological improvements. This should have resulted in an overall reduction in fuel consumption. However, counteracting effects occurred. As fuel consumption decreased, more car trips were made and longer distances were driven. Conversely, walking, cycling, or using public transportation were often foregone. In addition, more powerful engines could now be used, so that the fuel consumption per 100 kilometers driven ultimately increased. A classic rebound effect. Who would be satisfied today with a *VW Beetle* with 34 HP (25 kW) that took 30 seconds to accelerate from "0" to "100", consumed about nine to ten liters of gasoline at a top speed of 115 km/h—and with which I started my journey into automobility in 1976?

In addition to the **direct rebound** described here, other—environmentally negative—effects can occur. If driving costs less, for example, more people can afford an additional flight, which further increases emissions. In this case, we speak of an **indirect rebound.**

A direct rebound effect can also occur when **purchasing an electric vehicle**. When a customer buys such a vehicle, they have chosen a supposedly more sustainable product—at least if the electricity used comes from renewable sources. If the vehicle is now used much more intensively due to a "better feeling" (keyword **"Feelgood factor"**; see Sect. 1.3.4), and public transportation is foregone, a rebound effect also occurs. The positive effects of the "more sustainable" product are reduced by the buyer driving much more. This is also referred to as **backfire.**

A direct rebound effect also occurs when customers are increasingly drawn to products such as **plastic bottles due to their recyclability**. Then more plastic bottles are sold, even though the recycling issue has not been resolved. Avoidance would be a much better alternative in most cases.

4.4.1.2 Sustainable(er) Production Programs

After the development of eco-design for individual products and services, the question arises as to how the company's own offering can be made more sustainable overall. When we talk about a **production program**, we are on the **manufacturer's side.** Then the question is which products and services should be further developed or newly included in the company's own production program with regard to sustainability goals. When it comes to **assortment design**, we are on the **retail side.** Here the question is which offers are presented to customers in what form by the retailers (see in depth Sect. 4.4.1.6).

How a more sustainable orientation of the production program can be achieved is demonstrated below using selected examples. Almost every hour, more companies are joining in who want to make (parts of) their offerings more sustainable.

Activities in the Industrial Sector

Figure 1.33 clearly shows that the **industry** is already the second largest **source of harmful emissions** after energy production. Therefore, there are many initiatives to reduce the extent of production-related emissions—for example, by making the production program more sustainable. An example of this is the **steel industry,** which emits particularly many harmful greenhouse gases. Therefore, companies like ***Thyssenkrupp*** are working on solutions to produce **green steel**. The goal has been clearly formulated: Steel production at *Thyssenkrupp* is to become climate-neutral by 2045. This demonstrates the company's willingness to take on social responsibility. It also explicitly commits to the *Paris Climate Agreement* of 2015.

To achieve this major goal, emissions from production and processes at *Thyssenkrupp* as well as the emissions of the energy consumed are to be reduced by 30% by 2030 compared to the reference year 2018. For this purpose, *Thyssenkrupp* is building a facility for the production of green steel in Duisburg. *Thyssenkrupp* also aims to maintain as much as possible the high steel quality and the established production processes and value chains, even when reducing CO_2 emissions. The total cost of this project is estimated at two billion euros (see Thyssenkrupp, 2023).

In order to achieve this goal, the company is calling on politicians to create the necessary framework conditions for the upcoming transformation. This includes not only investment grants, but also operating aids for ongoing production. This is intended to at least partially offset the additional costs of a "green" production compared to conventionally produced steel (see Thyssenkrupp, 2023).

Thyssenkrupp joined the initiative ***Race to Zero*** supported by the *United Nations* in 2021. As part of this initiative, the companies involved are committed to limiting global warming to 1.5 degrees. In addition, *Thyssenkrupp Steel* has joined the global **Multi-Stakeholder Initiative*ResponsibleSteel*.** In this non-profit organization, companies, civil society groups, and associations work together to develop **standards for responsibly produced steel**. This involves looking at the entire value chain—from the procurement of raw materials to production and the recycling of steel after its use phase.

Another example—albeit much smaller—is provided by the Dutch company ***Vepa.*** The company has developed a multipurpose chair called *Blue Finn*, whose seat and backrest are made of plastic that consists of at least 85% recycled *Bluewrap* . *Bluewrap* is the packaging material for sterile instruments in hospitals. *Bluewrap* is made of 100% polypropylene and is easy to recycle. *Vepa* uses this material for a particularly innovative form of upcycling and makes chairs from it. This makes a significant contribution to directing the packaging flood caused by the pandemic into productive channels (see Vepa, 2023).

There are no limits to further creativity for **sustainable solutions in the industrial sector**. Figure 1.33 showed that the **transport sector** is the third largest source of harmful emissions. Therefore, the issue of sustainability in the **automotive industry** has been on the agenda for many years. The comprehensive focus on the **electromobility** is intended to make a decisive contribution to reducing harmful emissions. Efforts are now being made to put at least 30 million emission-free cars on European roads by 2030. This is the goal of the European Commission. At the same time, legislation is to ensure that the batteries of electric vehicles are designed to be more sustainable.

Vehicles with combustion engines currently generate about 82% of their emissions from the exhaust gases produced by the combustion of fuel. Another 18% of emissions are due to the production of materials for their manufacture. Since the usage phase accounts for the largest share of emissions, the solution for reducing emissions lies in the **electrification of the drivetrain.** However, the net-zero target defined for 2050 can only be achieved if the **sales of battery-powered electric vehicles** reach a market penetration of 100% by 2040. This explains why many countries have announced plans to ban the sale of combustion engines by 2035 (EU) or even earlier (see World Economic Forum, 2020, p. 7 f.).

In order to exploit the **potential of decarbonization** of vehicles, the focus after the electrification of the drivetrain shifts to material production. The values already mentioned—82% emissions during the usage phase, 18% during material production—will shift in 2040 to 40% emissions during the usage phase and 60% during material production. Consequently, the **decarbonization of material production** in the automotive sector then represents the greatest challenge in order to achieve a carbon-free value chain (see World Economic Forum, 2020, p. 7 f.).

To achieve such a **carbon-free value chain**, it is first necessary to build a comprehensive **transparency** about where the largest emissions occur in the production process. Then, it is necessary to win over and train partners—along the entire value chain—for more sustainable processes. At the same time, value creation teams need to be installed.

These must think and act beyond the boundaries of individual companies' areas of responsibility. Only through a cooperative approach in the **systems of value chains** described in Sect. 4.1 can ambitious emission reductions be achieved.

To promote this collaboration, various steps need to be initiated (see McKinsey, 2021):

- The partners in the value chain must agree on **industry-wide goals** and **milestones** for their achievement.
- The entire **value chain** (including the supply chain) as well as the **usage phase of vehicles** need to be rethought in their interplay. This is the only way to facilitate and promote further use of the resources installed in vehicles at the end of their usage phase (keyword "product or material passport").
- To **standardize processes and modules**, binding industry practices and standards aimed at sustainability need to be defined.

> **Note Box** To drive **decarbonization in the automotive industry,** it is important to recognize and utilize various **opportunities for pre-competitive collaboration.** This can pave the way for the necessary cross-industry changes.

It is interesting to look at how different car brands are currently rated in terms of their respective ESG score. The so-called **sustainability perception** was analyzed by *Brand Ticker* based on the ESG criteria of environment, social, and corporate governance. In this ranking, *Volvo* is at the top with a score of 53.6. The Swedish car manufacturer, which now belongs to the Chinese *Geely* group, aims to become a climate-neutral and circular economy company by 2040. This should reduce the CO_2 footprint per car over the entire life cycle by 40% compared to 2018. This is clearly reflected in the sustainability perception. How other brands score on the ESG scale is shown in Fig. 4.10 (see Reidel, 2022, p. 96).

The offer of **Product as a Service** (PaaS) can make a significant contribution to sustainability. In this business model, a company no longer sells its products, but offers them as a service. Instead of buying a product, customers pay for the use of the product, not for the product itself. PaaS contributes to sustainability because it shifts the focus from owning a product to **using a product.** Ideally, products are only provided when they are actually being used (see in depth Kreutzer, 2021, pp. 203–206).

The relevance of the usage phase is illustrated by looking at the actual **use of cars**. Today, it is assumed that cars in Germany are unused on the street about 96% of the time. After all, cars are driven on average just over an hour a day. If everyone only had access to a vehicle when it was really needed—for example through car sharing—gigantic resources for the production, maintenance, and parking of vehicles could be saved. However, this would also mean the loss of many well-paid jobs in the automotive industry. Since the responsibility for the care and maintenance of cars in car sharing lies in the

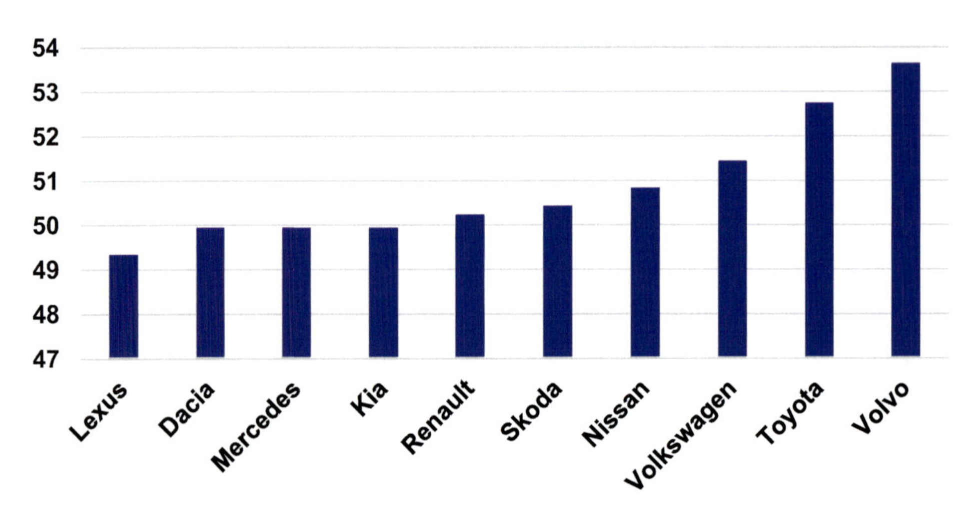

Fig. 4.10 Automotive brands with the highest ESG scores—2021

hands of the providers and not the customer, these tasks can usually be performed more efficiently.

> **Note Box** Product as a Service can contribute to sustainability in various ways.

Activities in Agriculture and the Food Sector

Figure 1.33 shows that the **agriculture** ranks fifth among the causes of harmful emissions after energy, industry, buildings, and transport. Therefore, this sector is also intensively discussing ways to achieve greater sustainability. The following measures are included:

- **Diversified cropping system**
  Instead of growing only one type of plant, several plant species are grown. This is intended to improve soil fertility and reduce the risk of crop failures due to diseases or pests.
- **Organic Agriculture**
  Organic farming avoids synthetic pesticides and fertilizers and instead relies on natural methods. This includes the use of plants that repel pests.
- **Water management**
  In many agricultural areas, the water requirement for growing crops is a significant factor. The use of modern irrigation methods, such as drip irrigation, can reduce water demand and conserve water resources.
- **Reduction of greenhouse gas emissions**

Agriculture is responsible for a large proportion of greenhouse gas emissions. The use of biogas plants or grassland management methods can reduce the CO_2 footprint of agriculture.

- **Promotion of biodiversity**
  Biodiversity in agricultural areas can be strengthened by growing plant species that are otherwise rarely cultivated. The protection of habitats for wildlife also strengthens biodiversity.

Many other measures are also being discussed and tested to improve sustainability in agriculture. This includes the use of artificial intelligence to improve resource efficiency. In addition, new plant species are being developed that require less water and fertilizer.

Significant changes are also evident in the food supply. In the 1980s, the company ***Rügenwalder Mühle*** advertised with the slogan: "The sausage that always tastes like it used to." Why? The company produced sausage specialties. As early as 2014, the range was expanded to include vegetarian and vegan products. A conscious realignment of the production program was carried out, also in line with the spirit of the times. Today the slogan is: "It tastes best together"—thus avoiding any product-related positioning (cf. Rügenwalder Mühle, 2023).

The following guiding principles form the basis for the **sustainability program of *Rügenwalder Mühle*** (Rügenwalder Mühle, 2023):

- "Sustainable economic, ecological and social action guides us in all areas of the company.
- We produce food in sustainable quality.
- We take responsibility for the supply chains.
- We contribute to more climate protection and reduce the greenhouse gas emissions caused by us.
- We use (natural) resources sparingly.
- We are a responsible, attractive employer and make a significant contribution to the region."

In the year **2020**, *Rügenwalder Mühle* developed a **code of conduct** and started its implementation. By the end of 2021, 100% of strategically relevant business partners had signed this code of conduct or were able to demonstrate compliance with the requirements of the code of conduct.

The company is concurrently working on the **transition from vegetarian recipes to vegan recipes.** The proportion of vegan recipes reached 85% by the end of 2022. In the procurement area, the goal was to increase the proportion of European plant proteins (measured by purchasing volume) to 80% by the end of 2022. For **employee development**, a **digital learning infrastructure** was implemented in **2021**. This is intended to provide continuous training for employees. In addition, the company has committed to capturing all relevant **Scope-3 emissions** by mid-2023 (see Rügenwalder Mühle, 2023;

see also on the topic of Scope Sect. 4.4.1.4). Through this corporate development, the company is consistently advancing on the path of sustainability.

The background against which the *Rügenwalder Mühle* and other food corporations have embarked on a more sustainable path is made clear by the following information. The **necessity to reduce meat production and meat processing** results from the fact that a reduced meat consumption would have a positive impact on the global climate and also on human health. The focus here should primarily be on **intensive agriculture** as a **driver of climate change**. In the so-called ***Meat Atlas,*** published by the *Heinrich-Böll-Stiftung,* the *BUND* and *Le Monde diplomatique*, the following results from a multitude of relevant studies can be found (cf. Heinrich-Böll-Stiftung, 2021, pp. 8 f., 10, 22, 24, 30 f., 36 f.):

- The **global meat production** continues to grow—driven by the increasing world population and the rising prosperity in many countries. Both factors drive growth in roughly equal parts. Overall, global meat consumption has more than doubled in the past 20 years. It reached a total of **320 million tons** in 2018. By 2028, meat consumption may grow by another 13%.
- However, **protection of the climate** and **biodiversity** can only be achieved if meat consumption is significantly reduced. Estimates suggest that a **halving of meat consumption** would be necessary to achieve a real turnaround.
- Today, about 70% of the total **agricultural land is used for livestock farming**. These areas include grasslands as well as pastures and fields for the **cultivation of feed crops**. Often, former peatlands are also used for animal husbandry. The fact is: If 3% of the agriculturally used peatlands in the European Union were to be rewetted, this could compensate for a quarter of the climate-damaging emissions from agriculture.
- If forests are also cleared for the cultivation of feed crops, the **natural habitat of wildlife** shrinks. The intensification of contact between humans and animals can promote the transmission of viruses and the emergence of new pandemics.
- In **agriculture**, there is an increasingly strong **structural change** towards fewer farms that keep more and more animals under industrial conditions. This is illustrated by one example: In 17 years, the number of broiler chickens has almost doubled. At the same time, almost three quarters of the fattening farms have given up. There is a clear concentration of animal husbandry taking place here.
- The use of **antibiotics in animal husbandry** leads to more and more resistant germs. As a result, the antibiotics that are indispensable in human medicine lose their effectiveness. 73% of all antibiotics sold worldwide today are used for animals—not for sick humans.
- Industrial animal husbandry is associated with a **routine use of antibiotics**. The **global market for veterinary drugs** has grown by 5 to 6% annually in recent years. Without legal regulations, a **rise in antibiotic consumption** in livestock is predicted to increase by 67% from 2010 to 2030. Antibiotics are sometimes also used to enhance performance in fattening—not to combat or prevent diseases.

- Moreover, so-called **reserve antibiotics** are already being used in the barn, which should actually be reserved for the treatment of humans.
- In the **feed crop producing countries**, a high amount of **pesticides** are used, which harm the groundwater and biodiversity. Some of the most dangerous pesticides have already been banned in the EU. However, the substances banned here are still used on a large scale in other parts of the world—especially for the cultivation of soy and corn. Often, these products are used to feed livestock, whose meat also ends up on the plates of consumers in EU countries.
- The five **largest meat and dairy corporations** emit as many **climate-damaging gases** as the oil corporation *Exxon*. According to the *Intergovernmental Panel on Climate Change (IPCC),* the share of the entire food sector in global human-induced greenhouse gas emissions is between 21 and 37%. Livestock farming contributes significantly to this with about 15%.

> **Note-Box** Agriculture in its current form is one of the main contributors to climate change. Changes can largely be brought about by the demand side—the consumers.

Activities in the Service Sector

Google has been making a (small) contribution to sustainability in Germany since 2022 through the further developed ***Google Maps App.*** In this app, **estimates of fuel savings** are now optionally presented for some routes. These are based on the selected engine type of the vehicle used, specifically gasoline, diesel, hybrid, or electric drive.

With the **fuel efficiency of a route**, the fuel consumption and consequently the associated CO_2 emissions are evaluated. The option is signaled in *Google Maps* by a **green leaf**. In the basic setting, fuel consumption is already taken into account when planning a route in *Google Maps*. Often, the fastest route is also the one with the lowest fuel consumption—but not always. To determine fuel consumption, *Google Maps* takes into account the **length of the route** and other factors, including road gradient and possible traffic jams (see Google, 2022). A small contribution from *Google* to reduce pollutant emissions.

Activities in the Financial Services Sector

Financial service providers are increasingly orienting their offerings towards sustainability. The relevant offers are referred to as impact investing. The term **Impact Investing** refers to investments in companies, organizations, and funds that aim to achieve a measurable positive social or ecological impact in addition to a financial return. This is why it is called "impact". With impact investing, the investor aims to contribute to the solution of social and/or ecological problems with the capital invested. An investment should therefore "pay into" the ESG criteria already presented in Sect. 2.1. **Impact investors** want to use their financial power to contribute to an improvement towards sustainability.

To this end, **impact investors** aim to invest capital in companies, organizations, and funds that are engaged in sectors such as renewable energy, healthcare, housing, education, and agriculture, and develop or implement product, service, and process innovations aimed at sustainability. It is essential that the measures initiated are also measurable, so that a "green label" on the investments does not turn out to be greenwashing afterwards.

The financial industry has already responded to this increasing demand and offers, among others, the following **products of Impact Investing** (see Commerzbank, 2023):

- **Corporate participations**
  Here, investors choose companies whose business strategy should be more focused on sustainability. For this purpose, the investors use their influence. Such participations can be subordinated loans or profit participation rights.
- **Microfinance funds**
  Such investments provide small loans to individuals in emerging or developing countries. The initiatives financed in this way are intended to combat poverty and hunger.
- **Funds with an impact-oriented approach**
  The managers of such funds often select the issuing companies based on ESG criteria. It is important that the orientation towards ESG requirements can also be proven.
- **Social Impact Funds**
  Such funds often invest in start-ups that pursue sustainable business models.
- **Green Bonds**
  Green Bonds are sustainable bonds that are issued to states, municipalities or companies. The funds provided are intended to finance sustainable projects. These include, for example, wind farms, photovoltaic systems or energy-efficient new buildings. The development of energy-efficient technologies can also be financed through such bonds.
- **Green Loans**
  Green Loans—also known as green loans—are also used to finance environmentally friendly, climate-protecting and resource-saving projects. In line with the EU taxonomy, so-called Green Loan Principles have been developed. These are intended to establish uniform standards for the use of and reporting on such green loans.
- **ESG-linked Loans**
  ESG-linked Loans are also known as Positive Incentive Loan or Sustainability-linked Loan. The special feature of these loans is that the loan conditions are linked to the achievement of defined sustainability goals. The use of the loan funds is not precisely defined here—unlike with Green Bonds and Green Loans. The money can be used for general corporate purposes. If a company's ecological balance improves in a targeted manner, the financing costs to be paid decrease. If the goals are missed, the financing costs increase.

- **ESG-linked promissory note**
  The margin of an ESG-linked promissory note is also linked to the company's sustainability performance.
- **Sustainability Supply Chain Finance**
  Here, the focus is on financing sustainable supply chains. The conditions for financing depend on the sustainability of the supplier. For example, one can orient oneself towards the eco-rating of the suppliers. In this way, corporate customers can provide financial incentives to their suppliers to make their activities more ecological and socially compatible. Orientation towards these criteria is becoming increasingly important for many companies. After all, their own sustainability and/or social balance is largely dependent on the performance of the companies in their own supply chain. In addition, financial institutions can also offer special conditions to suppliers who have a good eco and/or social balance.

> **Note Box** Providers must refrain from presenting their offers as "greener" than they are. Otherwise, the companies can be accused of the **prospectus fraud**.

At the same time, an exciting question needs to be answered:

> **Food for Thought**
> What happens if energy-intensive companies that cause high CO_2 emissions no longer receive financial resources for their business operations? These companies include, for example, manufacturers of concrete, steel and aluminium.
> Without concrete, steel and aluminium, however, no wind turbines and no high-quality, resource-saving buildings can be built. Such resources are also needed for the construction of solar panels.
> For fund providers, it would be easy to eliminate these providers from the funds. The effect on the CO_2 balance of the funds would be enormous. But how can an energy transition succeed if the companies that produce the necessary basic materials for this are cut off from the necessary financial resources?

The **European Central Bank** (ECB) already started in October 2022 to take into account the **climate protection in monetary policy** more strongly. For many years, the ECB has bought bonds from corporations that also cause a lot of CO_2 emissions for many billions of euros. The background to the realignment of bond purchases is the realization that climate change now also significantly influences economic development and thus inflation. Against this background, the ECB also defines climate protection as an important field of action. By orienting the purchase of bonds more towards sustainability criteria in the future, the climate risks in the ECB's balance sheet are to be reduced. In addition, the transition to a climate-neutral economy is to be supported (cf. Schnabel, 2022).

For this purpose, the ECB is abandoning the previously valid **principle of market neutrality.** In the future, the following selection criteria will also be taken into account when purchasing bonds (cf. Schnabel, 2022):

- Extent of the company's greenhouse gas emissions
- Development of further greenhouse gas emissions
- Quality of climate reporting, especially with regard to climate-related aspects

A shift towards climate neutrality is taking place step by step and not all at once. This is intended to motivate companies with high emissions to change their course.

In this context, the ***Dow Jones Sustainability Indices*** *(DJSI)* should also be considered. The group of these indices includes, among others:

- *Dow Jones Sustainability World Index (DJSI World)*
- *Dow Jones Sustainability North America Composite Index (DJSI North America)*
- *Dow Jones Sustainability Asia/Pacific Index (DJSI Asia/Pacific)*
- *Dow Jones Sustainability Europe Index (DJSI Europe)*

These indices list the "best companies in an industry" in terms of their sustainability performance. When being included in the indices, it is checked how sustainably the company's management is aligned. Global, identical for all industries, as well as industry-specific evaluation criteria are used. By taking into account ecological and social criteria—in addition to economic criteria—the *DJSI* distinguish themselves from both the traditional stock indices and purely ecology-oriented indices (cf. S&P, 2023).

With the ***S&P 500 ESG Index***, there is another broad-based and market capitalization-weighted index that aims for sustainability. The performance of securities is also measured by the extent to which the respective companies meet sustainability criteria (cf. S&P, 2023).

It is interesting that *Tesla* was kicked out of this index in May 2022. What happened? Why was *Tesla*, the pioneer of e-mobility, excluded from the index? After all, the CO_2 footprint of a *Tesla* vehicle is only a fraction of that of its combustion engine competitors. The reasons for the expulsion lie in other ESG criteria: racial discrimination, poor working conditions in the Californian factory, and the handling of the investigations by the US traffic safety authority into the fatal accidents with *Tesla's* driver assistance system "Autopilot" led to the downgrade. It was also emphasized that competitors from the automotive industry have now improved in terms of ESG criteria (cf. o. V., 2022b).

> **Food for Thought** Interestingly, there are already partial counter-movements when financial institutions act "too sustainably". This was experienced by the asset manager *Blackrock* in the USA in 2022. Republican-governed states withdrew investment amounts in the hundreds of millions because *Blackrock* was said to have focused too much on ESG criteria and too little on profitability in its

investment strategy. At the same time, *Greenpeace* criticized that *Blackrock* was doing too little to demand and promote sustainability in the business concepts of the companies in which *Blackrock* has invested.

Such attacks on more sustainable investment do not hit a small player. *Blackrock* is the most important asset manager in the world, with managed assets of over 8 trillion US dollars.

Is there a risk here that the supposed winner topic "sustainability" will ultimately become a business risk?

An analysis of the financial services sector should not overlook the **cryptocurrencies**. After all, the creation of cryptocurrency (so-called mining) as well as the execution of transactions involve enormous computational effort. A single transaction with *Bitcoin* produces 800 kg of CO_2 and approximately 375 g of electronic waste. The same amount of energy could be used to carry out 1.8 million credit card transactions. Alternatively, one could watch videos on *YouTube* for 133,000 hours or fly from Düsseldorf or Cologne to Mallorca. Even the consumption of 65 kg of beef would produce less CO_2 in terms of meat production than a single transaction (see Nestler, 2022, p. 24).

▶ **Food for Thought** As long as the electricity for mining and for transactions of cryptocurrency does not come from sustainable sources, but mining perhaps even takes place where electricity is particularly cheap—and possibly even particularly dirty—cryptocurrencies remain pure climate killers.

Activities in the Construction Sector

In 1990, the **building inventory** was still ranked third among the emitters of greenhouse gases. Since 2007, the transport inventory has pushed buildings to fourth place (see Fig. 1.33). Their continued importance becomes understandable when considering the resources necessary for the construction and maintenance of buildings. The construction industry is not only one of the **largest CO_2 emitters.** The **construction and demolition waste** also accounts for about half of all waste in Germany.

Therefore, **paths to climate neutrality** must also be pursued in the building sector. To reduce resource and land use, the entire life cycle of a building must be considered: **Build, Use, Dismantle.** For this purpose, the ***Cradle-to-Cradle principle*** is increasingly being used (see Engelfried, 2021, pp. 168–172).

The **resource use in the construction of buildings** is considerable. The production of concrete is extremely energy-intensive. In addition, non-renewable resources such as sand and gravel are used to a large extent. The fact that there is now a global shortage of construction sand illustrates the challenge. To remedy this, institutions such as the ***German Society for Sustainable Building*** *(DGNB)* or the ***ESG-Circle of Real Estate*** *(ECORE)* have developed special **rating systems for sustainable real estate**. These are intended to create the necessary transparency and thus comparability of construction projects. The *DGNB* offers an **international certification** to document the sustainability of buildings based on an objective description and evaluation. This considers the entire life cycle of the building (see DGNB, 2023). *ECORE* has launched an **initiative for ESG compliance** in real estate portfolios (see ECORE, 2023).

How can the ***Cradle-to-Cradle principle*** **be implemented in the construction industry?** To apply this principle to buildings, recyclable and chemically harmless building materials must be used in their construction. Only then can it be ensured that at the end of a building's life cycle, no non-recyclable waste is produced, but that the materials used could be reused. In an ideal world, **buildings** would become **long-term raw material storage** and **material storage,** whose contents could be repeatedly combined and used. This would not only preserve valuable raw materials such as concrete, wood, steel, and plastic, but also avoid huge construction waste dumps in nature.

The challenge is to capture and classify these quantities. Then it would be known which materials in buildings and roads are essentially "only" stored for later use. To create this transparency, a **digital material register** would need to be created. While the material recording for new buildings could be relatively simple due to the digital availability of data, old buildings pose completely different challenges. Here, the materials used would first have to be recorded through on-site inventories—object by object.

In order to enable further **resource use in the demolition of buildings** and to return the materials obtained from demolition to the cycle, appropriate concepts are being worked on. For this purpose, it is necessary to establish **registers for materials and products**, in which data on the materials and products used are recorded. The structures can be both buildings and bridges. Only if all components are recorded, information about the **separability of materials**, their **toxicity** (poisonousness) as well as about the total **bound** CO_2 in a structure is available. Based on the information to be stored in a register, it will also become apparent whether materials and products can be reused. Only such a register creates the **conditions for circular construction.**

The platform ***Madaster*** helps to build such a **register for materials and products**. Based on *BIM-(IFC)*- or Excel documents, a so-called **building resource pass** can be created. The abbreviation *BIM* stands for *Building Information Modeling,* i.e., a building information model. *BIM* is understood as the digital representation of the physical and functional characteristics of an object. Such an information model is the result of the digitization process of a building—a **digital twin** of the building is created. This digital twin contains all information about the entire building life cycle—from planning through the construction phase, administration/maintenance to disposal. To ensure easy collaboration between all parties involved, a standard format for data exchange is required. *IFC* stands for *Industry Foundation Classes* and defines this global, open standard for **data exchange in the construction industry** (see Biblus, 2020).

As a result, Madaster (2023) offers property owners the opportunity to store data of their own property and enrich it with further information (e.g., raw material values). **Material passports for buildings** can make an important contribution to the sale, but also to the recovery of materials. This allows companies like Heidelberg Cement to reduce their emissions by processing recovered cement (see Schmale, 2022, p. T3).

Through the platform *Madaster*, such a building resource pass can be created for both new buildings and existing objects. The installed material thus receives an identity. If a regular supplement of the data of the digital twin is ensured, this **resource pass** becomes a valuable digital data source to permanently support the process of construction, use, and recycling. Such a resource pass makes **new buildings** into (long-term transparent) **material storage** and allows for **old buildings** a **material mining,** to actually recognize the material treasures to be lifted there. This is referred to as **Urban Mining**, akin to mining in the city. In addition, the *Madaster Foundation* promotes the dissemination of know-how about the circular economy (see Madaster, 2023).

Comparable solutions are also offered by Concular (2023). Companies like iPoint (2023) support businesses with the ***iPoint Suite*** in the digitalization of the entire product lifecycle—from the design phase through production and use to recycling and reuse of raw materials.

An example of such a project in Heidelberg shows the achievable dimensions. It involves the former US-American residential settlement Patrick-Henry-Village with a total of 325 buildings. This complex contains 465,884 tons of material. Of this, 237,216 tons are concrete, 91,112 tons are bricks, and 3,881 tons are metals. This type of resource recovery is intended to help Heidelberg massively reduce its CO_2 emissions in the coming years. For Germany as a whole, the raw material substance of buildings amounts to approximately 15 to 16 billion tons. This corresponds to 190 tons per person. If civil engineering (including roads) is also taken into account, the total amount increases to approximately 29 billion tons (cf. Schmale, 2022, p. T3).

In addition to the reusability or recyclability of the raw materials used, renewable energies must be used in construction to reduce CO_2 emissions. Furthermore, construction measures must promote cultural and biological diversity. However, these requirements must be taken into account not only in new construction projects but also in renovation and refurbishment measures.

To document the use of environmentally safe, healthy, and also recyclable materials, so-called **Cradle-to-Cradle certifications** are used. The following aspects are considered (cf. Inventio, 2023):

- Material health
- Recyclability of materials
- Use of renewable energies
- Responsible water management
- Consideration of aspects of social justice

Depending on the evaluation, **Cradle-to-Cradle certificates** can be issued in Basic, Bronze, Silver, Gold, and Platinum.

On April 20, 2022, further requirements of the ***Quality Seal for Sustainable Buildings*** (QNG) came into effect. These concern new construction projects as well as the complete modernization of non-residential buildings. This **state quality seal** for buildings is awarded when proof of compliance with general and special requirements for the ecological, sociocultural, and economic quality of buildings can be provided. Compliance is determined by an independent examination after construction completion based on the completed planning and construction processes and by an inspection of the buildings (cf. BMWSB, 2022).

To reduce the resource use in the management of buildings, there is an increasing reliance on **photovoltaic systems** and **heat pumps**. The aim is for buildings to become energy self-suppliers, which are autonomous in terms of the energy consumed or can even feed energy back into the public grid. A prerequisite for this to succeed is particularly well-insulated buildings—and simple legal framework conditions.

> **Food for Thought** These examples make it clear: There is no shortage of ideas to get the circular economy moving. The only important thing is to take the customers on this journey.

4.4.1.3 Sustainable Supplier Management

Through the already discussed supply chain law, a **professional supplier management** is becoming increasingly important. The ecological footprint of a company is largely determined by the integrated supply chain. The integrated suppliers have always significantly contributed to the **economic success of the company**. Today, suppliers are increasingly required to also secure the **ecological and social success of the company**. Therefore, the goal of supplier management today is not only to build a **balanced supplier structure**. This is still indispensable to ensure a stable, resilient supply and thus a high resilience of one's own company. In addition, the requirements of the supply chain law defined in Sect. 2.4 must be taken into account when selecting and integrating suppliers.

Today, supplier management must increasingly orient itself towards sustainability factors. The following **drivers of sustainable supplier management** can be distinguished (see Fig. 4.1):

- Companies consistently implement the legal requirements (especially of the supply chain law) and/or respond to the expectations of other stakeholders in the sense of a reactive concept.
- Companies act proactively and creatively based on their own values and a purpose oriented towards sustainability.

> **Food for Thought** Today, it is indispensable for all companies that want to operate successfully in the long term to design supplier management in compliance with ecological, social, and economic criteria. After all, a company can only act comprehensively sustainable if its suppliers also act sustainably.

To meet the requirements of sustainable supplier management, the following **step-by-step concept for implementing the supply chain law** is recommended (see Inverto, 2022, pp. 7–13):

1. Stage: Analysis of the status quo

As part of a status quo analysis, it is necessary to check to what extent the company already meets the requirements of the supply chain law in order to identify specific areas of action. The following questions can help:

- Which **duties of the supply chain law** are already anchored in the company's **codes of conduct**? What gaps still need to be filled?
- Does the company have **standards for environmental protection and human rights?** Are these already uniformly defined and coordinated across all company functions?
- To what extent are standards for environmental protection and human rights already taken into account in **supplier management**? Are all suppliers checked for LkSG compliance—initially and in ongoing operations? Is such an evaluation also carried out for each product group to specifically identify high-risk groups?
- Are social, environmental, and ethical **risks in supplier relationships** systematically identified and classified? Are powerful **risk analysis tools** used for this—or do they still need to be installed? Are the insights gained from this communicated in a LkSG-compliant manner—in terms of content and timing?
- Are **qualitative and quantitative KPIs** defined to systematically determine compliance with the LkSG requirements among suppliers? Are **escalation mechanisms** defined when significant deviations from the guidelines are detected?
- Are **contacts for LkSG risk management** defined in purchasing and known throughout the company?
- Is the risk management prescribed by the supply chain law **Risk management** (including tools for risk analysis) integrated into the **training plan** for purchasing?
- Is the **achievement of sustainability goals** (e.g., CSR reports) and **emergency plans** communicated internally and externally?

To build the necessary transparency, methods of **Advanced Analytics** can be used. It is important to determine which products and services are purchased from which suppliers. At the same time, it is indispensable to inform the suppliers about the requirements defined in the supply chain law and to collect the necessary data.

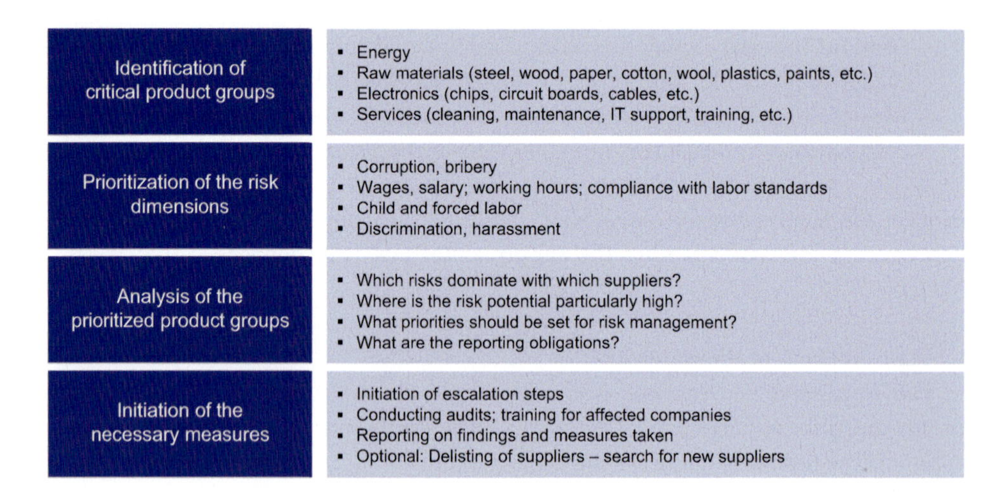

Fig. 4.11 Supply chain law-related product group analysis

A **supply chain law-related product group analysis** can be carried out as shown in Fig. 4.11.

Stage 2: Optimization of Risk Management

The challenge for companies is to use the enactment of the supply chain law as a stimulus to supplement their own risk management with the ESG criteria already presented. To this end, a **risk management audit** should be used to examine how the **risk management** has been organized so far:

- Does a **structured risk management** system exist in the company, or have crises been reacted to reactively and ad hoc so far?
- Are **risk management tasks** possibly still distributed throughout the entire company?
- Are there predefined—and generally known—**escalation steps**?
- Do the existing **risk management measures** cover the requirements of the supply chain law?
- Has a **supplier radar** been set up to identify emerging risks early and comprehensively?
- Is there a **priority setting** for particularly critical suppliers and/or product groups (keyword “risk suppliers” and “risk product groups”)?
- Are **preventive measures** being used to prevent violations of human rights or the environment?
- Are effective **control mechanisms** in place to detect rule violations early?
- Have **immediate measures** been developed to immediately rectify acute deficiencies in suppliers?

- Is there a balanced **supplier portfolio**—without major dependencies on individual suppliers or individual countries?

Such a **risk management audit** should be carried out at regular intervals. The insights gained here should immediately trigger optimization measures.

In order to better assess the performance of their own suppliers, supplier scorecards-**Supplier Scorecards** have been used in the past. Today, these scorecards increasingly include requirements derived from the ESG criteria, in addition to criteria such as cost, product quality, reliability, flexibility, etc. The more such supplier scorecards also take into account and weight **sustainability criteria**, the more a pursuit of sustainability permeates the entire supply chain. Unfair trading practices and the involvement of exploitative companies in the value chain are becoming less and less accepted—even if this can be associated with cost increases. At the same time, demands are growing louder to reduce the use of raw materials and energy and to avoid waste.

To support this process, a **supplier portfolio** can be used. This is based on the dimensions "relevance of the supplier" and "probability of a violation against LkSG" (see Fig. 4.12).

The **relevance of a supplier for one's own company** can be determined based on the following questions:

- What is the **supplier's share of one's own purchasing volume**—nationally, regionally, internationally?

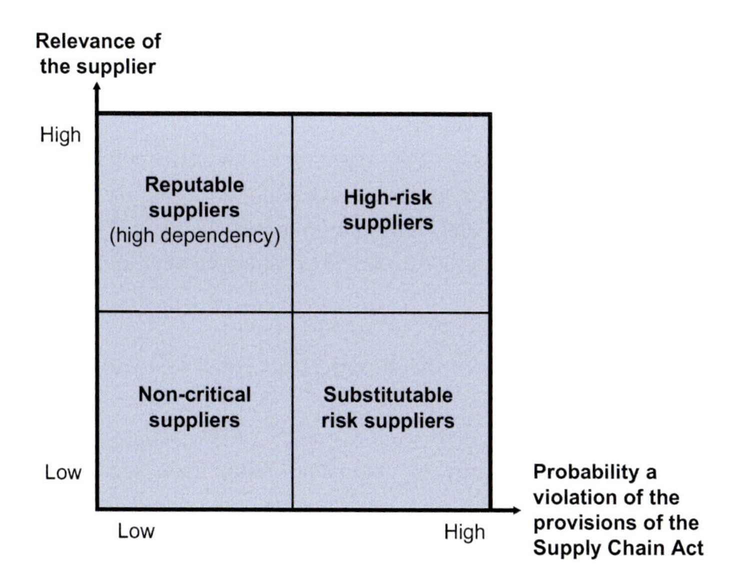

Fig. 4.12 Supplier portfolio for risk assessment according to the supply chain law

- What is the **share of one's own purchasing volume in the supplier's total sales**—nationally, regionally, internationally?
- How high are the **switching costs** (Switching Costs) when changing suppliers?
- What **share of one's own value creation** do the products or services of the supplier have?
- How large is the **visibility of the products or services** of the supplier to various stakeholders (customers, politics, general public)?

The probability of a violation of the provisions of the Supply Chain Act can be determined by the following questions:

- What is the **proportion of human labor in the value creation** of the supplier?
- Who is involved in the **provision of human labor**?
- What do the **type and duration of service provision** look like?
- What **risks to humans and the environment** are associated with the provision of services (processed materials, production and logistics processes)?
- In which **countries** or in which **regions** is the supplier active?
- What **labor and environmental standards** apply in these countries and regions?
- To what extent are **legal requirements** actually followed in these countries and regions? How are **controls** carried out? What is the **risk of bribery, corruption** etc.?

Companies are increasingly demanding detailed product information in their purchasing decisions in order to justify their own actions. For this, it is important that suppliers ensure **traceability of materials**. The transparency required here refers not only to the origin of the materials, but also to the conditions of their production. By seamlessly linking all data along the supply chain, a **360-degree view** is to be created. This requires as comprehensively networked supply chains as possible (see Sect. 4.1). In an expansion stage, an **supply chain ecosystem** can emerge.

To build such a supply chain ecosystem, *Apple* no longer relies solely on the supply of raw materials that are mined. The company is increasingly relying on reusable and recycled materials such as copper, tin, and aluminum in its supply chain. This not only makes the supply chain more resilient, but also allows for cost advantages and the acquisition of "sustainability points".

> **Note Box** Transparency about the supply chain should be used as a trust-building measure!

While many years ago the majority of companies relied on extensive **offshoring** to **reduce manufacturing costs**, today there is increasing talk of so-called **reshoring**. This process describes the relocation of production facilities back to the manufacturer's home country—or at least to neighboring countries. With this regional procurement, not only can the CO_2 emissions associated with long transport routes be reduced. Companies also become more independent of global supply chains and thus increase their own resilience.

Due to the tense political world situation, attempts are now increasingly being made to shape reshoring as **friendshoring**. This involves striving to locate production facilities in countries with which friendly relations exist—and which remain reliable partners even in crises. In this way, dependencies on uncertain countries are to be reduced and supply chain disruptions avoided.

> **Note Box** The trend is moving from offshoring to reshoring and friendshoring!

> **Food for Thought** One thing we must be clear about in this development, which dispenses with previously used globalization advantages: Such steps are associated with **welfare losses**—due to higher prices.

Building the necessary transparency about the supply chains poses a major challenge for the companies involved. How can these companies obtain the information necessary for sustainable supplier management? Does all of this have to be done in-house? Or are there relevant **service providers** who can assist in the **collection and analysis of supplier data and sustainability risks**? Relevant service providers have now established themselves who offer their services in these areas. A selection of such service providers can be seen below:

- ***EcoVadis***
  The company *EcoVadis* supports businesses through a **sustainability ranking.** This ranking currently includes more than 90,000 companies in over 160 countries with over 200 industry examples. With the services of *EcoVadis,* companies do not have to collect the relevant data themselves. This is done by *EcoVadis* in a standardized form—worldwide (see EcoVadis, 2023).
  The database of *EcoVadis* allows a **risk analysis of the suppliers,** without each company having to contact its suppliers individually. The risk profiles are continuously updated through 360-degree monitoring from a multitude of sources. Reporting is done via a dashboard that includes the relevant KPIs for the supply chain law. In addition, preventive measures and training are offered. Support is also provided in the development of plans to correct identified deficiencies.
  EcoVadis offers the opportunity to also evaluate one's own company based on international standards. For this purpose, a questionnaire tailored to the industry, company size, and respective country is used. The collected data are reviewed by sustainability analysts. The results can be compared internationally using **sustainability and car-**

bon scorecards. Based on this, suggestions for improvement are developed to eliminate identified weaknesses.

- ***Prewave***
  Prewave conducts **AI-based risk and sustainability monitoring**. The company monitors more than 250,000 direct and indirect suppliers by evaluating information from millions of sources in more than 50 languages. The risk radar captures CSR incidents (e.g., environmental pollution), cyber risks, governance issues (e.g., a change in management), product recalls, labor unrest, calls for boycotts, as well as political, financial, and legal dangers (see Prewave, 2023).
  By using artificial intelligence (keyword **"Predictive Analytics"**), *Prewave* can report on risks and issue **sustainability warnings** even before critical developments have occurred. **Risks in the supply and logistics chain** are identified early and can be proactively addressed. In addition, reliable **information for supplier selection** and for overall **supplier management** is available.
- ***Sustainalytics***
  Sustainalytics assesses the sustainability of over 4,000 listed companies based on their performance in the areas of environment, social, and corporate governance (ESG criteria). The **ESG risk assessments** are based on the **relevance of ESG topics** to a company and the **competence of management** to address these issues (see Sustainalytics, 2023).
  The ESG risk assessments are used by companies to check compliance with the requirements of the supply chain law. Banks access this data to determine credit risk. Investors can use the risk assessments to determine whether relevant ESG requirements are being considered.
- ***CSR Hub***
  CSR Hub provides **ESG ratings** that can be used for benchmarking. One's own performance can thus be compared with that of competitors, but also with companies from other industries and regions. Based on such data, **optimization strategies** can be developed. In addition, it can be recognized how stakeholders perceive the ESG performance of one's own company (see CSR Hub, 2023).
  In addition, the supply chain can be checked based on the ESG ratings of one's own suppliers. By accessing the data of the *CSR Hub*, **risks in the supply chain** can be identified without having to conduct comprehensive audits of the suppliers oneself.

Stage 3: Further Development of Supplier Management

An important fixed point of supplier management is a **selection of suppliers,** that is guided by human rights and environmental criteria. Only suppliers who meet these criteria are integrated into the supply chain. The next step is the **commitment of suppliers to comply with a code of conduct**. Subsequently, it is necessary to work together to ensure the **sustained compliance** with the criteria defined by the supply chain law. For this purpose, concrete **supply chain goals** are formulated together with the suppliers.

These should be quantified and defined in terms of time and region so that they can fulfill their steering function.

Based on this, concrete **measures** are developed to ensure compliance with the LkSG. These measures include the **development of milestones and escalation plans**. The **compliance with the LkSG standards** is to be checked based on own research and/or with the involvement of the presented service providers (such as *EcoVista, Prewave, Sustainalytics, CSR Hub*).

For this, a **regular exchange** between the own company and its suppliers is indispensable. This includes, for example, **supplier forums,** to exchange best practice examples. In addition, there should be an exchange with the direct suppliers about where **risks with indirect suppliers** could lie. Here too, it is important to identify possible human rights or environmental violations at an early stage. For this purpose, it can be helpful to support the direct suppliers in achieving LkSG compliance through **training and education measures**—also with regard to their direct suppliers. If central requirements are not met, the business relationship must be terminated (cf. Inverto, 2022, pp. 7–13).

This **three-stage process** is necessary for the **initial alignment with the requirements of the supply chain law**. Subsequently, a continuous review of the integrated suppliers must be carried out. Each potential supplier must be evaluated based on the measures described here before being included in the supply chain.

4.4.1.4 Sustainable Production Processes

To make your own manufacturing processes **more sustainable, a step-by-step concept for more sustainable production** can be used. This includes the following seven stages and takes into account the ***OECD* indicators for sustainable manufacturing.** Fig. 4.13 visualizes the individual stages (cf. OECD, 2022a).

Phase: Preparation

Step 1: Capture impacts and set priorities

In this stage, the **impacts of the current mode of production** are recorded. Where are the so-called "low-hanging fruits" to be found—optimization fields that can be easily tapped? In addition, the areas where there is a particularly high need for action because the biggest "sins" are present are to be identified. This is simply about setting priorities in order to develop a clear compass for the further steps.

To initiate this phase and consistently carry out all further steps, the installation of a **sustainability team** equipped with qualified personnel and sufficient budgets is recommended. This can be supported by external experts if the internal resources are not sufficient for this.

Step 2: Select indicators and determine data needs

The task at this stage is to identify the relevant indicators to determine the **impacts of production activities** on sustainability requirements. The **18 *OECD*indicators** listed in

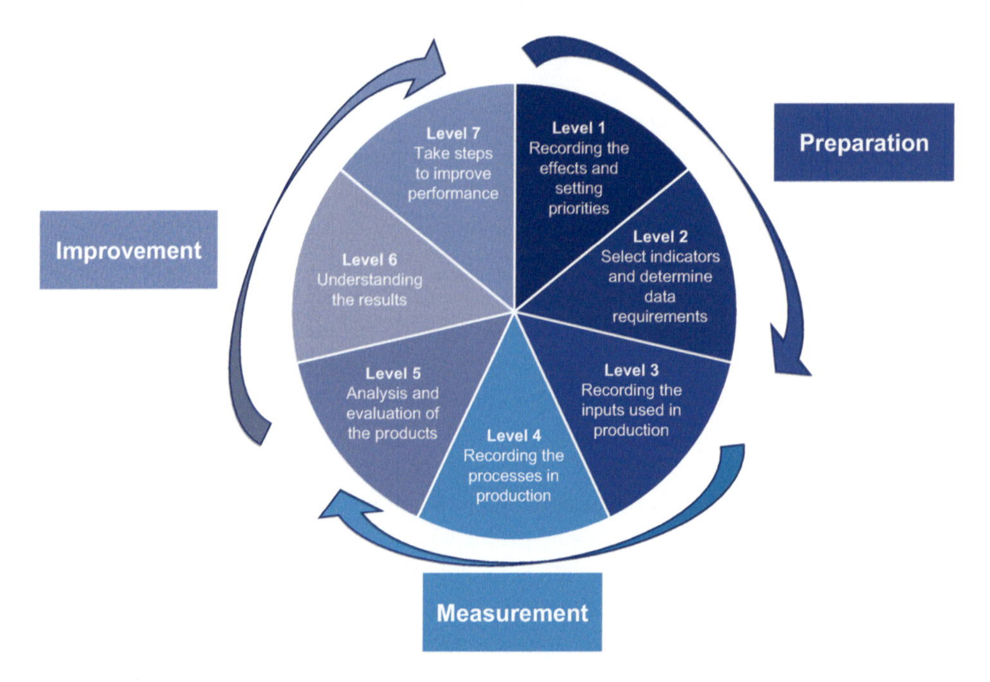

Fig. 4.13 OECD step-by-step concept for achieving sustainable production

Fig. 4.14 are a good starting point for this (see OECD, 2022b). These indicators also allow the successes of the measures initiated to be determined later.

These **18 *OECD*indicators** are briefly introduced below (see OECD, 2022b):

- **Input 1: Extent of the use of non-renewablematerials**
  This extent can be captured by the weight of the consumed, but non-renewable resources. The focus on this criterion is important because the supply of non-renewable resources is finite and therefore a particularly responsible handling of them is necessary. In addition to the finiteness of the materials themselves, further resources are used and emissions are caused for their extraction, processing and provision. Part of the materials used becomes part of the product itself, another part will end up as waste from the production process.
  This indicator measures the **consumption of non-renewable materials in relation to output** (e.g., number of products produced). The recording is done in tons.
- **Input 2: Extent of the use of limitedmaterials**
  This indicator measures the **consumption of legally restricted materials in relation to output** (e.g., number of products produced). The recording is also done in tons.
- **Input 3: Recycled/reused proportion of non-energeticmaterial use**
  To determine this indicator, the total weight of the recycled material and the total weight of the reused material are expressed as a percentage of the total weight of the

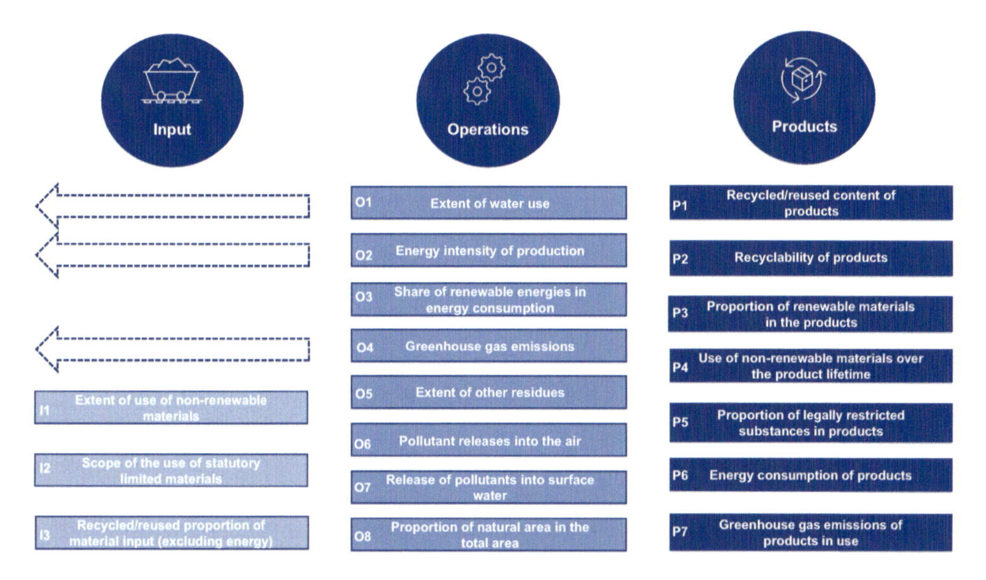

Fig. 4.14 18 *OECD* indicators for capturing sustainability in production

material use. This makes visible what proportion of the total material use in percent is accounted for by recycled or reused material. This indicator can be determined separately for all relevant input materials (such as paper, PET plastic). Here too, the recording is done in tons.

- **Operations 1: Extent ofwater use**

  To determine this indicator, the **water consumption in the manufacturing process** is related to the production quantity. Water is one of the renewable resources. However, there are increasingly local shortages and quality problems. Therefore, even in industrial processes, a resource-saving approach is necessary, for example, to avoid a lowering of the groundwater level.

  Water is a precious resource. Therefore, it is also important to handle this raw material carefully. It is important to know that about 1000 L of water are required to produce one liter of wine or one liter of milk. For the production of one kilogram of cheese, about 4000 to 5000 L of water are already required. For one kilogram of beef, it is often more than 15,000 L of water. For one kilogram of paper, about 10 L of water are needed—for the production of a computer chip about 35 L (see Diemand & Finsterbusch, 2022, p. 22). This list could be continued indefinitely and shows only one thing: water is an indispensable raw material for a variety of production processes.

  In addition to the water consumption itself, it is also necessary to record what percentage of the water used is recycled or returned to the cycle after use in the manufacturing process. The recording is done here in cubic meters.

- **Operations 2: Energy intensityof production**
  To calculate this indicator, the energy consumption for production and for the other company areas is related to the output. The recording is done in *megajoules*. The energy intensity of product use is calculated separately. These indicators are particularly important because every form of energy generation—renewable or non-renewable—also consumes non-renewable resources. These include habitats, fossil fuels and uranium, but also metals, plastics and other materials for the construction of hydroelectric power plants, solar collectors and wind turbines. In addition, greenhouse gases are produced when using fossil fuels.

> **Note Box** Many materials can be recycled, but one cannot: the consumed energy!

A review of the detailed data shows which processes contribute most to the facility's energy consumption. These should be primarily examined for potential savings. The energy intensity can be related to the value of the output.

- **Operations 3: Share of renewable energiesin energy consumption**
  This indicator is determined by reporting the consumption of renewable energy as a percentage of total energy consumption.
- **Operations 4: Emissionof greenhouse gas**
  To determine the emission of greenhouse gases, the greenhouse gas emissions of the following areas must be added:
  - Emissions from energy consumption in production
  - Emissions from the production area itself
  - Emissions from other company areas
  - Emissions caused by business trips

  This indicator combines the following defined Scopes 1, 2, and 3, which were established in the so-called **Greenhouse Gas Protocol**:
  - **Scope 1**
    Scope 1 includes the direct release of climate-damaging gases in the company, e.g., through its own production activities or its own fleet. This is also referred to as **direct emissions**.
  - **Scope 2**
    Scope 2 captures the indirect release of climate-damaging gases by the included energy suppliers. Specifically, it is about the **indirect emissions** from the consumption of purchased electricity, heat, or steam.
  - **Scope 3**
    Scope 3 targets the indirect release of climate-damaging gases in the upstream and downstream supply chain. This includes, among other things, the **indirect emissions** associated with the extraction and production of purchased materials and fuels, transport-related activities in third-party vehicles (e.g., business trips and

commuting by employees), as well as other outsourced activities and waste disposal.

The unit of this indicator is tons of CO_2 in relation to the achieved value of the output. Here, price effects can occur. For example, rising prices of the goods produced alone would improve the value of the indicator, as this would reduce the reported greenhouse gas intensity. To avoid such unwanted effects, the total amount of greenhouse gas emissions should also be recorded.

- **Operations 5: Extent of furtherresidues**

  To determine the amount of further residues, this indicator subtracts the weight of the manufactured products from the total weight of all inputs plus the weight of the consumed fuels. The waste quantity determined in this way can be related to the output. These residues can be released into the air, surface water, soil, and the sewer system. Waste can also be disposed of via recycling facilities, incinerators, and/or landfills.

  Here too, recording is done in tons in relation to the output. The indicator can be determined separately for the various forms of waste disposal. After all, recycling is preferable to landfill, for example.

- **Operations 6: Pollutant Releasesinto the Air**

  The intensity of pollutant releases into the air is determined by the weight of these releases in tons from production processes and possibly from other business areas—in relation to the output. Releases into the air are already included in the indicator extent of further residues (Operations 5).

  An isolated consideration of these releases is important, as such discharges into the air are often associated with environmental or health problems that occur in the vicinity of production facilities. This indicator should be determined separately for individual air pollutants.

- **Operations 7: Pollutant Releasesinto Surface Water**

  The intensity of pollutant releases into surface water is determined by the weight of the releases from the production processes and possibly other business areas into the surface water and is related to the output. The reduction of discharges is important due to the associated impairments of human health and the environment as a whole.

  It is recommended—as with the release into the air—to also record this value separately. However, this indicator does not take into account the effects of the mixture of released pollutants.

- **Operations 8: Proportion of natural areaof the Total Site**

  This indicator shows the percentage of natural areas in the total site of a production facility or a company. However, this indicator does not say anything about the sustainability of the plant or the products produced there. But the indicator shows the intensity of land use.

  The "natural coverage" on the site includes both unused land (such as natural forests, grassland, shrub or wetlands) and greened areas (including green roofs). In total, the indicator shows how extensively industrial plants interfere with existing ecosystems.

- **Products 1: Recycled/reused content of Products**
  The amount of the recycled or reused portion of products is determined by determining the percentage by weight of each product that is made up of recycled or reused material. This value can be determined—overall and for each category (recycled or reused). Input 3 already recorded which resources flow into production. A separate identification for the products is therefore important because the product design significantly determines the achievable proportion of recycled or reused materials. Therefore, it is also important to determine these indicators separately for each product—even if the data initially based on estimates. The value of these indicators can range between 0 and 100%.
  It is important to note that these indicators do not capture possible effects of recycled or reused materials on the quality of the final product. An increase in the proportion of recycled or reused materials can improve or deteriorate the marketability of the products. The proportion of recycled or reused materials can also have an increasing or reducing effect on the price.
- **Products 2: Recyclabilityof Products**
  The indicator for the recyclability of products indicates the percentage by weight of a product that can be reused. Ideally, products are developed so that they can be recycled, reused, reprocessed, composted, or biodegraded. This indicator shows to what extent this is the case.
  For this purpose, it is recommended to determine the indicator per product for the categories recycling, reuse, reprocessing, composting, and biodegradation. The indicator can assume a value between 0 and 100% in each category. An indicator that only captures these values for the entire product range is less meaningful.
  Recyclability at the end of a product's life cycle can be achieved through the choice of materials and the modularity of the products. Some materials are easier to recycle than others. Steel and aluminum, for example, are highly recyclable and are often reused.
  Products can also be designed modularly—without being glued. Then components can be replaced when they are defective. This is particularly relevant for components that wear out quickly. It is crucial that these do not become desired breaking points that are supposed to motivate a new purchase (see Sect. 4.4.1.1).
  Important: The indicator does not measure the extent to which products are actually recycled, reused, reprocessed, composted, or biodegraded. Here, only a potential is identified. Crucial for the further use of products is an infrastructure to open up such fields of use.
- **Products 3: Proportion of renewable Materialsin the Products**
  The proportion of renewable materials is expressed as a percentage of the total weight of a product. The more renewable materials (such as from plants or animals) are used, the less the need for non-renewable resources. Here too, it is useful to determine the indicator per product and not summarily for the entire product range. The indicator can again assume a value between 0 and 100%.

This indicator also does not take into account possible effects on product quality due to the substitution of non-renewable materials.
The extent to which such renewable raw materials should be accessed depends on the possibilities of producing such renewable materials without high negative economic, social, and ecological impacts.

- **Products 4: Use of non-renewable Materialsover the Product Lifespan**
  To calculate this indicator, the proportion of non-renewable materials of a product is divided by the expected lifespan of the product. The unit of the indicator is grams, kilograms, or tons/year. The value is to be determined per product and for the total amount of products produced per year.
  This indicator can be improved by two measures: extending the lifespan of products and reducing the use of non-renewable materials.
- **Products 5: Proportion of legally limited Substancesin Products**
  To indicate the proportion of legally limited substances in products, their weight proportion is expressed as a percentage of the total weight. Certain products can still only be produced by using materials whose release is harmful to humans and the environment. Often, the release of these substances should be avoided during the construction of the product and at the end of the life cycle (such as disassembly and recycling). Which substance use is limited is defined by national or regional legal regulations.
  The quantities of limited substances can be indicated once per product and also for the total production quantities. From the determined quantities, the necessity may arise to set up special return or recycling programs or to support their establishment.
  Possible effects on product quality due to a reduction of substances with restricted use (such as the renunciation of lead in solder for electronic devices) are not taken into account by this indicator.
- **Products 6: Energy Consumptionof the Products**
  To determine the intensity of a product's energy consumption, the average annual energy consumption of a product is determined. Additionally, the energy consumption of the total units produced in a year can be determined. The recording is done in megajoules. This value is part of a company's total "energy footprint" caused by the production of their products. By reviewing the detailed data, it is determined which products and which phase (materials, production, or use) contribute most to the total "energy footprint".
  The "Energy Consumption" indicator is of great importance for many products. These include buildings, airplanes, cars, household appliances, and other electronic devices. Here, the energy consumption in the usage phase is significantly higher than in their production or in the extraction and processing of the materials used for their production. To reduce energy consumption over the entire development, production, and usage duration of products, all three phases must be considered.
- **Products 7: Greenhouse Gas Emissionsof the Products in Use**
  To determine the greenhouse gas emissions of products in the usage phase, the average annual greenhouse gas emissions per product unit are determined. This indicator

measures the annual greenhouse gas emissions of the products produced in the reference year per production unit or the total product program. For the annual production volume, this amount is reported in tons of CO_2. This indicator quantifies the part of a company's total "carbon footprint" that is caused by the production and use of their products.
To reduce greenhouse gases overall, it is important to focus efforts on the phase that contributes most to the creation of greenhouse gases. This indicator focuses on the greenhouse gas emissions that occur during the use of the product. To reduce the intensity of this energy consumption, constructive changes in the products or their usage patterns must be achieved.
Greenhouse gases released during the production of materials and in the production process are or can be captured by the "Operations 4" indicator. The "Greenhouse Gas Emissions of the Products" indicator covers part of Scope 3 of the supply chain.

These quantitative indicators have proven themselves many times in defining, tracking, and improving performance. These indicators can support decision-making processes towards sustainability and can be adapted for all types of production. If necessary, additional company-specific indicators can be added. It is important that the entire processes are aligned with the collection of the data necessary for the calculation.

Phase: Measurement

Stage 3: Recording of the inputs used in production
The first group of indicators presented in stage 2 refers to the raw materials and intermediate products that flow into the production processes. Here, it is necessary to record what impact this material use can have on the environment.

Stage 4: Recording of the processes in production
At this stage, the processes that transform a multitude of input materials into end products for delivery and sale are to be examined more closely. The focus is on the main processing and manufacturing steps. This involves the energy intensity, the extent of greenhouse gas production, and other emissions into air and water.

Stage 5: Analysis and evaluation of the products
Here, the focus is on the impacts that the manufactured products have on the environment. For example, the energy consumption during use, the recyclability, and the use of hazardous substances are collected using the presented indicators. After all, these determine the degree of sustainability of the end product.

Phase: Improvement

Stage 6: Understanding the results
This stage involves interpreting the collected data. Here, it is important to identify possible trends in the indicators. The indicators provide important clues for further priority setting. These priorities document the need to develop appropriate measures to reduce undesirable impacts on the environment.

Stage 7: Initiating steps to improve performance
In the previous stages, the data for the relevant indicators were determined and prepared. The status quo is now precisely described and decisions for performance improvement have been made. Now, the adopted measures are to be implemented. For this, precise goals need to be defined, which are to be achieved through an action plan.

These seven stages are to be applied continuously in a **cyclical management process**. This enables the company to deeply understand the environmental effects of the procurement, production, and usage processes of their own offerings. Based on the values of the relevant indicators, the processes can be continuously improved.

> **Note Box**
> The pursuit of sustainable production is not a project with a defined start and end. Rather, it is a continuous process of learning, employing new insights, technologies, and processes to continually reduce negative environmental effects. Therefore, the seven stages should be gone through annually.
> The same applies here: **The beginning is the most important step!**

The example of ***Philips*** illustrates how comprehensively the company has already made the transition to sustainable production processes and thus to the circular economy (see Philips, 2023). *Philips* is intensively engaged with the topic of a more sustainable supply chain in order to achieve responsible and environmentally friendly material procurement and to build more sustainable supplier relationships. To build a **sustainable supply chain**, *Philips* not only aims for **decarbonization,** but also strives for circular procurement. The manufactured devices and their materials should be reusable or recyclable. The circular economy begins with such a holistic view not only with the disposal of the devices, but already with the design of the products (see Sect. 4.4.1.1).

What the **circular economy** looks like at *Philips* is shown in Fig. 4.15. Here, products, components, and materials are kept in their highest utility and thus in the inner cycle as long and as comprehensively as possible. This means that once created products circulate as long as possible among customers before they are subjected to less value-preserving use (decomposition, recycling, etc.). Closed cycles in this way maintain the value of a product as permanently and economically as possible while simultaneously reducing resource consumption and waste quantities (see Philips, 2023).

To this end, *Philips* is transforming its business processes—away from linear processes based on an **Take-Make-Break model**. The aim is to achieve circular processes with an **Make-Use-Return model.** This model includes the five return cycles defined in Fig. 4.15 **return cycles** (see Philips, 2023):

- **Dematerialization/Optimization**
  Dematerialization is about the question of whether physical products can be replaced by digital alternatives. An example of this is the substitution of a printed manual with an online file.
- **Service provision**
  In circular business models, customers only pay for the use of products and solutions—not for the products and solutions themselves. These are specific **subscription pricing models,** which replace a "payment for purchase" with a "payment for use" (keyword "Pay per Use"; see "Product as a Service" Sect. 4.4.1.2). These business models are based either on the length of a usage period, on each individual use and/or on the basis of defined results (see Kreutzer, 2021, p. 203 f.). In addition, on-site or remote services are offered to keep the aggregates running.

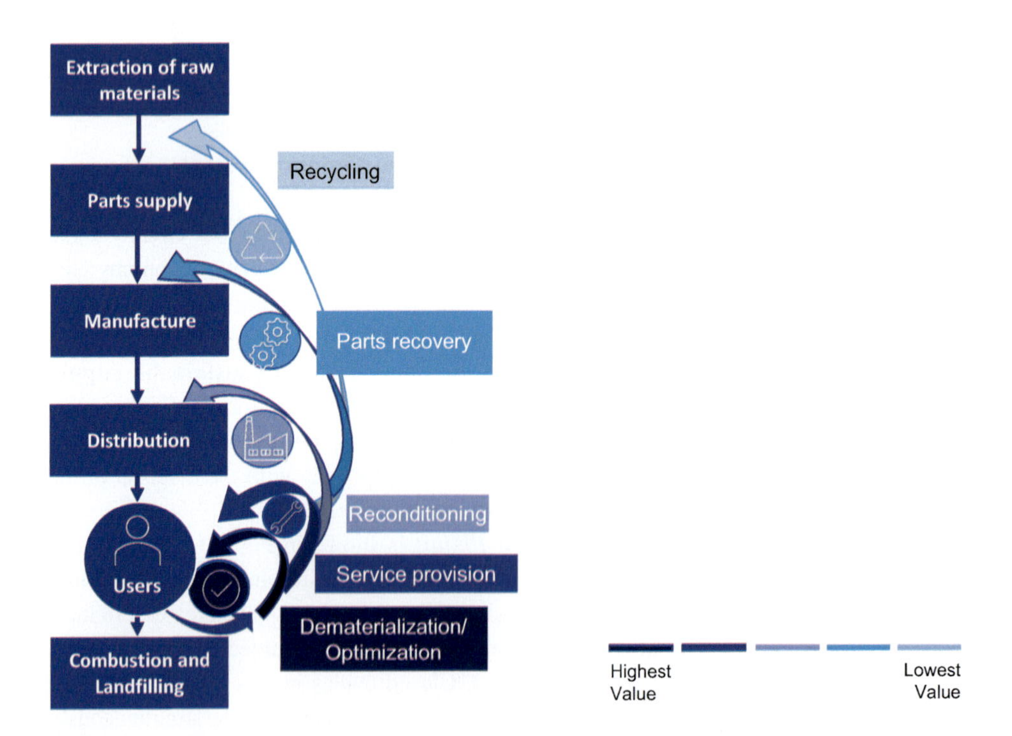

Fig. 4.15 Circular economy using the example of *Philips*

- **Refurbishment**
  For the refurbishment stage, *Philips* offers **product recycling services for medical imaging devices**. Through refurbishment or repair of individual parts, the **lifespan of products** is continuously extended.
- **Recovery of parts**
  Through the process of **decomposition**, parts of the products are to be recovered for reuse. This can at least maintain a high value for these parts, as further use takes place at the same process level.
- **Recycling**
  Products that can neither be repaired nor refurbished are sent for **recycling**. To facilitate recycling, *Philips* issues so-called **recycling passes**. These include important safety information for the recycling companies. This ensures that the products can be safely and environmentally friendly dismantled and recycled.

Philips has set ambitious goals for the **circular economy** for the year 2025 (see Philips, 2023):

- 25% of sales are to be generated with products, solutions and services based on the processes of the circular economy.
- Through trade-in programs for medical devices and solutions, cycles are to be closed and reuse ensured. The refurbishment should be carried out either by Philips itself or on site according to Philips' guidelines.
- Circular practices are to be comprehensively anchored at the company's locations. Disposal of waste via landfills is to be excluded.

To support these processes, ***Philips*** is a member of the **StEP initiative.** Philips also collaborates with the PACE initiative (see Sect. 4.4.5).

Philips and many other companies are trying to establish a **closed supply chain (Closed-loop Supply Chain)**. The **Closed-loop Supply Chain Management** includes the entire **forward logistics in the supply chain.** This includes material procurement, production and distribution. The **reverse logistics in the supply chain** focuses on the collection and processing of returned (used or unused) products and/or product parts to achieve economically and ecologically sustainable recycling. Only when this is achieved can the cycles really be closed.

Food for Thought

Sustainable corporate management is only achieved when a **responsible waste management** is ensured. In this context, the avoidable production of waste still represents the most important lever when companies implement environmentally friendly processes. Waste not only has negative effects on the environment, but also on the public perception of the company. Therefore, every company should strive for proper disposal of hazardous waste—and not export it to third world countries for cost reasons.

The range of such exports still extends today from plastic waste to computer scrap to ships that are dismantled on the coasts of African or Asian countries under inhumane and environmentally degrading standards. Even if avoiding such behavior may seem too costly at first, it can pay off later if a positive profile is achieved as a result.

4.4.1.5 Sustainable Logistics Processes

A important **"resource guzzler"** and **emission driver** are the **logistics processes.** The use of transport means on the road, rail, water and air consumes a large amount of resources and causes emissions. Estimates suggest that transport causes about a quarter of the **greenhouse gas emissions** in the EU. Figure 1.33 shows that transport ranks third in these emissions.

The high **interconnection of global value chains** is a central driver of national and international logistics. This interconnection has been and continues to be a central driver of prosperity in many countries of the world, but also a driver for the associated negative impacts on resources and the environment. The growth seen here was only partially slowed down by the Covid pandemic and will pick up speed again in the future. Consequently, the logistics industry is called upon to provide **climate-, environment- and socially compatible logistics solutions**.

Sustainable logistics solutions aim to minimize the environmental impact and resource consumption of the logistics chain, while at the same time improving the efficiency and performance of logistics processes. There are a number of **requirements for logistics processes,** that need to be considered:

- To increase the **satisfaction of B2C and B2B customers**, transparency about logistics processes in real time should be achieved—combined with a high degree of flexibility and reliability.
- At the same time, **resilience in the supply chain** is becoming increasingly important for companies (including multiple sourcing, warehousing, parallel structures in the supply chain).
- The **energy efficiency** should be increased by using more efficient means of transport and/or by optimizing routes and transport paths.
- A **reduction in emissions** is aimed at by using low-emission means of transport and by avoiding empty runs.
- A **conservation of resources** is aimed at by using recyclable or biodegradable packaging materials or by avoiding overproduction and excess stocks.
- A **social compatibility** of logistics solutions should increase the working conditions of employees.
- At the same time, **financial viability** is indispensable to ensure the financial success of the company.

It is important that sustainable logistics solutions harmonize all these requirements to create a truly sustainable logistics chain. To make progress in sustainability as a **logistics service provider**, the issue of sustainability must also be linked to the company strategy. This includes consistent and meaningful **sustainability reporting** to identify the most important levers for sustainability and to be able to monitor them continuously. For this, it is necessary to develop a **holistic view of logistics along the value chain** across all suppliers. This is partially enforced by the supply chain law (see Sect. 2.4). It can be implemented through a sophisticated value chain analysis and the system of the value chain (see Sect. 4.1).

A look at the **logistics service provider*Deutsche Post/DHL*** shows what steps have already been taken towards sustainability and what is still planned. For example, the offer of **parcel pick-up by couriers** at private customers contributes to the reduction of harmful emissions. This avoids unnecessary routes. In addition, private customers can choose **climate-friendly rail transport** for selected routes and shipments. This saves on average more than 30% of emissions compared to truck transport. The **climate-friendly reception** is also intended to reduce emissions. By coordinating the drop-off location, delivery day, and parcel redirection, costly and emission-intensive unsuccessful delivery attempts can be avoided. This is also supported by **addressing to branches and packing stations**, some of which are already equipped with solar panels. Paper consumption can be further reduced by a **digital delivery notification**; however, this also involves electricity-induced emissions (see DHL, 2023a).

The service ***GoGreen Plus*** from *DHL* is intended to further promote climate-friendly shipping. The goal of *GoGreen Plus* is the **avoidance of CO_2 emissions** in the shipping of goods within Germany. This concept is in a pilot phase with selected corporate customers in 2022. The core aim is to avoid **transport-related CO_2 emissions** on site in Germany. This is to be achieved through investments in alternative fuels and energy sources (such as biogas or green electricity). In this so-called **insetting**, in contrast to the **offsetting**, investments are made directly in measures within the value chain to avoid CO_2 at the place of origin (see DHL, 2023b).

The **mission** of *Deutsche Post/DHL* as part of the strategy ***"Mission 2050"*** is to reduce all logistics-related emissions to zero by 2050. For this, a nationwide sustainability program has been launched (see DHL, 2023b).

4.4.1.6 Sustainable Design of Wholesale and Retail

Many measures for a more sustainable design of logistics processes (see Sect. 4.4.1.6) also contribute to higher **sustainability in wholesale and retail**. In addition, the trading companies bear a great responsibility with regard to the following activities:

- **Use of environmentally friendly packaging materials**
  The use of recyclable or biodegradable packaging materials can reduce resource consumption in trade.
- **Introduction of environmentally friendly products**

If trading companies include environmentally friendly products in their assortments, this directly affects purchasing behavior. Motto: Supply creates demand. Through intensive cooperation with suppliers of sustainable products, the sustainability of the entire supply chain increases.

- **Avoidance of food waste**
  Trading companies still destroy large quantities of food—unless they are given away free of charge to food banks.
- **Use of energy-saving technologies**
  Trading companies can contribute to reducing energy consumption by using energy-saving technologies such as LED lighting or energy-efficient cooling systems. This also includes keeping the store doors closed in winter.
- **Use of solar or wind energy**
  Many trading companies have large roof areas that are suitable for the use of solar collectors. By generating their own electricity, CO_2 emissions in trade could be reduced.

A **sustainable trade strategy** includes that all measures are coordinated and support each other. The extent to which a **sustainability in food retailing** (LEH) is actually already being implemented is shown by a study by the Federal Environment Agency. This study analyzed the **environmental commitment of the eight highest-grossing LEH companies** in Germany. These companies include—sorted by turnover—*Edeka Group (Edeka, Netto Markendiscount), Rewe Group (Rewe, Penny), Schwarz Group (Lidl, Kaufland)* and the *Aldi Group (Aldi Süd, Aldi Nord).* The focus of the study was on the **supply chains** (purchasing and cooperation with suppliers), the respective **locations** and the **communication with customers.** The recording and evaluation of the environmental performance of the retailers took place in a total of 22 fields of action. For this purpose, 43 indicators and 112 sub-indicators were used. The processed data comes from publicly available sources. In addition, company-internal information was obtained through questionnaires. The **central results of this study** are shown here (see Federal Environment Agency, 2022a).

- Often, the **sustainability strategies of corporations** are insufficiently anchored organizationally and procedurally and therefore only guide actions to a limited extent. The public statements on environmental protection and sustainability then do not match the purchasing strategies, the pricing of different offers, and the advertising appearance.
- Companies perform well in **reporting on environmental goals**. The **increase in energy efficiency** in branches and production facilities is also specifically pursued. To this end, retailers invest in measures for more efficient resource use.
- Also, good results are sometimes achieved in **environmental campaigns** and in further **measures to raise awareness of consumers**.

- However, food retailers could make much more use of their scope for action. This applies in particular to **product range design** and **measures to reduce food waste**. In addition, communication to **raise consumer awareness** could be increased.
- With the **product range design** there are pros and cons. Over the past few years, food retailers have expanded their range of own **organic brands** and **vegetarian and vegan products**. Now, 62% of sales of organic food are made in conventional retail. For traditional risk raw materials—these include coffee, cocoa, and fish—large **proportions of certified goods** are already included in the range.
- When it comes to **(sustainable) purchasing of goods**, retailers could be much more committed. In **product range design, environmental protection aspects** can be taken into account more strongly. For example, the offer of air-freighted pineapples or mangoes could be dispensed with. A stronger **regionalization of suppliers** could also contribute to emission reduction—coupled with the question of whether fresh strawberries and fresh asparagus really belong in the range almost all year round. In addition, the **offer of plant-based alternatives** to animal foods could be expanded. In the context of *animal welfare,* for example, *Aldi* announced in 2021 that it will completely convert its fresh meat range to products with animal welfare farming methods 3 and 4 by 2030.
- Instead of leading the way through a more sustainable design of the product range, companies are driving the **development of a new environmental label**. The responsibility for a more environmentally friendly selection is thereby delegated from the trade to the consumer. As a result, retailers—at least in part—give up the opportunity to make the product range more environmentally friendly of their own accord.
- A more sustainable design of the product range also includes greater **transparency about the structure of the own product range**. However, hardly any company has key figures such as "sales share of meat and dairy products" and "sales share of plant-based (substitute) products" in its product range controlling. Half of the retailers could not provide any information on this—some only qualitative assessments. Reliable data, however, are indispensable for a sustainable product range orientation.
- **Campaigns to reduce food waste** have been launched. This is mainly done with fruit and vegetables. There is often also a **commitment to the sustainability goal** of halving food waste per capita at the retail and consumer level by 2030. However, seven out of eight companies were not yet able to quantify the reduction of food waste at their own locations. The companies are now working on determining the reduction quantity within the framework of the dialogue forum ***"Too good for the bin"***. But there is still a lot of room for improvement here.
- In addition, the **sensitization of consumers for more sustainable offers** should be significantly intensified. This area includes measures of **store design** and the **placement of sustainable offers**. In addition, the **advertising activities** of retailers could motivate customers much more strongly to make **more environmentally friendly purchasing decisions**. The study showed, for example, that animal, environmentally

harmful products were advertised much more extensively than more environmentally friendly plant-based alternatives. To be precise: almost 20 times more often.

In summary, this study recommends that retailers should manage their **sustainability** much more systematically overall. This includes, above all, the definition of verifiable **goals for ecological sustainability.** Specifically, this involves verifiable climate goals and goals for deforestation-free supply chains. Reliable data is needed to increase the **transparency of purchasing decisions.** This transparency is often still lacking. In addition, it is recommended to link the activities aimed at sustainability much more strongly with the management, purchasing, and the management of the product groups (cf. Federal Environment Agency, 2022b).

> **Note Box** Sustainability management is a team sport!

In parallel, the *Federal Environment Agency* calls on **politics** to take accompanying measures. These include **financial incentives.** By realigning the value-added tax, more sustainable foods could be taxed at a lower rate. This could guide purchasing behavior through price incentives. In addition, **regulatory measures** are recommended to achieve the internalization of external costs (cf. in depth Sect. 1.5). Furthermore, **minimum standards** could be defined for the purchase of raw materials (e.g., for palm oil and soy). It is also necessary to ban harmful fishing methods (cf. Federal Environment Agency, 2022b).

Overall, it becomes clear what challenges—not only in food retailing—need to be mastered. Every **trading company** should therefore check how—oriented towards legal requirements and the expectations of stakeholders—its own **product range design** can be trimmed towards sustainability. Many retailers and wholesalers prominently communicate on their websites that they want to align their product ranges more sustainably. Until now, sales and profit goals were primarily in focus when designing product ranges. Now—in addition—**criteria of ecological and social sustainability** must be taken into account. However, a product range oriented towards sustainability is an indispensable success factor to help sustainable action break through.

> **Note Box** Trading companies should not wait for the legislator before a stronger focus on sustainability takes place. Companies that only follow legal requirements miss the chance to profile themselves as a sustainable company—before everyone else does!

A **sustainable product range design** can start at various points:

- **Selection of suppliersbased on sustainability criteria**
  A first important step towards a more sustainable product range planning is the selection of suppliers, which is oriented towards the ESG criteria and the requirements of the supply chain law—even if there is (still) no obligation to do so. According to the already quoted motto "Supply creates demand", wholesalers and retailers can thus influence purchasing behavior—within limits—towards sustainability.
- **Selection of offersoriented towards sustainability criteria**
  Closely linked to the selection of suppliers is also the choice of offers that are included in the range. The following questions can help with range design:
  - Are the products organically produced, fairly traded, and recyclable?
  - Were the products produced in an environmentally friendly or carbon-neutral way?
  - Should products be marketed that "give up the ghost" after a short period of use?
  - To what extent should products from a fast-fashion collection be marketed?
  - Should products be marketed that cannot be repaired and therefore have to be disposed of in the event of a defect?

  Among the questions to be clarified here is also the decision on which **form of packaging** is chosen:
  - Are chargeable plastic bags offered?
  - Is paper used for packaging?
  - Are the goods handed over to the buyer unpackaged?
  - To what extent should extensively shrink-wrapped food be offered as convenience food?
  - Is completely unpackaged goods included in the range?
  - Can customers bring their own containers to fill goods?
  - Are refillable packaging offered?
- **Inclusion of upcycling productsin the own offer**
  An important step towards a sustainable range is the inclusion of upcycling products. *Lufthansa* offers an **Upcycling Collection** of lifestyle products. These unique items are made from materials from decommissioned aircraft as well as from onboard utensils, uniforms, and other items from the *Lufthansa* world. This results in furniture, bags, and other utensils that breathe new life into original waste (see Lufthansa, 2023a). Corresponding products can also be included in the range by third-party providers to use finite resources longer.
- **Pricing positioningof sustainable offers**
  At what prices are sustainable offers positioned? Is a **price skimming strategy (Skimming Pricing)** pursued, i.e., high prices? After all, customers who focus on sustainability are also willing to pay more. However, this will only slowly develop sustainable consumption. This strategy is often seen in traditional retailers.

  Or is a **price penetration strategy (Penetration Pricing)** pursued with low prices to convince as many customers as possible of the sustainable offer as quickly as

Fig. 4.16 Unpackaged goods and elaborately packaged goods in the same organic store. *Source* Photo by the author

possible? Here, discounters like Aldi and Lidl have taken the lead, who developed an organic offer across the entire range earlier than others. A significant increase in demand for organic products was therefore triggered by the discounters.

- **Spatial positioningof sustainable offers**
  Retail companies have a significant influence on purchasing behavior through the spatial presentation of goods. Among other things, the following questions arise:
  - Where are the sustainable offers located—on the floor or in the online shop?
  - Are sustainable products and services aggressively "put forward" so that they are perceived by as many customers as possible and thus also purchased?
  - Or are such offers positioned rather in "back corners" of the offline and online sales space? Perhaps also to be able to say afterwards: Nobody wanted to buy the sustainable products!
- **Packagingof sustainable offers**
  Especially organic markets have now started to offer **unpackaged goods**. For this purpose, unpackaged stations **Unpackaged Stations** are placed in the stores. This is the case at *Bio Company* and Tegut. However, it can happen that unpackaged goods are offered in the direct vicinity of elaborately packaged convenience products. The photos shown in Fig. 4.16 were both taken in the same organic supermarket in Bonn. The flood of packaging on the one hand and the reduced "unpackaged offer" is not really consistently aligned with "organic" and "zero waste"—quite the contrary.

> **Food for Thought**
> It is exciting—and also frustrating—to see that in some organic stores not only the use of **unpackaged stations** is increasing. Often there is a simultaneous massive

increase in elaborately packaged food (so-called convenience food), while fresh food counters are losing importance. Here, some forms of distribution—despite "organic claim"—are taking an increasingly less sustainable path.

Then the rule is: Instead of waiting at the counter for freshly sliced goods, you can also buy—comprehensively packaged—six slices of salami. Here, the packaging sometimes weighs more than the "meat filling".

- **Advertising Positioningof sustainable offers**
  Retailers can also steer demand through advertising presentation. The following questions arise:
  - How extensively are sustainable offers advertised?
  - Do these—possibly analogous to the presentation of offers—lead a rather neglected existence?
  - Or are sustainable offers aggressively put forward—also to position the company as sustainable?
- **Handling of perishable Goods**
  A major challenge continues to be the handling of perishable goods. Here, large volumes of valuable food are still being lost. Therefore, the following questions should be asked:
  - How is perishable goods handled—destruction or passing on to food banks?
  - For example, is it demanded from the integrated distribution partners when selling bread and rolls that they must offer the complete range until closing time?

> **Food for Thought**
> Is it not reasonable for a customer today that especially fresh food can also be sold out? Especially shortly before closing time? The necessity for this is not only derived from sustainability, but simply from an appreciation for food that should not be thrown away.
>
> Here, the handling of best-before and expiry dates also needs to be checked. Sustainable retailers offer such products at a reduced price to sell them before they spoil. Most products can still be consumed well after these dates have expired.

Lidl has launched a campaign **"Save me"** to counteract the destruction of about eleven million tons of food per year in Germany. Of this, 59% is attributable to private households and 7% to trade. Against this background, *Lidl* has introduced the so-called ***Saver Bag*** in all 3200 branches nationwide as part of a **"Save me" concept.** In this bag, less perfect looking but edible fruit and vegetable products are offered at a discounted flat rate of 3 € per bag (cf. Lidl, 2022).

The *Rescue Bag* is, in addition to the donation of food to food banks, another component of **Lidl's Sustainability Strategy 2030.** In this, the company has committed to reducing food waste and organic waste by 30% by 2025. Since 2020, *Lidl* has been offering **products with a short shelf life** at reduced prices in **"I am still good" boxes** throughout Germany. The final step is the utilization of no longer edible food in **biogas plants** (see Lidl, 2022). All these measures fit perfectly into a circular economy.

With regard to sustainability in trade, **fashion** is often at the center of heated discussions—and rightly so. A study by ***Kearney*** analyzed the **environmental pollution of the fashion industry.** European consumers spend over 200 billion euros per year on fashion. A large part of this is spent on **fast fashion.** According to estimates, the **CO_2 emissions of the fashion industry** amount to 1.2 to 1.7 billion tons. Production and distribution cause 94% of these emissions. The fashion industry thus emits more harmful emissions than air and sea transport combined. The fashion industry's share is about 3 to 5% of global CO_2 emissions. In addition, the fashion industry accounts for about 11% of the fresh water used in the entire industry (see Kearney, 2020).

As part of this **fashion study**, 8000 Germans were also asked about their consumption behavior. The following findings were obtained (see Kearney, 2020):

- On average, Germans own 97 pieces of clothing.
- On average, 17 new pieces are bought per year.
- After six years, the clothes are discarded because…
  - the clothing is damaged or worn out (36%),
  - they no longer fit (28%),
  - they are no longer liked (20%),
  - space is being made for new clothes (11%),
  - the clothing was bought for a one-time purpose (5%).

In the fashion sector, a **systemic cause for the high environmental pollution** can also be identified. The vast majority of fashion manufacturers produce blindly—not knowing whether the produced goods, which often travel from the Far East for several months, will meet with interest from customers. If a cut, a color, or a combination does not catch on, the goods are usually destroyed. The companies need much more data to avoid such overproduction (see the digital value chain in Sect. 4.1). To gain more data and evaluate it in a targeted manner, *Boss* has set up a Digital Campus in Portugal. There, 300 employees are trying to answer the question: "Where, what, when does the customer want? We give this to the designer as a manual."—so says *Boss* CEO Grieder (2022, p. 20).

The ***Circular Fashion Index*** by *Kearney* determines how sustainably Europe's 100 fashion brands operate and whether or how they extend the lifecycle of their clothing. Eight criteria are used to determine this index. These capture both the **primary market** (with new products) and the **secondary market** (with second hand and recycling). According to these results, only three of the 100 largest European fashion brands are on the path to a sustainable, **circularly designed business model:***Patagonia, The North Face* and *Levi's.* These three companies emphasize in their communication that their products should last longer (cf. *Patagonia* Sect. 5.4.3). In addition, the companies encourage their customers to think about the environmental impact of purchasing another piece of clothing. Some of these brands are already working with recycled clothing or rely on **Cradle-to-Cradle materials.** These are, for example, biological nutrients that can be returned to the natural cycle (cf. Kearney, 2020).

Other companies strive to highlight the topic of "sustainability" at least occasionally in **range design** and advertising. Thus, ***H&M*** promotes a product category in its online shop called ***Conscious Choice.*** The "more sustainable products presented under this label are made with particular consideration for the planet: They consist of at least 50% more sustainable materials such as organic cotton or recycled polyester. The exception is recycled cotton, which for quality reasons makes up a maximum of 20% of a product" (H&M, 2022).

It is interesting here how cautiously—and thus transparently—*H&M* communicates. They speak of "more sustainable" products. In addition, it is clearly communicated that only a lower recycling content can be achieved with recycled cotton for quality reasons. This achieves transparency and credibility in communication. However, it must be noted that the proportion of *Conscious Choice* in the overall range at *H&M* is still relatively small.

On the website of ***C&A***, under the heading **"Wear the Change"**, it is postulated (C&A, 2023):

> "Let's wear fashion responsibly. For today and for tomorrow."

This is followed by the appeal to wear new outfits that also contain recycled materials. Sustainable fashion should become a matter of course. *C&A* is now even making the corporate motto "Wear the Change" the **brand slogan** for its commitment in the area of sustainability. This is intended to place a stronger focus on the topic of sustainability. The associated changes not only affect the collections. In addition to better products, better working and living conditions are also to be offered to their own employees and partners in the manufacturing countries.

To this end, *C&A* (2023) states:

> "Especially now, when our culture, environment, society, traditions and fashion are undergoing a strong change and the desire for a more sustainable lifestyle is becoming clear, C&A presents itself as a progressive advanced brand that goes one step further and leads a traditional family business into a greener future. That's exactly why we continuously adapt our business and our operations. We focus on our customers and their needs, which are changing just like the world. At the same time, we try to reduce the footprint that arises in our company and along the supply chain. … C&A is the perfect brand for lively, authentic and modern people with a positive attitude—for open and relationship-oriented people who value values and make no compromises in terms of sustainability. 'Wear the Change' stands for all this—for a better world—today and especially tomorrow."

In addition, *C&A* (2023) makes an important promise:

> "Of course, our sustainable products are still recognizable at first glance—just look for the green symbols that indicate the respective sustainability aspects."

In this way, the **seriousness of one's own approach** can be conveyed and important **orientation in the purchasing process** can be provided.

The extent to which various **fast-fashionretailers** in Europe have already transitioned to the circular economy was determined in the countries of France, Germany, Italy, Spain, and the United Kingdom. This involved the inclusion of **sustainable fashion collections** in their own range. In the first quarter of 2021, ***H&M*** had almost 25% of the highest proportion of sustainable collections among the marketed clothing lines (see Statista, 2022c, p. 28). At ***Mango***, a significant increase can be seen between 2020 and 2021. Interestingly, the proportion of sustainable clothing at *Zara* and ***C&A*** has significantly decreased from 2020 to 2021. This clearly shows that words and actions do not always align (see Fig. 4.17).

Many other companies are also working to make their ranges more sustainable. This is regularly reported on the websites of the respective retail companies, such as *dm Drogeriemarkt* or *Kaufland.*

4.4.2 Emissions Trading and Compensation

To provide compensation for environmental pollution, compensations can also be used. The **philosophy of compensation** is that it is not crucial for the global climate whether

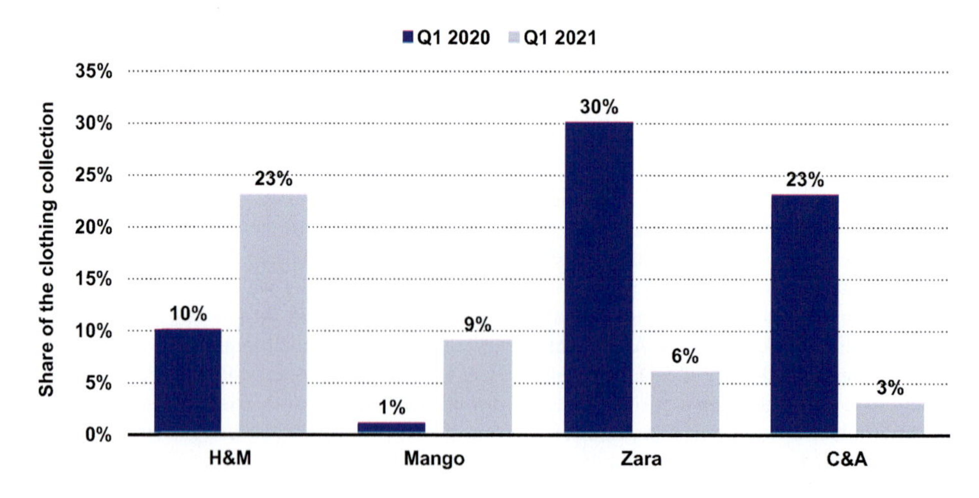

Fig. 4.17 Proportion of sustainable clothing at Fast-Fashion Retailers in Europe—1st quarter 2020 and 2021. (Data source: Statista, 2022c)

emissions are offset at the place of origin. Therefore, the emissions caused by a major event in Berlin can also be compensated by measures in the Amazon region.

The following **forms of compensation** can be distinguished:

- Companies provide compensation due to legal obligations.
- Companies provide compensation on a voluntary basis.
- Customers provide compensation on a voluntary basis.

Until green electricity, green hydrogen, green aviation fuel, and green steel are available, and all other production and consumption steps can be set to "green", companies and consumers cannot avoid dealing with the **emissions trading** and **compensation of harmful entries into the environment**.

For the foreseeable future, many forms of service provision will continue to have external effects and associated external costs. These will often be reduced, but often not completely avoidable. Here, for example, we are thinking of **production processes** where CO_2 emissions occur. If you do not want to completely stop production, other forms of coping are required. For this, the so-called **emissions trading** is used. Through the required purchase of **"pollution certificates""**, the costs of external effects are internalized. This gives pollution a price, which is borne by the company and passed on to customers through prices. This overcomes the **market failure** described in Sect. 1.5.

> **Note-Box** Through emissions trading, an **internalization of external costs** takes place. This overcomes **market failure**.

When companies compensate for the emissions associated with their services due to legal obligations or their own initiative, market failure is—at least partially—overcome. However, such acting providers may suffer **disadvantages in price competition**. After all, the **internalization of external costs** leads to higher costs and possibly also to higher prices. Whether such disadvantages come into play depends on the clientele of the companies—and on the company's communication strategy.

> **Note-Box** Compensation is, in a sense, an ecological indulgence trade! Companies have to pay for their sins in the form of climate-damaging emissions.

The **emissions trading** is a **market-based instrument,** to reduce the emissions of CO_2 and other greenhouse gases. To this end, governments determine how many tons of CO_2 a certain group may emit in total. Anyone who belongs to the group—e.g. certain companies—and emits CO_2 must purchase an emission allowance for each ton of CO_2 emitted. These allowances can be purchased by the members of the respective group, for example, at state-organized auctions. Anyone who emits CO_2 without such an allowance is acting illegally and must pay penalties. If companies emit only a small amount of CO_2, only a few allowances need to be purchased. Those who burden the climate less are therefore rewarded (see BMUV, 2022).

> **Note-Box** If "pollution certificates" are sold too cheaply, the incentive for companies to vigorously work towards reducing emitted greenhouse gases is lost. If such certificates are sold too expensively, the affected companies may relocate to other countries with less stringent environmental regulations.

The **European Emissions Trading System** (EU **ETS/Emissions Trading System**) was introduced as early as 2005. All major power and heat generation facilities are required to participate. Large industrial plants (e.g. steel mills, refineries, and cement plants) and since 2012 also aircraft operators must purchase allowances for their emissions. In Europe, currently about 11,000 facilities—in Germany about 2,000—and several hundred aircraft operators are subject to emissions trading. In 2021, an additional emissions trading system was introduced in Germany for almost all other CO_2 emissions resulting from the combustion of fossil fuels. This also applies to road traffic and heating. However, the corresponding **obligation to surrender emission allowances** does not affect car drivers or homeowners. The obligation lies with the so-called "introducers". These are the companies that first sell diesel, gasoline, etc. in Germany (see BMUV, 2022).

Companies can also **voluntarily** undertake a **compensation of negative entries into the environment**. At the tour operator ***Chameleon,*** such an **internalization of external costs** for trips is achieved by including the **compensation measures** in the travel price. Anyone who books here can travel with a good feeling. It is not left to the traveler's decision whether to compensate or not. The decision has already been made by the organizer. From this, each guest receives a certificate for a personal piece of forest. Its size depends on the emissions of the respective trip (see Chameleon, 2022, p. 5, 393, 450; see also the contribution by *Chameleon* in Sect. 5.4.5).

A **voluntary compensation** for private customers is carried out by ***Deutsche Post/ DHL***. Since 2011, all national parcels and packages of ***DHL*** **private customers** have been transported CO_2-neutral with ***GoGreen***. This is achieved through investments in global climate protection projects. Since July 2022, *GoGreen* is also automatically included in international shipments from private customers. There are no additional costs for private customers (cf. DHL, 2023a). In addition, *DHL* offers companies an **optional compensation** for a climate-neutral shipping of **national advertising and press mailings**. For this, a special *GoGreen* contract must be concluded, which regulates the **costs of compensation**. From a minimum quantity of 50,000 shipments per year, companies can contribute to climate protection when shipping national advertising mailings and press products (cf. Deutsche Post, 2023).

If companies do not take action themselves—as in the case of *Chameleon*—then **CO_2-credits** can be purchased directly from project operators via consulting firms. However, a **government seal of approval for the certificates** is often lacking. To ensure that a **high-performance certificate** is obtained, it is recommended to orientate oneself towards the ***WWF***-co-developed ***Gold Standard***. By meeting the following criteria, the compensation projects are certified to actually lead to a **reduction of greenhouse gases** and at the same time make a **contribution to the sustainable development of the participating countries** (cf. WWF, 2023; Gold Standard, 2023):

- The supported projects must have a **supplementary character**.
- The selected projects focus on **renewable energies** and/or serve to **increase energy efficiency**.
- **Sustainability criteria** must be observed.
- Through clear consultation procedures, the **participation of the local population** in the projects is ensured. This should ensure a balance between environmental requirements and practicality.

Only when **all four requirements are met** do the compensation projects comply with the ***Gold Standard***. Anyone interested in compensation projects should check whether the requirements of the *Gold Standard* have been reviewed in the selection of projects. By promoting such projects, significant amounts from the economy can flow into climate protection. Therefore, it is important to pay attention to partners who have earned a special credibility.

One relevantly active company is ***atmosfair***. This non-profit organization supports the compensation and reduction of climate-damaging greenhouse gases. The company, founded in 2005, offers climate-damaging emitters (e.g., airlines, hotels, organizers) a **Climate compensation**. Companies and private individuals can determine climate-impactful emissions of their own activities at *atmosfair* and purchase **emission certificates** in the required amount. *atmosfair* promises that for each purchased certificate, one ton of greenhouse gas emissions will be saved in a climate protection project in the global south (cf. atmosfair, 2023).

Atmosfair reported revenues from donations and from economic business operations amounting to 21 million euros in its 2021 annual report (cf. atmosfair, 2022, p. 20). Such an amount is less than a drop in the bucket in terms of climate change. How companies can communicate the topic of compensation is discussed in more depth in Sect. 5.3.8.

The ***TÜV NORD CERT*** and other institutions certify climate protection projects according to the specifications of the ***Gold Standard*** in voluntary emissions trading (cf. TÜV, 2023). Other standards include the ***Verified Carbon Standard*** (cf. Verra, 2023), the ***Fairtrade Climate Standard*** (cf. Fairtrade, 2023) and the ***Climate, Community and Biodiversity Standard*** (cf. CCBA, 2023).

As already indicated, the **customers** themselves can also be motivated to make a **compensation on a voluntary basis**. This approach will be illustrated using the example of air travel. Anyone who embarks on a **flight** burdens the environment. How can travel providers deal with this? On the one hand, providers—as shown in the example of *Chamäleon*—can make the compensation independently and include it in the price.

On the other hand, travel providers can offer customers a **chargeable offer for compensation**. Such offers can already be found with many airlines today. During the booking process, the customer is asked whether they want to offset the emissions of their flight.

The ***Lufthansa*** offers to make the flight "directly in the booking process" CO_2-neutral. The following options are offered to the passenger, although not all of them represent real compensation (cf. Lufthansa, 2023b):

- **Promotion of innovative fuel technology**
  Sustainable Aviation Fuel (SAF)—so-called sustainable aviation fuels—are not made from fossil raw materials, but from used cooking oils and fats as well as from agricultural waste. This can save about 80% of CO_2 emissions compared to conventional kerosene. To achieve a complete CO_2 reduction for the passenger, Lufthansa uses the required amount of SAF. This specifically avoids CO_2 emissions.
- **CO_2compensation through climate protection projects**
  Lufthansa supports climate protection projects worldwide together with *myclimate*. These include the renaturation of moors in Germany, the construction of biogas plants in Brazil, the use of energy-saving stoves in Kenya, Peru or Burundi, and the protection of forests in Tanzania. The passenger contributes to climate protection, promotes biodiversity and improves living conditions for people on site. The focus here is partly on real compensation of CO_2 emissions. However, the use of energy-saving stoves does not compensate for Lufthansa's emissions, but reduces emissions elsewhere. In total, the overall amount of emissions is reduced.
- **Combination of measures**
  Here, investments in climate protection projects and the use of Sustainable Aviation Fuel are combined.

Figure 4.18 shows an example of how the compensation options are displayed in the booking process of *Lufthansa*. The amounts to be invested for compensation for a one-way flight from Berlin to Vienna range from €1.51 to €55.04. All options can be selected voluntarily—or not.

Even already booked or completed flights can be quickly and easily offset by integrating the offer of ***Compensaid*** operated by the *Lufthansa Innovation Hub* (cf. Compensaid, 2023).

According to *Lufthansa*, more passengers are now willing to make such a compensation. However, the values are still—depending on the route—only between 4 and 9% of all flights. In previous years, only one percent of customers were willing to make a compensation payment (see Kotowski, 2022, p. 18).

Advertising campaigns themselves also cause a CO_2 footprint. But how big is it? Here, we talk about the ***Green GRP***. GRP stands for Gross Rating Point and serves as a measure of advertising pressure. Specifically, the GRP measures the gross reach in per-

Fly CO_2 -neutral

The CO_2 emissions of one person for the Berlin-Vienna flight amount to 75 kg for the entire flight.

1.51 €	12.22 €	55.04 €
100 % climate projects	80 % Climate projects	100 % sustainable aviation fuel (SAF)
	20 % sustainable aviation fuel (SAF)	CO_2 emissions are reduced immediately
Add to shopping cart	Add to shopping cart	Add to shopping cart

Fig. 4.18 Options for compensating CO_2 emissions at *Lufthansa*

cent within a predefined target group potential. The *Green GRO* is about the **ecological footprint of an advertising campaign**.

To make progress here, ***Mediaplus*** has established an interdisciplinary working group on sustainable media. In addition to economic sustainability, ecological and social sustainability should also be considered in media use. The goal is the **development of sustainability standards** for the market and the **introduction of sustainable media products.** After its introduction, the *Green GRP* should be available to agencies, marketers and advertisers. After all, only knowledge of the CO_2 footprint of an advertising campaign enables companies to book their advertising in a climate-neutralized way in the future. *Mediaplus* offers its advertising customers the option to compensate for the CO_2 emissions caused by a campaign. For *Mediaplus*, the Green GRP is an important part of its own sustainability initiative (see Mediaplus, 2023).

But how is the *Green GRP* determined? If a customer decides for the **climate neutralization of an advertising campaign**, *Climate Partner* determines the CO_2 footprint of the campaign based on the planned consumption data. This footprint depends on the circulation, the paper used, the power consumption for digital processing, etc. From this, the costs for a corresponding **compensation** are derived. In addition to the advertising campaign, the customer receives a report on the CO_2 footprint and an invoice for the compensation through investments in the selected climate protection projects. The advertising customers can advertise to their customers with the label "climate neutral" or more correctly "climate neutralized" campaign.

The label awarded by ***Climate Partner*** for climate-neutral products, services and companies confirms that the remaining CO_2 emissions after a reduction has already been initiated have been offset through certified climate protection projects. The "climate neutral" label is intended to provide transparency and ensure traceability of climate protection measures. The **certified offset process** documented by the label is reviewed annually by *TÜV Austria*. For this, *Climate Partner* offers a system of ID numbers. Stakeholders can trace the offset of emissions via these ID numbers. Customers can see in the tracking system which climate protection projects have compensated for which product. For this, the "climate neutral" label contains an ID number and a tracking ID. This process is also certified by *TÜV Austria* (see Climate Partner, 2023). The extent to which companies can actually operate "climate neutral" is discussed in Sect. 5.3.8.

The company NOAH (2022) has found for 60 large companies from the DACH region that these companies have spent approximately 211 million US dollars on climate certificates since 2011. Through this payment, the companies were able to credit themselves with the avoidance of 17 million tons of carbon dioxide. This results in an average price of about 12 US dollars per ton. However, it was determined that only a part of the paid amounts is actually invested in climate projects. According to this study, up to 60% of the amounts are consumed for brokerage fees by brokers, but also by lawyers and consultants. On site—so the analysis—often less than 10% of the money arrives. The non-governmental organization Carbon Credit Watch (2023) even speaks of outright **garbage certificates** (see o. V. 2022a, p. 19).

> **Note Box** There are hardly any creative limits to the possibilities for compensation. However, it is important to work with reputable partners who, for example, adhere to the *Gold Standards*. Nevertheless, compensation remains the third best alternative after avoidance and reduction—and not always do the funds reach the projects to the desired extent.

4.4.3 Round off the existing business model with concepts of the circular economy

Companies can not only strive for greater sustainability of the existing business model and/or compensations, but also implement a **rounding off of the existing business model** with concepts of the circular economy. In this case, an existing company—in addition to its core business—establishes another business model that primarily aims at the goals of the circular economy. All companies are called upon to intensively examine in which form the existing business model can be supplemented with components of the circular economy. Selected concepts will be presented below.

4.4.3.1 Amazon Marketplace

In addition to its classic e-commerce activities, ***Amazon*** opened the Marketplace in 1999. Initially called Z-Shops, it was renamed to ***Amazon Marketplace*** in 2000. *Amazon Marketplace* is an e-commerce platform where third-party sellers can sell new and used products at fixed prices alongside *Amazon's* regular offerings. This gives third-party sellers—even for used products—access to *Amazon's* customer base. *Amazon* was thus able to significantly expand its range of services without having to invest in additional offerings itself. At the same time, used products can be kept in the market for longer.

For the **sellers**, the incentive is to generate revenue for unwanted or unused products. **Buyers** have the opportunity to purchase desired offers at lower prices than for new products. A **win-win-win-win**-situation—for sellers, buyers, *Amazon*—and for the environment! To be fair, it should be noted that for *Amazon*, the entry into the circular economy was probably not the main reason for the establishment of *Amazon Marketplace*, but simply the opening up of another source of revenue.

The companies active on *Amazon Marketplace* use the **network effects** of *Amazon's* platform. Through *Amazon Marketplace*, **two-sided markets** are connected via a platform. On one side are the suppliers, on the other side the demanders. **Positive network effects** occur because the presence of many suppliers is advantageous for the demanders. Consequently, demanders go where many suppliers can be found, because they expect a large selection and possibly high price pressure. The suppliers themselves are pleased about many demanders. Therefore, the suppliers (with their offers) go where they can reach many demanders via one channel.

What **proportion of e-commerce sales** from *Amazon* is attributable to the **sale of used items** cannot be determined from the company's figures (see Amazon, 2022). *Amazon* also wants to end the **mass destruction of new goods** criticized in Sect. 1.2. If *Amazon* itself or traders active on *Amazon Marketplace* cannot sell their products, these often end up in landfills or incinerators along with returned items. To avoid this, *Amazon* plans to involve **remainder dealers** and a specific ***Amazon*-Outlet** to sell these goods at a lower price. Through these third-party sellers, the traders could still achieve between 30 and 60% of the original price. In addition, *Amazon* wants to bring used items into sales in the future. For this purpose, new **shipping and sales options** are to be introduced (see Goebel, 2021).

In the context of *Amazon* and other shippers, there is frequent complaint that these companies often destroy returned and/or damaged but still usable goods, rather than donating them to charitable organizations. Why is this the case? The reason given for this is that it is cheaper for companies to destroy such goods rather than to donate them. The reason is that there has been **no tax exemption for in-kind donations** so far. This would mean that a tax payment of 19% would be due on an in-kind donation of 1000 € (190 €), even though the company itself has not made any turnover. The destruction of products, on the other hand, only costs between 100 and 200 €—per ton!

A change in this **VAT treatment of in-kind donations** would have to take place at EU level. However, such a change is not planned, although discussions on this continue. The *Federal Ministry of Finance* has **systematic concerns** against a **VAT exemption for in-kind donations** (with input tax deduction). A corresponding tax exemption would lead to an **untaxed final consumption** and thus run counter to the objectives of the **VAT system directive**. A VAT exemption for in-kind donations (with input tax deduction) would lead to a **disadvantageous treatment of private consumers**. Private individuals who buy items to donate them for charitable purposes would still have to pay VAT. In the case of a VAT exemption for in-kind donations for entrepreneurs, items could be handed over without VAT burden (cf. BFM, 2021).

However, in 2021 there was already a **equity rule,** to facilitate the reduction of the stockpiles accumulated by traders during the lockdown of the Corona pandemic. Therefore, in-kind donations were exempt from VAT until the end of 2021. An equity rule was also introduced in 2022 for donations to people affected by the war in Ukraine.

> **Food for Thought**
> **Tax regulations** stand in the way of **introducing functional products into the circular economy through donations**. The legislator prefers to accept that—even brand new—goods are destroyed, rather than providing for an adjustment of the legal framework conditions in order to conserve resources.
> How long can and do we want to afford this?

4.4.3.2 Zalando Pre-owned

The **clothing industry** requires large amounts of water for the production of cotton, is one of the largest producers of CO_2, and is also one of the largest waste producers. Estimates suggest that three out of five of the 100 billion garments produced each year are disposed of in landfills within a year, releasing toxic chemicals. In addition, approximately 10% of CO_2 emissions are caused by the fashion industry—more than air and sea transport combined. Furthermore, about 11% of the fresh water used in the entire industry flows into fashion industry factories. The vast majority—about 85%—of the negative environmental impacts occur in the producing countries. However, customers—for example in Europe—are largely unaware of this (see Fisch, 2022, p. 12).

It is clear that the dominant **linear fashion system** is not sustainable: fashion items are bought, worn by one person—or not—and in the worst case disposed of in landfills. This can happen after days, weeks, months or even years. Only a small portion of the no longer used clothing is reused through collections or appropriate clothing containers.

To address this issue—at least on a small scale—***Zalando*** has developed the ***Zalando Zircle***, a contribution to the circular economy in the fashion industry. With its sustainability strategy *do.MORE*, the company has set itself the goal of applying the **principles of the circular economy** by 2023 and extending the lifespan of at least 50 million fashion products. The focus here is on waste prevention and better resource use. The targeted 50 million is a commendable goal—but given the total of about 100 billion garments produced per year, it is just a drop in the bucket—to be precise, it is only 0.05%!

Specifically, *Zalando Zircle* aims to repurpose garments—even after the end of their individual use phase—for further, similar use. In this way, at least a small contribution can be made to extend the lifespan of fashion items and thus change the classic linear fashion system towards a circular economy.

How does the platform ***Zalando Zircle*** work? This platform offers "pre-owned" products—much more euphonious than "second hand"—in the *Zalando Fashion Store*. Since 2018, a free app has been making selling clothes as easy for *Zalando* customers as buying them. Users of the app can digitize their wardrobe. This gives them an overview of the current inventory. This can and should not only provide incentives for a new purchase, but also for selling rarely or never worn garments.

Users have the opportunity to offer their own garments to the community or *Zalando*. Customers can exchange used items for free shipping. With just a few clicks, garments can be exchanged for a *Zalando* voucher or a donation to the *Red Cross* or *WeForest*. In addition, products can be purchased directly from other users. *Zalando Zircle* sees itself not only as a marketplace. Here, a **digital community for fashion-interested people** is to be created who want to exchange ideas with others (see Zalando, 2022).

The following figures show what **action is needed in fashion**. Each year, according to estimates, new goods worth about €7 billion are destroyed in Germany alone. It is assumed that about a third of the garments produced are never worn. The destruction of new goods is already factored into the price. Wouldn't recycling be a preferable alternative to destruction? More than half of textiles no longer use natural fibers. The material is plastic! Its quality is often so poor for cost reasons that it cannot be processed into new products. What is the result? Unworn dresses and shirts become cleaning rags or filler material in the automotive industry (see Fisch, 2022, p. 12).

> **Food for Thought** 10% of the yarn used in the clothing industry is wool. More cannot be produced due to the limited number of animals. 30% of the yarn is cotton. More cannot be produced here either without increasing land use. This means that about 60% of clothing today is made of synthetic fibers. This is polyester, whose micro particles accumulate in the water. "If we continue like this, by 2050 we will have more plastic in the water than fish"—at the same time it applies: "No planet, no fashion"—so says *Boss* CEO Grieder (2022, p. 20).

Further exciting insights into the more or less sustainable actions in the textile industry are provided by the ***Fashion Transparency Index.*** Here, 250 of the world's largest fashion brands and retailers were evaluated in terms of the degree of their disclosure of human rights and environmental policies, as well as the associated practices and their impacts (see Fashion Revolution, 2022).

4.4.3.3 IKEA Second Chance

At ***IKEA Second Chance***, the motto is: "Do you have a well-preserved *IKEA* piece of furniture that you no longer need? We buy selected products from you so that they can get a second chance in our bargain corner and find a new owner" (IKEA, 2023). With this offer, *IKEA* wants to contribute to the circular economy in terms of its own products. For this purpose, *IKEA* defines the following conditions, all of which must be met before a product is taken back:

- **The piece of furniture is an original *IKEA*product.**
  Copies and counterfeits, as well as pieces of furniture not manufactured by *IKEA*, are not accepted.
- **The piece of furniture is complete and works perfectly.**
  Damaged products are not taken back.
- **Every piece of furniture to be returned must be brought back fully assembled.**
  Only if the piece of furniture is fully assembled can it be checked and valued by an *IKEA* employee.
- **The piece of furniture must be kept in its original condition without any changes.**
  If a piece of furniture has been painted, for example, or parts are missing, it is excluded from repurchase.
- **The piece of furniture must not show strong signs of wear.**
  Pieces of furniture that are heavily worn are therefore excluded from repurchase.

Due to these high **requirements for a repurchase**, it remains to be seen what proportion of products *IKEA* actually accepts. In addition, it is unclear how and whether customers should deliver larger pieces of furniture (such as cabinets, beds, or entire kitchens)—fully assembled—to *IKEA*. Because such assembly and return transport to *IKEA* always carries the risk that the pieces of furniture will not be accepted. Then all the effort would have been in vain.

The result is interesting when you search for "used products" on the *IKEA* website: "Results for 'used products': We found 0 products and 73 content results."

The **bargain corner** or the **Second-Chance Market** can only be found in the *IKEA* furniture stores on site. So, interested parties must first drive to the nearest *IKEA* market to possibly find a used product that meets their own wishes there.

> **Food for Thought** The question arises whether *IKEA*, due to the strict return conditions and a lack of online search option for used products, has perhaps only put on a too short green coat, without really wanting to make a decisive contribution to the circular economy.

4.4.3.4 Utilization of Waste Heat from Data Centers

A **rounding out of the existing business model** can also be achieved by **data centers** tapping into the **waste heat** generated by their operation as an additional source of revenue. Bitkom (2022a) has presented an exciting study on this. By connecting data centers to public and private district heating networks, they could make a direct contribution to the **basic supply of heat** in Germany. At the same time, the often critically assessed **energy balance of the growing data center industry** would be significantly improved.

After all, the heat generated in data centers today, which is CO_2-free, is usually released unused into the environment. Here, an interesting **district heating potential** could be used to supply municipal facilities (such as swimming pools, event rooms), but also private apartments and commercial operations. According to calculations by *Bitkom,* the waste heat from data centers in Germany could supply around 350,000 apartments annually. This number corresponds approximately to the housing stock of the city-state of Bremen.

Which **data centers** would be interested in these usage possibilities? The study assumes that especially medium and larger data centers with an annual **IT connection capacity of more than 5 megawatts** should tap into this additional source of revenue—for the benefit of the environment. Such data centers are mainly found in Germany in the metropolitan regions of Frankfurt/Main, Berlin, Hamburg, and Munich. Here, a connection capacity of 965 megawatts is achieved. About half of this could be integrated for the use of waste heat (see Bitkom, 2022b).

In regions without a district heating network, the waste heat from data centers could be released to surrounding buildings. However, the waste heat from data centers usually does not reach the necessary temperature of the district heating networks. Therefore, heat pumps would have to be used here to bring the temperature up to the level of the heating network. However, these are all manageable challenges (see Bitkom, 2022a).

> **Food for Thought** The waste heat from data centers in Germany could supply heat to approximately 350,000 apartments. How long do we want to continue to forego the use of the waste heat potential of data centers? This question is particularly urgent given that a growing demand for computing capacity is expected in the future.

4.4.4 Development of New Business Models for the Circular Economy

Following, **selected business models** are presented that fundamentally contribute to the goals of the circular economy. Every week, new companies introduce themselves that want to contribute to greater sustainability through the core of their business activities.

4.4.4.1 Ebay

The core idea of the founders of ***ebay*** in 1995 was certainly not to make a significant contribution to the circular economy. The focus was simply on a **consumer-to-consumer marketplace,** where old and new goods could be sold directly by consumers. Today, *ebay* is also active in B2B and B2C business with new goods. Nevertheless, *ebay* contributes to the circular economy by keeping products in the usage cycle for longer.

4.4.4.2 Rebuy

When developing **new business models** to advance the circular economy, companies like ***Rebuy*** come to mind *(rebuy.de).* Rebuy (2023) defines itself as follows:

"We are the first choice for second hand."

To achieve this, *Rebuy* strives for a **sustainability match** between a used product and an interested user. In doing so, the Re-Commerce company**Re-Commerce company**, founded in Berlin in 2009, contributes to **Zero Waste** by taking on an important function within the circular economy: The company trades in used electronics and used media. It positions itself as the European market leader with over five million customers (cf. Rebuy, 2023).

Especially with electronic devices, the company addresses the still existing lack of environmentally friendly or sustainable use—and offers a **platform for used products**. The need for action is particularly great here, as a significant proportion of electronic waste is still not properly recycled—even though the **recycling rates for household waste** from Fig. 1.12 suggest otherwise.

A representative survey of 5400 consumers in Germany documents a large untapped potential for reducing electronic waste. In 2021, repairs for all households in Germany prevented almost 200,000 tons of electronic waste. The repair rate for damage cases is currently 24% for electrical appliances. A moderate increase to just 30% would reduce the average amount of electronic waste per household to 7 kg. That would be 2 kg less than today (cf. Droste & Höwelkröger, 2022, p. 9).

The goal of *Rebuy* is to recover 450 tons of resources annually by 2025 and 1000 tons by 2030 through the **refurbishment and sale of used electronic devices**. To this end, the used products are refurbished using modern methods—keyword **remanufacturing** (cf. Rebuy, 2023). If one compares the goal of *Rebuy* with the electronic waste in Germany in 2020 of almost 800 million tons, it becomes clear that many more companies with a service like *Rebuy* are needed to manage the electronic waste from Germany alone in a circular manner (cf. Statista, 2021b). After all, the 1000 tons targeted for 2030 represent just 0.000125% of electronic waste in Germany.

Rebuy has now expanded its range of services to include the reuse of books. Through the **return of books to the usage cycle** and the **optimization of packaging**, at least 60,000 trees per year are to be saved by 2025.

4.4.4.3 Unpackaged Stores

A new—or very old—business model underlies the so-called **unpackaged stores**. Here, the use of packaging is consistently avoided and goods are offered for open sale as in the past—"at Aunt Emma's"—usually in a much more stylish way! These retail stores offer their range loose—without containers and therefore free of packaging.

Instead of reaching for packaged goods, buyers fill the desired products themselves. For this, either **own containers** are used or the retailers offer **biodegradable packaging** or **reusable containers** (sometimes on a deposit basis) (cf. Fig. 4.16, bottom left). To comply with hygiene regulations, some foods are offered in so-called **Gravity-Bins** (cf. Fig. 4.16, left).

These vertically mounted containers are often used at breakfast buffets as cereal dispensers. They allow the removal of goods in the desired quantity. Vegetables and fruits are offered in **wooden crates**, for example. Unpackaged (2023), Original Unpackaged (2023) and Freikost Deinet (2023) are among the stores specializing in "unpackaged". However, **unpackaged goods** are not only offered in special stores, but also supplement the standard range in the form of unpackaged stations, especially in organic markets (cf. also Sect. 4.4.1.6).

4.4.4.4 Reverse Logistics Group

The pursuit of comprehensive supply chain concepts also enables the development of new B2B business models. The globally represented ***Reverse Logistics Group***, for example, has developed return solutions and systems for products, components, and materials.

The goal is to connect the involved companies through a scalable platform within a **reverse logistics value chain**. In this case, "sustainability" is not a secondary condition of corporate activity, but its core (see Reverse Logistics Group, 2023).

4.4.4.5 Ecosia

A business model that combines various facets of a sustainable business model is the German-based search engine ***Ecosia.*** Unlike common search engines, it does not sell user data to advertisers. Also, no third-party behavior trackers are used. With this, *Ecosia* contributes to **social goals**.

The revenues generated through advertising are largely used to plant trees worldwide. According to the company, over 155 million trees have already been planted—and another tree is added every 1.3 seconds. In total, more than 36 million euros have been invested in over 13,000 planting locations for this purpose (see Ecosia, 2023). The planting of trees is not primarily intended as compensation for the company's CO_2 emissions, but is another core of the company's purpose. This contributes to **ecological goals**.

4.4.4.6 Wintershall Dea

Wintershall Dea is Europe's leading independent **natural gas and oil company.** The European company supports the EU's goal of becoming climate-neutral by 2050. Therefore, the goal has been set to reduce the greenhouse gas emissions of categories Scope 1 and 2 in all self- and third-party operated exploration and production activities to net zero by 2030. The company's own methane intensity is to be reduced to below 0.1% by 2025, and the routine flaring of associated gas is to be completely avoided. Emissions from the use of hydrocarbons are also to be reduced. *Equinor* is an international energy company focused on long-term value creation in a low-carbon future (see Wintershall Dea, 2023).

Wintershall Dea and *Equinor* joined forces in 2022 to develop a new business model. This involves a comprehensive and secure **value chain for the capture, transport,**

and corporate storage of CO_2. This process is also called **Carbon Capture and Storage (CCS)**. Through this project, CO_2 emitters from the European mainland are to be directly connected with offshore storage sites on the Norwegian continental shelf. This is intended to make the Norwegian-German CCS project *NOR-GE* contribute to reducing greenhouse gas emissions in Europe—and open up a new business field for *Wintershall Dea* and *Equinor* (see Wintershall Dea, 2023).

To achieve this, work is being done on a **cross-border CCS value chain** for Europe that meets regulatory conditions. This is intended to achieve a technical solution that enables the necessary **decarbonization of carbon-intensive industries**. The targeted **underground storage of CO_2** is intended to contribute in the future to reducing unavoidable emissions from industrial processes—such as in the production of aluminum, glass, and cement. During the production process, the carbon dioxide bound in the raw materials escapes. To make such companies emission-free, only one technology helps today: The greenhouse gases have to be captured and stored permanently. Depleted natural gas or oil fields under the sea are suitable for this (see Wintershall Dea, 2023).

Through the partnership between *Wintershall Dea* and *Equinor*, Germany, as the largest European CO_2 emitter, is to be connected with Norway. The country has the highest CO_2 storage potential in Europe. For this purpose, a pipeline approximately 900 km long is to be built from northern Germany to the storage sites in Norway. The pipeline is to be operational before 2032 and have a transport capacity of 20 to 40 million tons of CO_2 per year. This corresponds to approximately 20% of the total annual German industrial emissions. During the construction phase of the pipeline, transport to the underground storage sites could already be carried out by ship. To enable these steps, *Wintershall Dea* and *Equinor* are applying for **licenses for offshore storage of CO_2**. The aim is to store 15 to 20 million tons per year on the Norwegian continental shelf (see Wintershall Dea, 2022).

Food for Thought

The capture of carbon dioxide from industrial exhaust gases and its storage underground is recommended by the *Intergovernmental Panel on Climate Change IPCC*. The *German Energy Agency* suggests developing a roadmap for Carbon Capture and Storage. The *Foundation for Climate Neutrality*, the *Agora Energiewende* and the *Agora Verkehrswende* also demand that the topic be put back on the political agenda (see Hovius, 2021). A multitude of computer simulations on global CO_2 emissions provide a clear result:

Without Carbon Capture and Storagethere will be no climate-neutral economy. Because the world is failing in climate protection, it is now necessary to start extracting excess CO_2emissions from the atmosphere and storing them underground. Only negative emissions can still stop climate change.

Peatlands and reforested forests can also absorb excess CO_2 emissions. But this will not succeed as quickly and comprehensively as would be necessary to reduce the emissions as required.

What is the situation in the country of the energy transition? In Germany, Carbon Capture and Storage is currently still prohibited except for experimental facilities. Here, the fear of leaks, of unwanted earth movements, and of the release of pollutants underground dominates. This leads to citizen protests and the stoppage of experimental facilities.

The solution for Germany? Since the underground **storage of CO_2** is (still) prohibited in Germany, it is to take place in Norway. By the way: No country in Europe emits more carbon dioxide than Germany. This behavior fits into the picture that Germany rejects fracking, but at the same time imports gas and oil obtained through fracking in the USA on a large scale. Germany—or the government—rejects nuclear power and has been importing nuclear electricity from France for years. Plastic waste is still being exported from Germany to developing countries. We have relevant data for combating the Corona pandemic provided to us from Israel, because data collection in Germany is hindered by data protection.

In the case of (supposedly) risky and/or (supposedly) difficult to communicate technologies, Germany relies on **outsourcing**.

Responsible action by a still wealthy industrial nation would look different.

4.4.4.7 Loop

The purpose of the company *Loop* is *Eliminating the Idea of Waste*. The company wants to ensure that waste is no longer a consideration. To accelerate this process, *Loop* has developed a global **platform for reuse**. The company works with manufacturers to develop **refillable versions** of the conventional disposable products previously distributed (see Loop, 2023).

To this end, it collaborates with over 200 **consumer goods manufacturers** and more than a dozen major **retailers**. The jointly developed offerings are to be used both in brick-and-mortar stores and in online shops. The goal is to establish a **circular system for reuse.** Ideally, this system should cover thousands of different product categories—from coffee cups to shampoo bottles. The great challenge is to make reuse as simple and convenient as single use. The **platform for reuse** is currently available in the United States, the United Kingdom, Canada, Japan, and France (see Loop, 2023).

Loop Global Holdings is part of *TerraCycle. TerraCycle* collects and recycles hard-to-recycle waste via national **recycling platforms.** The range includes hundreds of different waste streams—from cigarette butts to dirty diapers. The goal is to switch from linear systems to circular systems in as many areas as possible. The focus here is on the development from single-use systems to multi-use systems. It remains to be seen to what extent the promise "Recycle non-recyclable material with TerraCycle" can actually be fulfilled.

4.4.4.8 H2Fly

The company ***H2Fly*** wants to contribute to **sustainable air traffic**. For this purpose, the company focuses on the development of a hydrogen-electric drive train for aviation. Ten years of research have gone into the development of an emission-free solution. The aircraft *HY4* used for demonstration purposes is powered by clean hydrogen. This is converted into electrical energy in the fuel cell drive train (see H2Fly, 2023).

The company *H2Fly* plans to use this drive system in various aircraft. In the planning are **air taxis,** which can cover ranges of well over 500 km. In the future, the system is also intended to power **business aircraft** with several seats or even **regional aircraft** with a range of 2000 km (see H2Fly, 2023).

Whether these visions will become reality will only become apparent in the coming years. But it is clear that many companies are working on solutions for a more sustainable economy and a more sustainable life.

4.4.4.9 Cirplus

The company ***cirplus*** sees itself as the **global marketplace for recyclates and plastic waste.** The team of this company consists of technology partners and digital experts. They have a high level of know-how in the production, processing, and recycling of thermoplastic plastics. The company strives for high neutrality towards all market participants. The mission of *cirplus* is to make the purchase and distribution of recycled plastics as simple as possible (see Cirplus, 2023).

For this purpose, *cirplus* has developed a **B2B online marketplace for the waste, recycling, and plastics industry** as an **end-to-end transaction platform**. On this platform, the company handles the essential steps of the transaction. This also includes logistics, credit insurance, and accompanying communication. The materials traded on this platform can come from the waste streams of consumers and companies (see Cirplus, 2023).

4.4.4.10 Circular Economy Accelerator

For business ideas in the circular economy, dedicated **accelerator programs** are now being launched. These are intended to help relevant start-ups achieve a significant growth step as quickly as possible. An example of this is the program ***Circular Economy Accelerator,*** which is based in Wuppertal. This **Circular Economy Accelerator** runs programs with 15 start-ups twice a year (see Circular Valley, 2023).

What is offered as part of the accelerator? An important basis for growth for start-ups is the existing **network** here with over 100 partners from industry, R&D, and government. The start-ups are offered tailored **expert workshops** and **training modules** as well as individual **mentoring** and **coaching** to specifically promote the growth process. This professional mentoring and coaching is accompanied by **financial support.**

This offer is primarily aimed at start-ups whose **concepts can contribute to waste prevention.** This includes production technologies with minimal environmental impact and sustainability-oriented labeling (e.g., product passports). The entire area of tracking

& tracing, collection, and separation of waste streams (including the entire waste logistics) are also promoted here. This also includes the development of novel recycling technologies and innovative service models around sustainability.

The accelerator focuses in accordance with the EU action plan for the *Green Deal* and the circular economy on the sectors of electronics, information and communication technology, vehicles (including batteries), packaging, plastics, textiles, furniture, construction and buildings, as well as heavily burdened intermediate products such as steel, cement, and chemicals (see Circular Valley, 2023).

4.4.4.11 Non-Profit Solutions

To the circular economy many business models also contribute, which do not pursue economic goals, but make a contribution to the common good. Most activities are financed by donations and the voluntary commitment of many people.

- **Food Banks**
  An example of this is the ***Food Bank Germany.*** This is a non-profit association as the umbrella organization of almost 1000 food banks. This association is financed by donations. The **core concept of the *food banks*** is that about 60,000 helpers collect food from retailers and manufacturers and distribute it to needy people. These foods would have been destroyed without the involvement of the *food banks,* even though they are still edible.
  Each year, about 265,000 tons of food can be made available to needy people. Today, over two million people are already being supported with food. In addition, many *food banks* offer additional services besides the food distribution. These can include a hot lunch, tutoring, or childcare (see Food Bank, 2023).
- **Open Bookcases**
  Another—based on voluntary commitment—"business model" are the **open bookcases.** In line with the circular economy, the idea is pursued that books should continue to live even if they are no longer needed by their owners. The **principle of open bookcases** is easy to understand. Anyone can put books into these bookcases and/or take books out for free. There are no deadlines or other regulations, except to provide books in good condition and to treat the bookcases with care. The range of available books covers the entire possible spectrum, from cookbooks to crime novels to specialist literature. To make it easier for children and young people to access books, the bottom shelf of a bookcase is usually reserved for children's books.
  The open bookcases are often voluntarily maintained by residents. The task as patrons is to regularly check the content and condition of the bookcase. After all, these bookcases are in public space.
- **Clothing Chambers**
  Clothing chambers also make an important contribution to a circular economy. Clothing chambers are social institutions that are financed by donations. Anyone who wants to donate wearable clothes and often other household items (dishes, toys, small

furniture, books, etc.) is welcome at the clothing chambers. The mostly voluntary employees accept the **in-kind donations** and display them in their premises. Socially disadvantaged people can choose the items they need. These are either given away for free or for a small fee.

> **Note Box** The **ideas for voluntary commitment** are limitless. All concepts that lead to the reuse of already produced products contribute to a sustainable use of the earth's limited resources.

4.4.5 Entrepreneurial Commitment in (International) Organizations for the Promotion of the Circular Economy

Companies can contribute to further advancing developments towards a circular economy through **participation in (international) organizations**. Here, a selection of relevant organizations is presented:

- ***PACE Initiative****(Platform for Accelerating the Circular Economy)*
- ***StEP Initiative****(Solving the E-Waste Problem)*
- ***#breakfreefromplastic***
- ***Alliance Against Plastic Waste in the Environment****(Alliance to End Plastic Waste, AEPW)*

The 2018 founded ***PACEInitiative (Platform for Accelerating the Circular Economy)*** is a global platform for collaboration between public and private decision-makers to further advance the circular economy for plastics, textiles, electronics, food, and capital goods. Nearly 100 leading representatives from governments, businesses, and civil society from all continents and sectors have joined the *PACE Leadership Group* to accelerate the transition to a circular economy worldwide (see PACE, 2023).

The ***StEP*-Initiative (Solving the E-Waste Problem)** is a global public-private initiative supported by the *UN*. Its founding members *include Cisco Systems, Dell, Ericsson, Hewlett-Packard* and *Microsoft*. Other governmental, non-governmental and academic institutions, as well as recycling and processing companies, are also involved. The main goal of *StEP* is the worldwide standardization of recycling processes to recover valuable components from electronic waste (e-waste). Markets for their reuse are to be established. In addition, the extension of product lifetimes is promoted. Another goal is the worldwide harmonization of legislation and policy regarding e-waste (cf. StEP, 2023).

#breakfreefromplastic is a global initiative that advocates for a future without plastic pollution. Since 2016, more than 11,000 organizations and individuals worldwide have joined the movement. The goal is to demand a massive **reduction of single-use plastic** and to push for permanent **solutions to plastic pollution**. This involves looking at the entire plastic value chain—from extraction to disposal. The primary focus is on avoiding single-use plastic (cf. breakfreefromplasti, 2023).

The initiative offers various forms of participation. Non-profit organizations can apply for a **core membership**. These members are the driving force behind the movement, developing and implementing the overall strategy of *#breakfreefromplastic*. The **associated members,** who have registered on the website, receive regular communications from the movement and are supported by it in various ways. **Individual supporters** are individuals who do not belong to any organization. They are also continuously informed by the initiative about activities (cf. breakfreefromplastic, 2023).

To curb the tide of plastic, the ***Alliance Against Plastic Waste in the Environment (Alliance to End Plastic Waste, AEPW,*** 2023) was founded. Its goal is to **avoid plastic waste** and ultimately eliminate it completely. Since 2019, the Alliance has brought together over 90 member companies, project partners, allies, and supporters to work towards this goal. The network combines resources and expertise to develop innovative solutions and disseminate them worldwide. It works with policymakers, non-governmental organizations, and local communities. The following **challenges of plastic waste** need to be mastered:

- Today, there is still a lack of a **powerful infrastructure** as well as **systems for collection** and **concepts and facilities for processing** domestic and urban waste. The plastic in the oceans mainly comes from waste generated on land. The plastic waste is first "collected" via rivers and then washed into the sea, as there is still no sufficient infrastructure for waste collection and recycling worldwide.
- In addition, there is a lack of **social awareness,** to perceive and utilize plastic waste as a valuable resource—or better yet, to avoid the use of plastic as much as possible from the outset.
- Moreover, newly manufactured **plastic raw materials** are often cheaper and more hygienic than recycled raw materials.
- For companies, it is usually cheaper today to leave the disposal of plastic waste to third parties. These **external costs of the plastic economy** are not captured in the companies' calculations, but are left to society and nature (cf. on this market failure Sect. 1.5).

Over the next five years, the ***Alliance Against Plastic Waste in the Environment*** wants to invest 1.5 billion US dollars in various projects and collaborations. The following measures are at the center of this:

- **Development of infrastructure** for waste collection, waste management, and increasing recycling
- **Development and scaling of innovative technologies** for recycling and recovery of high-quality plastics

- Innovative solutions for **minimizing and eliminating plastic waste** are to be developed. This includes solutions that allow the reuse of already used plastics. This is intended to enable a circular economy for plastic. One goal is to develop **industrial standards** for this and to establish them in the market.
- **Support in educating about plastic waste** involving governments, businesses, and communities
- The *Alliance Against Plastic Waste in the Environment* regularly informs about partner projects that are intended to prevent the entry of plastic waste into the environment in the ***AEPW Progress Report***.
- **Initiation of cleanup activities** in regions that are already heavily burdened by plastic waste (such as rivers and certain coastal regions)
- The *Alliance Against Plastic Waste in the Environment* initiated the World Cleanup Day in 2020. As part of this "All Together Global Cleanup" campaign, waste was collected worldwide from the environment (see AlltogetherGlobalCleanup, 2023).

Food for Thought

The activities of the *Alliance Against Plastic Waste* in the environment sound very good. However, it is necessary to check what successes have already been achieved. Journalists repeatedly uncover how few concrete measures have been implemented.

In addition, it is worth taking a look at the 1.5 billion US dollars that the Alliance intends to invest in various projects and collaborations over the next five years. Each year, approximately 8 billion tons of plastic are produced worldwide by over 60,000 companies. This generated a turnover of 608 billion US dollars in 2022 (see Statista, 2022a). If you put the Alliance's budget of 1.5 billion US dollars in relation to the turnover in five years, these 1.5 billion represent only 0.05% of this turnover. Not really much to come up with convincing solutions worldwide (!). This suggests the suspicion of greenwashing!

Additional exciting and sobering results on the topic of plastic waste are provided by a study by ***Greenpeace USA*** (see Greenpeace, 2022):

- Of the 51 million tons of plastic waste that US households generate each year, only 2.4 million tons, and thus just under 5% of the total amount, are recycled.
- In principle, **recycling facilities** only accept two types of plastic: polyethylene terephthalate **(PET),** which is often used for water bottles, and high-density polyethylene density **(HDPE),** which is used, for example, for shampoo bottles. These plastics carry the numbers 1 and 2 (see Fig. 5.12). The recycling rate for PET is 20.9%, for HDPE it is 10.3%.
- The **plastic types 3 to 7** are recycled at less than 5%. These types are used for coffee cups, children's toys, and to-go containers.

- Products of plastic types 3 to 7 often carry the **recycling symbol**, although they are **not recyclable** according to a classification by the *Federal Trade Commission* (see Fig. 5.12). The reason for this is that products are only classified as recyclable if corresponding recycling facilities are accessible to a substantial majority (here approx. 60% of the population). In addition, the collected products would have to be used for the production of new items—the very core of recycling. However, this is not the case with plastics of groups 3 to 7.
- At the same time, the **production of new, non-recycled plastic** continues to increase. Why? Because new plastic is easier to produce and therefore cheaper than recycled plastic—while being of higher quality at the same time.

Food for Thought

The vast majority of plastic products are not recycled. Even if a **recycling symbol is found on products**, this does not mean that the products are actually recycled, nor that they are recyclable at all. This realization is something to ponder!

For the companies that use plastic, the **recycling solution**—or more honestly the **recycling lie**—is perfect. After all, by pointing out a potential for recycling—to be exploited by third parties—they evade any responsibility of their own.

Greenpeace (2022) succinctly summarizes why the **concept of plastic recycling** has failed:

- Plastic is produced in huge quantities and is extremely difficult to collect.
- Plastic waste cannot be recycled mixed, but must be sorted in advance.
- The recycling process is harmful both to the environment and to the personnel involved.
- Due to the contamination of plastic waste, recycled plastic cannot be processed back into food-grade material. This is not allowed by hygiene regulations.
- The process of plastic recycling is very expensive—often more expensive than the production of new, high-quality plastic.

Movie Tip **"The Recycling Lie"**—available in the ARD media library.

Also in the **fashion industry**, various initiatives have been formed to guide the fashion industry as a whole towards a circular economy. Here is a small selection of these initiatives:

- *Better Cotton Initiative*
- *Fashion for Good*
- *Global Fashion Agenda*

The ***Better Cotton Initiative*** aims to improve the situation for farmers who produce cotton through its ***2030 Strategy***. Today, about a quarter of the world's cotton is produced according to the ***Better Cotton Standard***. 2.4 million cotton farmers have been trained in sustainable farming methods and have a license to grow ***Better Cotton.*** The *Better Cotton Initiative* is working towards a sustainable world where farmers know how to better cope with climate change and environmental threats while earning a decent living. To enable this, the global demand for sustainable cotton is to be promoted. To this end, suppliers, manufacturers, retailers, and brands are encouraged to use *Better Cotton* more intensively (see Better Cotton, 2023).

Fashion for Good is a global initiative. It aims to stimulate change and make fashion a force for good through a collective movement. The initiative works directly with the fashion industry to develop innovative solutions. At the heart of *Fashion for Good* is an **Innovation Platform.** Through this, manufacturers, brands, retailers, and funders are brought together to work collectively on sustainable innovations. The goal is to bring new ideas and technologies from the niche to mass production (see Fashion for Good, 2023).

The vision of the ***Global Fashion Agenda*** is to advance the fashion industry on its way to becoming an industry positive for people and the planet. The ambitious goal is to become an industry that gives back more (positive) to society, the environment, and the global economy than it takes out. To drive such a change and take action, the non-profit organization is committed to mobilizing, inspiring, influencing, and educating all stakeholders. The ***Global Circular Fashion Forum*** is a global initiative to stimulate local actions in the countries of textile production to accelerate and expand the recycling of post-industrial textile waste. This is an attempt to contribute to the establishment of a long-term and scalable circular economy in the fashion industry (see Global Fashion Agenda, 2023).

▶ **Food for Thought** Every company is called upon to consider active participation in the relevant organizations that advocate for the expansion of the sustainable economy. This networking not only promotes the exchange of ideas on the circular economy but can ideally also unleash the informational and financial power to achieve sustainable changes.

4.5 SheIn—An Extreme Example

A particularly exciting example of how closely light and shadow can lie together in a business model is provided by the company ***SheIn*** (*shein.com*, originally started as *SheInside*), founded in 2008 by *Chris Xu*. This is one of the largest **online shops from China.** It currently focuses on the **fashion segment.** The company's target groups are primarily women of Generation Z worldwide (see Sect. 1.3.1) as well as young parents. Men's and children's clothing are now also on offer. *SheIn* has a high proportion of 18-

to 29-year-old customers in Germany (see Statista, 2021a, p. 3). What is special about this company is not the use of fashion bloggers or the bonus points program, which rewards regular use of the app. What is exciting is that *SheIn* has been able to reduce the **design and production time** for fashion items from 21 days to three days, as achieved by ***Zara,*** one of the inventors of **Fast Fashion**. Today, it manages to bring about 1000 new styles to the market—per month. This was achieved through shorter production cycles, just-in-time logistics, and high investments in flagship stores in many city centers around the world. The speed and variety of the product range were the most important drivers of success for the fast-fashion companies.

The increasing digitization among consumers and providers (manufacturers as well as retailers) has now enabled a further leap in development towards to **Ultra Fast Fashion** and **Realtime Fashion** or **Realtime Retail.** The networking allows the company *SheIn* to bring 3000 to 7000 new products to the market—per day. Assuming an average of 5000 new products per day, this can be over 1.8 million new products per year. This is also made possible by the fact that *SheIn* requires significantly less time for product tests than *Zara* (see Verhaeghe, 2021).

▶ Note Box **Ultra Fast Fashion** and **Realtime Fashion** or **Realtime Retail** shorten the lead time for new fashion products from a few weeks to a few days! This business model runs counter to all sustainability initiatives in the fashion industry.

In contrast to fast-fashion retailers, who still have to invest in logistics and warehousing as well as physical retail stores, the **Realtime-Fashion providers** shift their business to a large extent to the Internet. Here the focus—apart from logistics and warehousing—is on online presence and especially the online shop (see SheIn, 2023).

The **Realtime-Fashion company** *SheIn* strives with the mission "Everyone can enjoy the beauty of fashion" to offer not only fast but also stylish quality products at attractive prices to customers worldwide (SheIn, 2023). To achieve this, *SheIn* hired designers from different parts of the world. Today, the company has a design team of about 800 people. *SheIn* uses *Google Trends* and various social listening tools to anticipate new clothing trends in fabrics, colors, and styles.

To evaluate this **data stream in real time**, **sophisticated AI systems** are used. These automatically derive new product ideas from the search and social media behavior of users in individual countries. The insights gained from this are implemented by designers closely tied to the company in a *SheIn* community. These designs are automatically forwarded to the suppliers via a central *SheIn* ERP system. There, the products are initially produced in limited quantities and offered in the respective region via the online shop. For the products that are particularly successful in terms of viewing and buying behavior in the webshop, the order volume for production is increased in real time via the ERP system. *SheIn* is thus strongly guided by consumers in its offering. It is important to know:

▶ *SheIn* does not have its own style!

The company rather produces the goods that resonate the most in different regions of the world. This way of working can be compared to the **engagement algorithm** of ***TikTok.*** On *TikTok,* new videos also automatically appear in the timeline of more and more people who have previously proven themselves to a limited number of viewers. This creates a **data-based feedback loop,** which continuously improves the digital customer experience and promotes the production of customer-relevant products. At the same time, warehousing is reduced and the production of hard-to-sell offers is avoided. *SheIn* is able to reflect local fashion taste in real time—and to generate a different but always relevant offer for over 200 countries. This does not require traditional market studies, but "only" an **evaluation of user habits in real time** (see Verhaeghe, 2021).

Intensive collaboration with various manufacturers enables the implementation of the concept **Consumer-to-Manufacturer (C2M).** The special feature of this **C2M concept** is that it leads to a **direct relationship between customers and manufacturers** (see Fig. 4.19). As a result, the stages of intermediate trade are eliminated in C2M, for example, the involvement of wholesalers.

What is the difference between **C2M (Consumer to Manufacturer)** and **D2C (Direct to Customer)?** In D2C, intermediaries are also eliminated. However, D2C is only a specific distribution concept—also called **direct sales**—where there is direct contact between the customers and the manufacturer. Orders are placed there and delivery is triggered from there. Many traditional manufacturers (like *Adidas, Montblanc, Nike*) use their own online shops for this direct sales. This is usually done in parallel with indirect sales through wholesalers and retailers.

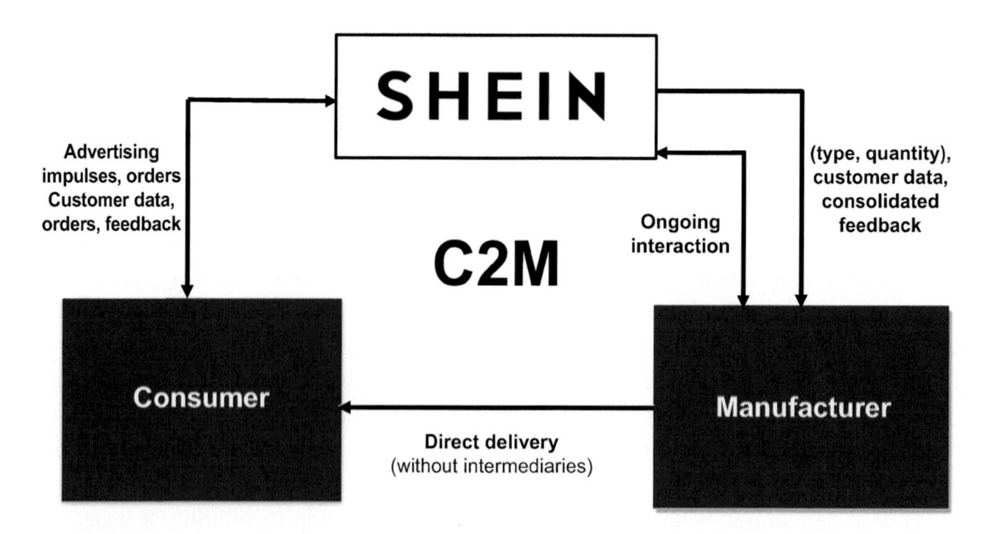

Fig. 4.19 Consumer-to-Manufacturer concept

In contrast to the D2C approach, the consumer in the **C2M concept** triggers specific production processes. Here, customers have a high degree of influence on what products look like and in what quantities they are produced.

> **Remember Box** In contrast to the **supply-driven system Direct-to-Customer** (D2C), **Consumer-to-Manufacturer** (C2M) is a **demand-driven system.** Ideally, only the pieces that actually meet demand are produced—and only in the required quantity. In addition, customers are supplied directly from the manufacturer—and not from the company responsible for marketing.

To achieve this, customers in the **C2M model** are directly or indirectly connected to manufacturers via online-platforms (like *SheIn* or *Pinduoduo*) (see Fig. 4.19). Interposed distribution stages are eliminated, enabling more intensive and direct exchange of information between customers and producers. Demand is bundled via the platforms and forwarded to the production sites. There, the desired items are produced in the ordered quantities and delivered directly to the customers after a few days. At the same time, **feedback loops** are created that allow comprehensive feedback from consumers to manufacturers. Only comprehensive networking and the digitization of information make such a business model possible—combined with AI-supported evaluation options. This promotes the trend towards **individualization** in the fashion sector (see Scherrer, 2021).

A convincing **appearance on social media**—involving various influencers—and an online shop with a good customer experience are prerequisites for the **points system** used by *SheIn* to have its effect. This points system rewards users for a variety of actions—whether it's verifying the email address, writing a product review, or adding a photo with a *SheIn* item. Additional points are earned by daily log-in, participating in outfit competitions or live events. The earned points are worth money and can be used to pay for a purchase. Thus, **gamification elements** are used for systematic customer loyalty—and comprehensive customer engagement is promoted (see Verhaeghe, 2021).

In evaluating the business concept of *SheIn*, it should be noted that the company often faces **accusations of plagiarism**. Both large brands and smaller designers are affected, whose designs are often almost identically copied. Due to production in China, deliveries abroad can often result in **long delivery times**. Also, **Realtime Fashion** has nothing to do with sustainable fashion. Another point of criticism is the **high intransparency**—regarding the company and the acting persons at *SheIn*. A detailed analysis of the business model, profitability and especially sustainability is therefore not possible. Nevertheless, investors now value *SheIn* at 100 billion US dollars—more than *Inditex (Zara)* with about 70 billion US dollars and *H&M* with about 20 billion US dollars combined.

Food for Thought It is interesting that *SheIn* has become the **shooting star of fast fashion** and is making billions with it. At the same time, the core target group **Generation Z** seems to have no problem reconciling ***Friday for Future*** and **Fast Fashion**.

Questions you should ask yourself

- How do we respond to ecological, social, and economic challenges—pathologically, reactively, proactively, or creatively?
- How comprehensively do we regularly analyze our value chain—for benefit enhancement, cost reduction, and identification and exploitation of sustainability potentials?
- Who is responsible for analyzing sustainability potentials in our company?
- Is our own value chain integrated into a system of value chains?
- Where are there still areas for optimization and growth in our own value chain or in the system of value chains? Who takes care of this?
- Has a digital value chain already been integrated? How well does it work? Who is driving its further development?
- How comprehensively do we use the potentials of the Internet of Everything with regard to sustainable corporate management? Where does the responsibility lie for this?
- Which options for sustainable corporate management are we already exploiting or do we want to exploit?
- Which facets of the circular economy do we focus on to advance the issue of sustainability?
- How sustainable are our product and service designs (including packaging) already?
- Do our designs enable repair, remanufacturing, and/or upcycling?
- Can the concepts used lead to direct or indirect rebound effects?
- How professionally is our supplier management oriented towards sustainability?
- How consistently do we implement the current and future requirements of the supply chain law?
- What reporting obligations are coming up for our customers that we ourselves must respond to with data?
- Which reporting obligations can we only fulfill by collecting more information ourselves—also from our own suppliers?
- Which companies can support us in determining our own Corporate Carbon Footprint or the Product Carbon Footprint?
- How do we organize the data management of the required information?
- Who bears the overall responsibility for meeting various legal requirements in our company?
- Are sufficient personnel and financial resources available for this?

- Is there a training agenda for the internal communication of relevant information? Who is responsible for this?
- How consistently is our production program oriented towards sustainability? What further steps can be initiated here?
- How sustainably are our production processes oriented?
- How consistently do we focus on avoidance, reduction, and recycling of non-renewable resources?
- Do we manage to keep products in the usage cycle as long as possible (e.g., through services)?
- Can we develop new business models based on service here?
- Can products and/or services be digitized to reduce resource use?
- What necessities result from emissions trading for our own company?
- Do we carry out proactive compensation that is reflected in the prices?
- Can we take care of the compensation ourselves, or should service providers take over the compensation?
- Do we motivate our customers to compensate? If yes, how extensively are such offers used?
- What possibilities exist to supplement the existing business model with concepts of the circular economy?
- Can we develop new business models ourselves that have certain concepts of the circular economy as their content?
- Do we engage in (international) organizations to advance the circular economy?
- What opportunities and risks can be derived from the concept of *SheIn* for our company in terms of sustainability?

References

Alliance to End Plastic Waste. (2023). We are working together to end plastic waste. https://end-plasticwaste.org/. Accessed 4 Jan 2023.

AlltogetherGlobalCleanup. (2023). Every little piece counts. https://alltogetherglobalcleanup.org/. Accessed 2 Jan 2023.

Amazon. (2022). 2021 Amazon. Annual report. https://ir.aboutamazon.com/annual-reports-proxies-and-shareholder-letters/default.aspx. Accessed 28 July 2022.

Atmosfair. (2022). Jahresbericht 2021. https://www.atmosfair.de/de/ueber_uns/jahresberichte/. Accessed 29 July 2022.

Atmosfair. (2023). Atmosfair 2.0. https://www.atmosfair.de/de/. Accessed 2 Jan 2023.

Better Cotton. (2023). Better Cotton Annual Report. https://bettercotton.org/. Accessed 2 Jan 2023.

BFM. (2021). Fragen und Antworten zur umsatzsteuerrechtlichen Behandlung von Sachspenden. https://www.bundesfinanzministerium.de/Content/DE/FAQ/2021-03-23-FAQ-sachspenden.html. Accessed 4 Aug 2022.

Biblus. (2020). IFC- und Open BIM-Format, alles was man wissen sollte. https://biblus.accasoftware.com/de/ifc-und-open-bim-format-alles-was-man-wissen-sollte/. Accessed 30 Sept 2022.

Bitkom. (2022a). Abwärme von Rechenzentren für Heizung und Warmwasser einsetzen. https://www.bitkom.org/Presse/Presseinformation/Abwaerme-Rechenzentren-fuer-Heizung-Warmwasser-einsetzen. Accessed 2 Sept 2022.

Bitkom. (2022b). Studie "Rechenzentren in Deutschland". https://www.bitkom.org/Bitkom/Publikationen/Rechenzentren-in-Deutschland-2022b. Accessed 2 Sept 2022.

BMUV. (2022). Emissionshandel. https://www.bmuv.de/themen/klimaschutz-anpassung/klimaschutz/emissionshandel. Accessed 1 Aug 2022.

BMWSB. (2022). Qualitätssiegel Nachhaltiges Gebäude (QNG). https://www.nachhaltigesbauen.de/austausch/beg/. Accessed 29 Sept 2022.

breakfreefromplastic. (2023). Who we are. https://www.breakfreefromplastic.org/about/. https://www.breakfreefromplastic.org/Zugegriffen. Accessed 2 Jan 2023.

C&A. (2023). Wear the Change, https://www.c-and-a.com/de/de/corporate/company/nachhaltigkeit/wearthechange/. Accessed 4 Jan 2023.

Carbon Credit Watch. (2023). Introducing CMW. https://carbonmarketwatch.org/about/. Accessed 2 Jan 2023.

CCBA. (2023). CCB Standards. https://www.climate-standards.org/ccb-standards/. Accessed 2 Jan 2023.

Chamäleon. (2022). *Seele*. Liebe. Herz.

Chui, M., Löffler, M., & Roberts, R. (2010). *The Internet of Things*. In: McKinsey Quarterly 2/2010

Circular Valley. (2023). Circular Economy Accelerator Program. https://circular-valley.org/about_the_program. Accessed 3 Jan 2023.

Cirplus. (2023). Wir sind cirplus. https://www.cirplus.com/de/about-us. Accessed 2 Jan 2023.

Climate Partner. (2023). Klimaschutz im Unternehmen—jetzt starten. https://www.climatepartner.com/de. Accessed 2 Jan 2023.

Commerzbank. (2023). Impact Investing einfach erklärt. https://www.commerzbank.de/investieren/wissen/impact-investing-einfach-erklaert/. Accessed 2 Jan 2023.

Compensaid. (2023). Make your air travel CO2 neutral now. https://compensaid.com/. Accessed 1 Jan 2023.

Concular. (2023). Werte schaffen durch Zirkuläres Bauen. https://concular.de/. Accessed 3 Jan 2023.

CSR Hub. (2023). Consensus ESG Ratings, Big Data Technology Creates Corporate and Investment ESG Solutions. https://www.csrhub.com/. Accessed 2 Jan 2023.

Deutsche Post. (2023). Gehen Sie mit uns den grünen Weg. https://www.deutschepost.de/de/g/gogreen.html. Accessed 3 Jan 2023.

DGNB. (2023). Europas größtes Netzwerk für nachhaltiges Bauen. https://www.dgnb.de/de/. Accessed 1 Jan 2023.

DHL. (2023a). Unser Weg zu null Emissionen in Deutschland. https://www.dhl.de/de/privatkunden/kampagnenseiten/gogreen.html. Accessed 2 March 2023.

DHL. (2023b). GoGreen Plus—die nächste Stufe des klimaneutralen Versands. https://www.dhl.de/de/geschaeftskunden/paket/leistungen-und-services/services/service-loesungen/gogreen-plus.html. Accessed 2 Jan 2023.

Diemand, S., & Finsterbusch, S. (2022). Wasser marsch. *Frankfurter Allgemeine Zeitung*. 6.8.2022. p. 22

Droste, M., & Höwelkröger, R. (2022). Reparieren statt Wegwerfen: Wertgarantie sensibilisiert für Elektroschrott-Problem. https://www.sozialeverantwortung.info/unternehmen-fuer-mensch-und-umwelt/reparieren-statt-wegwerfen-wertgarantie-sensibilisiert-fuer-elektroschrott-problem/. Accessed 17 Jan 2023.

ECORE. (2023). ECORE—ESG Circle of Real Estate. https://www.ecore-scoring.com/. Accessed 1 Jan 2023.
Ecosia. (2023). Suche im Web und pflanze Bäume. https://info.ecosia.org/what. Accessed 2 Jan 2023.
EcoVadis. (2023). Die weltweit zuverlässigsten Nachhaltigkeitsratings für Unternehmen. https://ecovadis.com/de/. Accessed 2 Jan 2023.
Engelfried, J. (2021). *Nachhaltiges Umweltmanagement* (2. edn.). UVK.
Europäisches Parlament. (2022). Recht auf Reparatur: Warum sind EU-Rechtsvorschriften wichtig? https://www.europarl.europa.eu/news/de/headlines/society/20220331STO26410/recht-auf-reparatur-warum-sind-eu-rechtsvorschriften-wichtig. Accessed 9 Aug 2022.
Fairtrade. (2023). Fairtrade Klimastandard. https://www.fairtrade-deutschland.de. Accessed 2 Apr 2023.
Fashion for Good. (2023). Fashion for Good is here to make all fashion good. https://fashionforgood.com/. Accessed 2 Jan 2023.
Fashion Revolution. (2022). *Fashion Transparency Index*. 2022 Edition. London.
Feinkost Deinert. (2023). Herzlich willkommen bei Freikost Deinet! https://www.freikost.de/. Accessed 2 Jan 2023.
Firouzi, F., Chakrabarty, K., & Nassif, S. (Hrsg.). (2020). *Intelligent Internet of Things: From Device to Fog and Cloud*. Springer.
Fisch, I. (2022). Lagerkoller. *Frankfurter Allgemeine Zeitung*. 15.5.2022. p. 12
Global Fashion Agenda. (2023). About Global Fashion Agenda. https://globalfashionagenda.org/about-global-fashion-agenda/. Accessed 2 Jan 2023.
Goebel, J. (2021). So will Amazon die massenhafte Vernichtung von Retouren beenden. In: Wirtschaftswoche. https://www.wiwo.de/unternehmen/dienstleister/outlet-statt-vernichtung-so-will-amazon-die-massenhafte-vernichtung-von-retouren-beenden/27480108.html. Accessed 3 Aug 2022.
Gold Standard. (2023). A higher standard for a climat secure and sustainable world. https://www.goldstandard.org/. Accessed 2 Jan 2023.
Google. (2022). Kraftstoffsparende Routenplanung in der Google Maps App. https://support.google.com/maps/answer/11470237?hl=de. Accessed 10 Sept 2022.
Greenpeace. (2022). New Greenpeace Report: Plastic Recycling Is A Dead-End Street—Year After Year, Plastic Recycling Declines Even as Plastic Waste Increases. https://www.greenpeace.org/usa/news/new-greenpeace-report-plastic-recycling-is-a-dead-end-street-year-after-year-plastic-recycling-declines-even-as-plastic-waste-increases/. Accessed 22 Dec 2022.
Grieder, D. (2022). "Wir haben keine Stoffe auf Xinjiang". *Frankfurter Allgemeine Zeitung*. 13.6.2022. p. 20
H2Fly. (2023). Zero-emission aviation. Powered by hydrogen. https://www.h2fly.de/. Accessed 2 Jan 2023.
H&M. (2022). Conscious Choice. https://www2.hm.com/de_de/damen/nachhaltigkeit/our-products.html. Accessed 24 March 2022.
Heinrich-Böll-Stiftung. (2021). Fleischatlas. Daten und Fakten über Tiere als Nahrungsmittel 2021. https://www.boell.de/de/fleischatlas. Accessed 8 Sept 2022.
Hovius, N. (2021). Eine Forschungspause, die dem Klima nicht hilft. https://www.faz.net/aktuell/wissen/erde-klima/ccs-technologie-in-deutschland-eine-forschungspause-die-dem-klima-nicht-hilft-17408711.html. Accessed 5 Sept 2022.
IKEA. (2023). IKEA Zweite Chance. https://www.ikea.com/de/de/zweitechance/. Accessed 3 Jan 2023.
Inventio. (2023). Nachhaltigkeit weitergedacht: "Cradle to Cradle" in der Immobilienbranche. https://inventio.de/magazin/cradle-to-cradle/. Accessed 2 Jan 2023.

Inverto. (2022). *Wie Sie das Lieferkettengesetz (LkSG) erfolgreich umsetzen und für sich nutzen.* Köln.
iPoint. (2023). Entdecken Sie die iPoint Suite. https://www.ipoint-systems.com/de/software/. Accessed 4 Jan 2023.
Kearney. (2020). Can circularity save the fashion industry? https://www.kearney.com/consumer-retail/article/-/insights/can-circularity-save-the-fashion-industry. Accessed 2 Aug 2022.
Kotowski, T. (2022). Lufthansa verschärft Klimapläne. *Frankfurter Allgemeine Zeitung*. 20.9.2022, p. 18
Kreutzer, R. T. (2021). *Toolbox für Digital Business. Leadership, Geschäftsmodelle, Technologien und Change-Management für das digitale Zeitalter*. Springer Gabler.
Kreutzer, R. T., & Klose, S. (2023). *Metaverse kompakt*. Springer Gabler.
Lidl. (2022). Lidl gibt Obst und Gemüse eine zweite Chance. https://unternehmen.lidl.de/verantwortung/csr-news/2022/220801_rettertuete-national. Accessed 9 Aug 2022.
Loop. (2023). Designed for Reuse. https://exploreloop.com/. Accessed 3 Jan 2023.
Lufthansa. (2023a). Upgrade your life. https://www.worldshop.eu/de/upcycling/?p=7yQGmmy2evE. Accessed 3 Jan 2023.
Lufthansa. (2023b). CO_2-neutrales Fliegen. https://www.lufthansa.com/de/de/flug-kompensieren. Accessed 1 Jan 2023.
Madaster. (2023). Mehrwert für Materialien. https://madaster.de/. Accessed 3 Jan 2023.
McKinsey. (2021). This surprising change can help the auto industry tackle emissions goals. https://www.mckinsey.com/business-functions/sustainability/our-insights/sustainability-blog/this-surprising-change-can-help-the-auto-industry-tackle-emissions-goals. Accessed 12 Sept 2022.
Mediaplus. (2023). Green GRP. https://www.mediaplus.com/de/landingpages/green-grp1.html. Accessed 21 July 2023.
Nestler, F. (2022). Dreckige Minen. *Frankfurter Allgemeine Zeitung,* 24.6.2022. p. 24
NOAH. (2022). Europe's #1 Growth Capital Fundraising Platform. https://www.noah-conference.com/. Accessed 6 Dec 2022.
o. V. (2022a). Frust über Klimazertifikate. *Frankfurter Allgemeine Zeitung,* 5.12.2022. p. 19
o. V. (2022b). Rauswurf erzürnt Musk. *Frankfurter Allgemeine Zeitung,* 20.5.2022. p. 23
OECD. (2022a). Action steps for sustainable manufacturing. https://www.oecd.org/innovation/green/toolkit/actionstepsforsustainablemanufacturing.htm. Accessed 10 Aug 2022.
OECD. (2022b). OECD sustainable manufacturing indicators. https://www.oecd.org/innovation/green/toolkit/oecdsustainablemanufacturingindicators.htm. Accessed 30 Aug 2022.
Original Unverpackt. (2023). Willkommen. https://original-unverpackt.de/. Accessed 2 Jan 2023.
PACE. (2023). Who we are. https://pacecircular.org/about. Accessed 2 Jan 2023.
Philips. (2023). The circular imperative. https://www.philips.com/a-w/about/environmental-social-governance/environmental/circular-economy.html. Accessed 3 Jan 2023.
Porter, M. E. (2004). *Wettbewerbsvorteile*. Campus
Prewave. (2023). Lieferketten Risiko Monitoring. https://www.prewave.com/de/. Accessed 2 Jan 2023.
Rebuy. (2023). Wir sind rebuy. https://company.rebuy.com/. Accessed 2 Jan 2023.
Reidel, M. (2022). Die Marken mit dem höchsten ESG-Score 2021. *Horizont, 12–13*, 96.
Reverse Logistics Group. (2023). Ein innovatives Pfandsystem für Schottland. https://rev-log.com/de/. Accessed 2 Jan 2023.
Rügenwalder Mühle. (2023). Unsere vegetarischen und veganen Produkte. https://www.ruegenwalder.de/vegetarische-und-vegane-produkte. Accessed 3 Jan 2023.
S&P. (2023). Invited companies. https://www.spglobal.com/esg/csa/invited-companies#DJSI. Accessed 4 Jan 2023.

Sapriel, C. (2019). How to Continually Build Your Crisis Resilience: A checklist. https://www.iabc.com/crisis-resilience-checklist/. Accessed 20 Oct 2020.

Scherrer, A. (2021). Das C2M-Modell von Shein erklärt. https://blog.carpathia.ch/2021/06/02/das-c2m-modell-von-shein-erklaert/. Accessed 18 Nov 2021.

Schmale, O. (2022). Ein großer Schritt zur Wiederverwertung. *Frankfurter Allgemeine Zeitung*, 1.7.2022. p. T3

Schnabel, I. (2022). "Anreize für Klimaneutralität setzen". https://www.tagesschau.de/wirtschaft/finanzen/interview-schnabel-ezb-101.html. Accessed 29 Sept 2022.

SheIn. (2023). Über uns. https://de.shein.com/About-Us-a-117.html. Accessed 3 Jan 2023.

Statista. (2021a). eCommerce: Shein.com in Deutschland 2021a Brand Report. https://de-statista-com.ezproxy.hwr-berlin.de/statistik/studie/id/97520/dokument/ecommerce-sheincom-in-deutschland-brand-report/. Accessed 18 Nov 2021.

Statista (2021b). Elektroschrott. https://de-statista-com.ezproxy.hwr-berlin.de/statistik/studie/id/101889/dokument/elektroschrott/. Accessed 27 July 2022.

Statista. (2022a). Market size value of plastics worldwide from 2021 to 2030. https://www-statista-com.ezproxy.hwr-berlin.de/statistics/1060583/global-market-value-of-plastic/. Accessed 26 July 2022.

Statista. (2022b). Number of Internet of Things (IoT) connected devices worldwide from 2019 to 2030. https://www.statista.com/statistics/1183457/iot-connected-devices-worldwide/. Accessed 26. Juli 2022.

Statista. (2022c). Fast Fashion in Europe. https://www-statista-com.ezproxy.hwr-berlin.de/study/70830/fast-fashion-in-europe/. Accessed 28 July 2022.

StEP. (2023). Who we are. http://www.step-initiative.org/. Accessed 2 Jan 2023.

Sustainalytics. (2023). Company ESG Risk Ratings. https://www.sustainalytics.com/esg-ratings. Accessed 2 Jan 2023.

Tafel. (2023). Die Tafeln: Lebensmittel retten. https://www.tafel.de/. Accessed 2 Jan 2023.

Thyssenkrupp. (2023). Mit grünem Stahl in die Zukunft. https://www.thyssenkrupp-steel.com/de/unternehmen/nachhaltigkeit/klimastrategie/. Accessed 3 Jan 2023.

TÜV. (2023). Goldstandard. https://www.tuev-nord.de/de/unternehmen/zertifizierung/goldstandard/. Accessed 2 Jan 2023.

Umweltbundesamt. (2022a). Ökodesign-Richtlinie. https://www.umweltbundesamt.de/themen/wirtschaft-konsum/produkte/oekodesign/oekodesign-richtlinie#umweltfreundliche-gestaltung-von-produkten. Accessed 2 Aug 2022.

Umweltbundesamt. (2022b). Wie nachhaltig sind die deutschen Supermärkte? Systematische Bewertung der acht umsatzstärksten Unternehmen des Lebensmitteleinzelhandels in Deutschland aus Umweltsicht. https://www.umweltbundesamt.de/presse/pressemitteilungen/nachhaltigkeit-im-supermarkt-handel-schoepft. Accessed 27 Sept 2022.

Unverpackt. (2023). Unverpackt—lose, nachhaltig, gut. https://unverpackt-kiel.de/. Accessed 2 Jan 2023.

Vepa. (2023). Willkommen bei Vepa the furniture factory. https://vepa.nl/?lang=de. Accessed 2 Jan 2023.

Verhaeghe, S. (2021). How Shein is revolutionizing the fashion industry. https://www.dukeandgrace.com/en/insights/articles/how-shein-is-revolutionizing-the-fashion-industry. Accessed 19 Nov 2021.

Verra. (2023). Verified Carbon Standard. https://verra.org/project/vcs-program/. Accessed 2 Jan 2023.

Wiesner, K. (2016). *Faires Management und Marketing*. De Gruyter.

Wintershall Dea. (2023). Wintershall Dea und Equinor entwickeln gemeinsam CCS-Infrastruktur in der Nordsee. https://wintershalldea.com/de/newsroom/wintershall-dea-und-equinor-entwickeln-gemeinsam-ccs-infrastruktur-der-nordsee. Accessed 2 Jan 2023.

World Economic Forum. (2020). Forging Ahead. A materials roadmap for the zero-carbon car. https://www.weforum.org/reports/forging-ahead-a-materials-roadmap-for-the-zero-carbon-car. Accessed 12 Sept 2022.

WWF. (2023). Der touristische Klimaabdruck. https://www.wwf.de/aktiv-werden/tipps-fuer-den-alltag/umweltvertraeglich-reisen/klima-fussabdruck. Accessed 2 Jan 2023.

Zalando. (2022). Zircle—Der Kleiderschrank der Zukunft. https://corporate.zalando.com/de/magazin/zircle-der-kleiderschrank-der-zukunft. Accessed 28 July 2022.

Unlocking Sustainability Potentials in Marketing and Brand Management 5

It is no shame to take a step back, but it is a shame not to take a run-up afterwards. Anyone who wants to overtake must change lanes.

Wisdom of life

Abstract

Marketing and branding are not the starting point of sustainable business management, but rather important companions. Green marketing and green branding cannot exist without sustainable business management. Marketing can help answer the question of which target groups are particularly open to sustainable offers—and how sustainable action can contribute to the profiling of the brand. In addition to the elements of green brand management, particularly exciting examples of green brand management are also presented.

5.1 Classification of Green Marketing and Green Branding

Is it too simplistic to say that the term "green" in the sustainability discussion is an invention of marketing managers—to sell the whole thing a bit prettier? In fact, it needs to be examined whether "Green Marketing" and "Green Branding" is or must be nothing more than "sustainable marketing" and "sustainable brand management". Alternatively used terms are also Eco-Marketing or Sustainable Marketing. However, the question always arises whether the activities behind these contents actually aim at a Triple-Bottom-Line (see also Bauer & Sobolewski, 2022).

© The Author(s), under exclusive license to Springer Fachmedien Wiesbaden GmbH, part of Springer Nature 2024

R. Kreutzer, *The Path to Sustainable Corporate Management*,
https://doi.org/10.1007/978-3-658-43974-3_5

> **Note Box** **Green Marketing** cannot exist without sustainable corporate management. A **Green Branding** also cannot succeed without a company having already implemented a multitude of measures of sustainable corporate management. The activities in marketing and branding communicate the sustainability-oriented measures that a company has implemented or plans to implement in other areas of the value chain. In addition, sustainability potentials can also be tapped in marketing itself.

For marketing and brand management, it is crucial that the aspects and topics of sustainable corporate management are worked out and communicated, which were presented in the aforementioned chapters and are already being applied by the companies. Brands and brand management must not be misused for a **Greenwashing**—to merely cloak a company's "inaction in terms of sustainability" with a **green mantle**. This would have nothing in common with a strategic marketing orientation and responsible brand management (see also Weigand, 2020, p. 65 f.; further Peterson, 2021).

> **Note Box** Marketing and brand management can and must make "green" elements of actual corporate action visible in words and deeds. Such "green" elements must not be invented!

> **Food for Thought**
> Why should companies refrain from greenwashing? Quite simply. In the digital age, the rule is:
> **What can come out, will come out! Sooner or later!**

Therefore, green marketing and green brand management are not at the beginning of a corporate **Green Journey.** Marketing and especially **marketing research** are important informative companions at the beginning of this journey towards sustainability. After all, the measures to be initiated for sustainable corporate management must be reconciled with the expectations of the various stakeholders. In a later phase, it is then necessary to report regularly and honestly about measures that have already been taken and planned. This also includes reporting on activities that have not achieved the desired success and are therefore being stopped. Such an approach is also part of **honest communication.**

When talking about **sustainability in marketing** , the entire marketing management must be checked for sustainability potentials. Figure 5.1 shows the comprehensive **concept of marketing management** relevant for this.

For many decades, **marketers** have been advised to place an empty **chair** in their meeting rooms—representing the respective target group to be addressed. So that no one forgets who ultimately has to like all the marketing measures. Today, often **persona concepts** are used to describe these target persons more precisely (see Kreutzer, 2021, pp. 371–373).

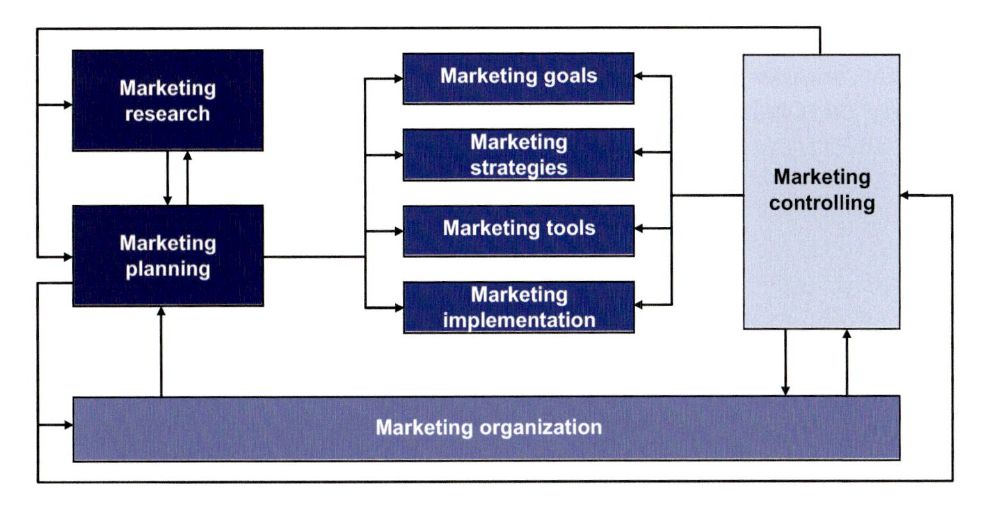

Fig. 5.1 Concept of Marketing Management

In addition, another symbol should be introduced in the meeting rooms today—ideally in all of them. This is the **photo of the key stakeholder: our blue planet** (see Fig. 5.2). After all, we only have one! And without a healthy living environment, humanity cannot survive in the long term. Therefore, in all corporate decisions—far beyond marketing—a look should be taken at the blue planet and the question should be asked what this important stakeholder would say about the respective decision.

Fig. 5.2 Equipping the meeting rooms with a photo of the key stakeholder. *Source* NASA

> Food for Thought How does the saying go? The Earth meets another planet. This one asks: "You look quite worn out, what happened to you?" The Earth replies: "Oh, I think I have Homo sapiens." The reaction of the other planet: "That's not so bad! It will pass!"

The following will show how the different areas of marketing management should be designed to leverage the **sustainability potentials in marketing**. The first to be mentioned is marketing research.

Here we deliberately speak of **marketing research**. This includes, in addition to the **market research**, the other fields of the micro and macro environment of a company that go beyond the "market" and cover the other fields shown in Fig. 5.3.

Only such research will meet the information needs of sustainable corporate management appropriately. **Marketing research** is defined here as the comprehensive collection, processing, analysis and interpretation of information to support the overall process of marketing management with information. An important task of marketing research is first to determine the openness of customers to sustainable offers. This also includes the willingness of customers to possibly pay higher prices for more sustainable solutions. It is also necessary to determine through which form of storytelling customers can be motivated to consume more sustainably. A particular challenge is to identify a possible attitude behavior gap and ideally to overcome it.

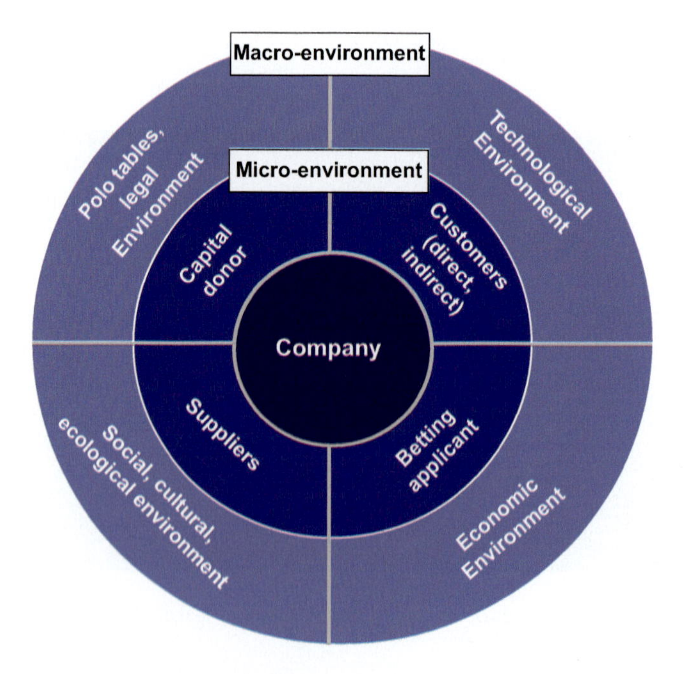

Fig. 5.3 Micro and macro environment of the company

Zalando derives concrete ideas from the study results already discussed in Sect. 1.3.2 to close the **Attitude Behavior Gap** (cf. Zalando, 2021, pp. 27–33). These **recommendations for action to overcome the Attitude Behavior Gap** are formulated in such a way that they are relevant for many industries—not just for the fashion industry:

- Focus on transparency in your communication and take your customers on the (exciting) sustainability journey!
- Talk about sustainability in a way that everyone can follow your explanations!
- Help your customers understand the relevance and urgency of your company's sustainability mission!
- Help your customers to buy right, not more!
- Use data and technology in such a way that discounts that are not sustainable are avoided!
- Increase the sale of sustainable products by convincing your customers through factors such as "quality in a broader sense"!
- Use your influence in a meaningful way by giving a voice for sustainability to your customers and employees as well as influencers!
- Apply the principles of the circular economy throughout the lifecycle of your products or services!
- Invest in second-hand solutions—whether through your own platforms or by supporting corresponding concepts!
- Help your customers to use, maintain and repair your products and services correctly for a longer period!
- Consistently turn your customers into fans of sustainable approaches!

The **market research** has the important task of identifying ideas for and acceptance of such approaches in order to continuously develop them.

In addition to this, **marketing research** focuses on further fields (cf. Fig. 5.3). The initial question is which **legal obligations** regarding "sustainability" already exist today and will be faced by companies in the future. It is also necessary to check which (new) **technological possibilities** are available to promote the circular economy. The **economic conditions**—for example in the form of global competition—must also be taken into account. However, a particular focus is on the **social, cultural and ecological environment.** Here, the corporate solutions must meet with acceptance and contribute to the avoidance or reduction of harmful effects.

5.2 Target Groups of Green Marketing

A important question of green marketing relates to the particularly interesting **target groups:**

- For which customer groups are sustainable marketing and sustainable brand management particularly important?
- Which segments should be primarily addressed—as long as sustainable demand has not yet become mainstream?

For green **brand management in the B2C market**, one segment is particularly exciting: the target group of **LOHAS**. This acronym stands for **Lifestyle of Health and Sustainability.** The studies presented below show that their consumption and leisure behavior differs significantly in many dimensions from that of the rest of the population.

Note Box

LOHAS are not consumption refusers—quite the contrary! LOHAS are about full enjoyment—but with a clear conscience. It's not about renouncing consumption, but about conscious, responsible consumption, for which more money is gladly spent.

LOHAS often have "green" values and feel committed to the principle of sustainability. At the same time, they value a high standard of living and a convincing aesthetic of the offers.

A look at the **age distribution of LOHAS** in the total population in Germany shows that only from the age group of 40 to 49 years is the proportion of LOHAS larger than the proportion of this age group in the total population. With increasing age, the proportion of LOHAS in the total population also increases. This proportion also increases with higher **educational qualifications** (see Fig. 5.4; see Statista, 2021f, p. 3, 5 f.). These and the following information about the target group of LOHAS are based on a study with a representative sample for Germany of 22,563 respondents.

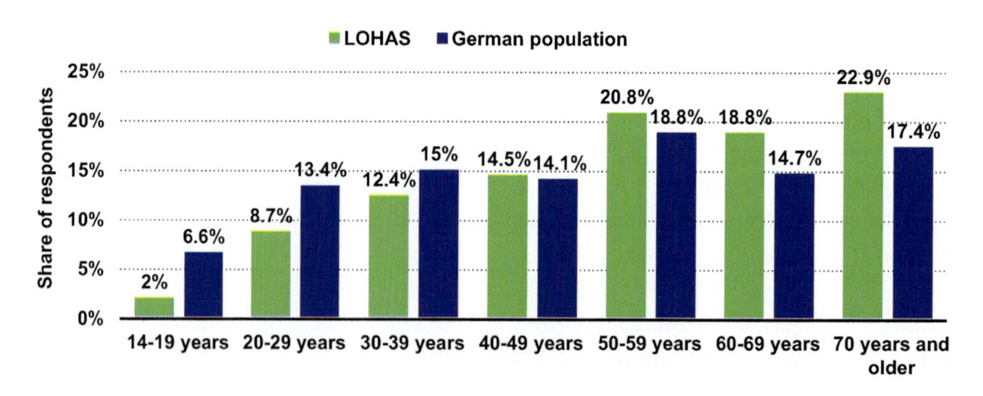

Fig. 5.4 Proportion of LOHAS in different age groups in Germany—2021. Data source: Statista, 2021f, p. 3, 5 f.

> **Food for Thought** If one keeps this fact in mind, the following can be determined with relative certainty: The **LOHAS proportion** will continue to increase due to the age-related population development in Germany. Therefore, companies would do well to examine the **"green potential" of their offers** today.

In 2021, around 36% of LOHAS were **men.** In the total population, their share was around 49.3%. The proportion of **women** among LOHAS is significantly higher at 64% compared to their share in the population at 50.7% (see Statista, 2021f, p. 2). A look at the **value orientations of the LOHAS** in Germany is also very informative (see Table 5.1). Here, compared to the total population, there is a significantly higher interest in social justice and close relationships with other people. Nature, family, but also a great deal of independence are important to LOHAS. This is also accompanied by fun and a curiosity to experience new things (see Statista, 2021f, p. 9). These aspects must be taken into account when communicating with this target group.

The **areas of interest of the LOHAS** are shown in Fig. 5.5. Here, it becomes visible again how differently the LOHAS group "ticks" compared to the general population. In total, one can observe a much greater openness and a greater interest in people and the environment among the LOHAS. This also corresponds with the higher educational qualifications of this group (see Statista, 2021f, p. 17).

The dominant values and areas of interest among LOHAS are also reflected in the **shopping behavior of the LOHAS.** In 2021, 86.8% of LOHAS agreed with the statement "When it comes to food, I mainly focus on quality and not so much on price". In the general population, this proportion was only about 38.8%. The other statements

Table 5.1 Value orientations of LOHAS in Germany—2021. (Data source: Statista, 2021f, p. 9)

	LOHAS	German population
Social justice	95.5%	64.2%
Having good friends, close relationships with other people	94.0%	85.2%
Experiencing nature, being a lot in nature	91.4%	43.3%
Being there for the family, committing oneself to the family	90.6%	80.5%
Independence, being able to largely determine one's own life	87.5%	70.8%
A happy partnership	85.5%	75.2%
Helping people in need	83.0%	54.0%
Always learning new things	74.0%	48.2%
Having a lot of fun, enjoying life	72.8%	62.9%
Dealing with questions about the meaning of life	70.6%	25.2%
Having children	68.6%	59.1%
Getting to know the world, other countries and cultures	67.8%	42.0%

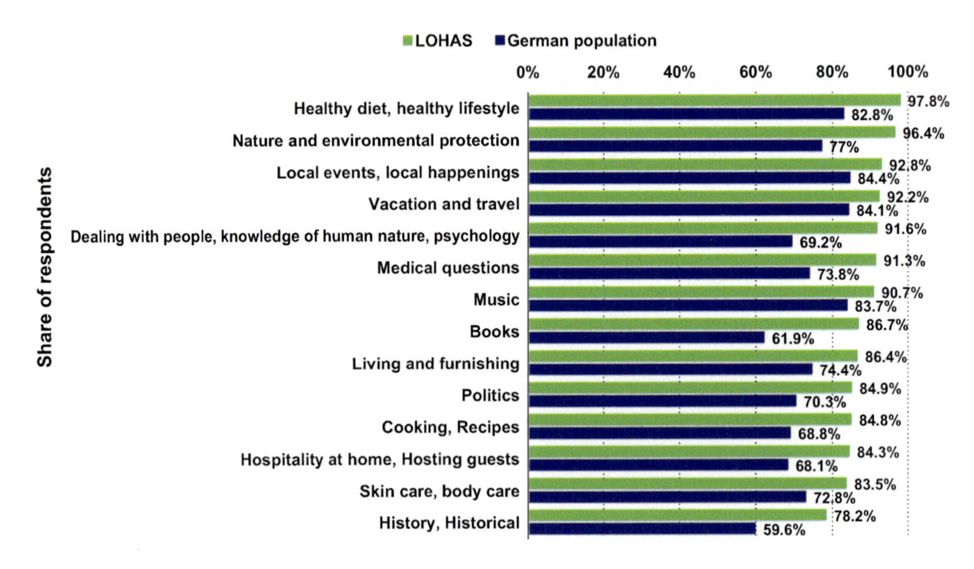

Fig. 5.5 Areas of interest of the LOHAS in Germany—2021. (Data source: Statista, 2021f, p. 17)

queried in Table 5.2 also show significant deviations between the information provided by the LOHAS and the general population (see Statista, 2021f, p. 10).

It is advisable to start exploiting the potentials shown here before corresponding legal regulations come into force. Because only those companies can credibly profile themselves through green brand leadership that have already started doing so when there was no compulsion for more sustainable marketing.

Table 5.2 Shopping behavior of the LOHAS in Germany—2021. (Data source: Statista, 2021f, p. 10)

	LOHAS	German population
I am willing to pay more for good quality	91.1%	68.7%
I prefer to buy regional products from my home country when shopping	88.5%	54.9%
When it comes to food, I mainly focus on quality and not so much on price	86.8%	38.8%
I am willing to pay more for environmentally friendly products	81.3%	37.6%
When buying food, I pay particular attention to the products coming from animal-friendly farming	76.6%	34.0%
When it comes to food, I value products from organic farming, organic products	75.7%	29.8%
When buying products, I pay attention to their durability, i.e., that I can use them as long as possible	75.4%	51.3%

However, green brand management should not be limited to the LOHAS segment. That would be a **resignation to the status quo.** Instead, companies and brands are called upon to actively take their customers on the **journey towards sustainability** and not just react (see Fig. 4.1). After all, the world and its natural resources are not waiting for an even greater **change in consciousness among customers** to take place.

At the same time, companies should consider how **customer expectations over time** have changed (see Fig. 5.6). Green brand management can address and serve some of the changes shown here, thus also helping to close the Attitude Behavior Gap.

In sum, the **expectations of prospects and customers** can today be characterized by the buzzwords **"me, everything, immediately and everywhere".** And increasingly, another requirement is added: **"and green"** (see Fig. 5.7).

A study by **Concept M** (2022) provides further exciting insights. Overall, it can be stated that **"sustainability"** is becoming an increasingly **omnipresent mainstream topic.** The triggers for this are the increasing weather disasters—which also claim victims in Germany. These issues are charged by the protest actions of ***Fridays for Future, Last Generation*** and ***Extinction Rebellion.*** Here, **horror scenarios with end-time narratives** are sometimes designed, which see a reversal—today and by everyone—as indispensable to save the planet.

As a result, even supposedly uninvolved and innocent customers develop **feelings of guilt**. There is a sense of a **moral pressure** to also display **more sustainable behavior.** The ideas revolve around avoiding plastic, possibly reducing meat consumption, and taking fewer or no flights (keyword "flight shame"). However, the vast majority still find it difficult to break away from **ingrained habits.** This is especially true when changes are primarily experienced as renunciation.

For the "normal" customer, the **supposed solutions** remain **contradictory,** confusing—and hardly achievable. After all, the **facts are also confusing**—and even the experts hardly reach a consensus. Moreover, it is often said:

Formerly	Level	Today
Rather passive, accepting customer	Customer	Active, demanding, qualified customer
Use of a few sources of information	purchasing behavior	Often extensive research before purchase (online and offline)
Supply purchases often dominate	Purchase type	Also pure "pleasure purchases"
Customer reports on his experience in his own environment	Communication	Often extensive offline and online communication about experiences
Good price-performance ratio required; standard solutions are accepted	Expectations	Criteria such as sustainability etc. are gaining in importance; search for individual offers

Fig. 5.6 Changes in customer expectations

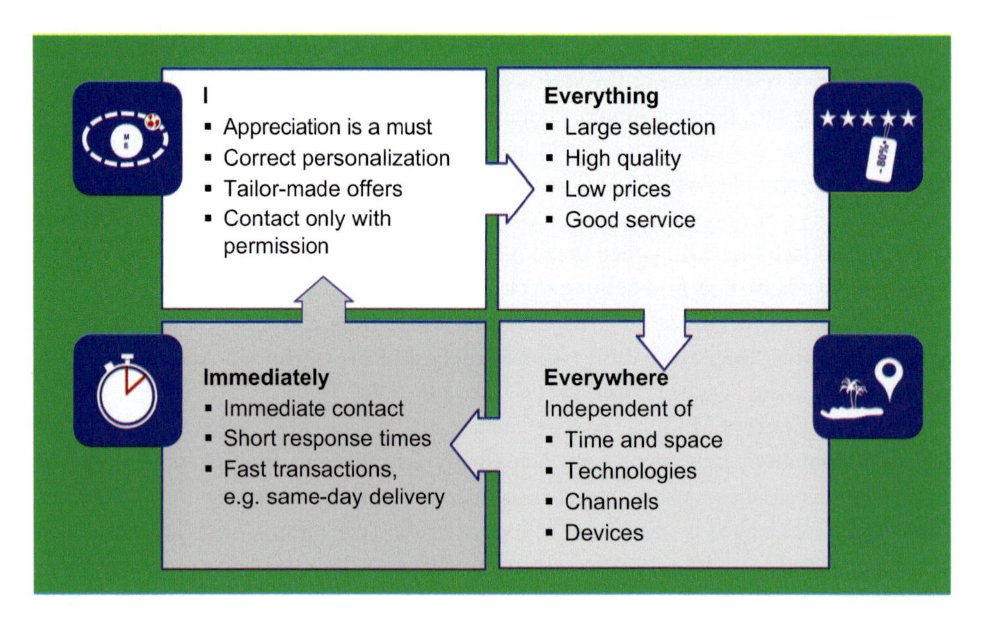

Fig. 5.7 Me-everything-immediately-everywhere-and-green-expectation on the customer side

▶ Sustainability is something you have to be able to afford first!

The uncertainty diagnosed here can be eliminated by responsible and honest green brand management. To do this, good arguments must be communicated to buyers in an easily understandable way—and also symbolically. Then **green brands** can firmly establish themselves in consciousness and—step by step—lead to a **more sustainable consumption.** The necessary change processes will not be achievable without intrinsic motivation. It is about achieving a deep understanding of why change is necessary today and also personally for each individual. It is about understanding that the need for change cannot—alone—be delegated to companies, politicians, etc. Such a **delegation attitude** needs to be overcome—through convincing narratives in brand communication. Otherwise, companies will become scapegoats because as an individual customer, you seemingly can't change anything (see in depth Sect. 1.3.4).

▶ **Food for Thought** To win customers for sustainability, communication of renunciation should not dominate. It's about conveying to buyers through a communication of gain what they can "gain" if they behave more sustainably. Communication should highlight a win-win situation:

- "Win" for the customer, because they feel better when buying and consuming.
- "Win" for nature, because it will be preserved a little longer.

Also, **companies** should be addressed as **customers of sustainable offers.** The need for more sustainable leadership in companies results from the expectations and demands of a multitude of stakeholders. Companies can identify these stakeholders and their respective expectations through the Stakeholder-Onion-Model (see Fig. 8.4). For the **providers of more sustainable solutions,** the following questions arise:

- Which companies are already acting more sustainably out of **their own initiative** and are looking for partners to continue and develop this path?
- In which companies is the **pressure to change** so great today that a switch to more sustainable solutions is already necessary?
- In which companies is there still a **ignorance towards more sustainable action,** so that the relevance must first be communicated before corresponding orders can be expected?

Companies that proactively act on their own initiative belong to the particularly exciting target group of **innovators** and **early adopters** of sustainable solutions. Here, the potential partners to jointly develop innovative concepts are predominantly found. Companies that primarily act out of a pressure to change can often only be convinced to opt for more sustainable solutions through a stringent **cost-benefit argument.** With regard to the **segment of "ignorants",** comprehensive communication is needed to wake them from their slumber. Certainly, this is a target group that should not be addressed at the beginning of a marketing process.

5.3 Elements of a Green Brand Leadership

The term **"green brand leadership"** is appearing in more and more discussions today. This term has also been used several times in this work. Is this just a new fashion trend that will fade away after a short time? Or is this a development that poses new challenges to brand leadership in the long term?

5.3.1 Basic Concepts of Brand Leadership

To approach the task, the terms "brand leadership" and "green" need to be clarified first. **Brand leadership** is the process of defining the purpose, the underlying values, and the long-term identity of a brand by the responsible managers. This **brand identity** is intended to position the company's own service offering (product, service, company) in the market and differentiate it from competitive offerings and competitors. At the same time, this brand identity should increase the desirability in the eyes of customers and other stakeholders—such as investors and (potential) employees.

The **process of brand leadership** also includes the communication of the brand identity through various channels internally and externally. Within the company, the aim is for all employees to behave "in line with the brand" and to actually live the defined brand values. In external relations, a certain **brand image** should be built—a convincing brand image in the eyes of the relevant stakeholders. This brand image should correspond as closely as possible to the brand identity defined as the target.

Figure 5.8 shows how a brand identity is built overall and how a brand image is created. The **brand identity** represents the "self-image of the internal target group". It describes how the brand should look from the perspective of the employees responsible for building the brand within a company. Guided by their own **competencies** (What can we do?), **values** (What do we believe in?), **personality** (How do we appear?) and **vision** (Where do we want to go?), the **services** are defined (What do we do?). The statement "The future needs the past" plays an important role here. The **origin** (Where do we come from?) therefore forms the basis of all decisions. The thus defined **self-image of the brand** is then transmitted into the market via the 5 Ps of marketing (cf. Burmann et al., 2018, p. 15).

In the course of building a **brand-customer relationship**, a influenced **brand image** is created among the external target groups (here especially among the customers, but also among other stakeholders). This represents the "external image of the external target group". A prerequisite for building a brand image is the **brand awareness**. Through communication, **brand attributes** are conveyed as **functional** and **symbolic benefit associations** of the brand. All these elements are reflected in the **brand expectations** of the external target groups (cf. Fig. 5.8).

In **brand management in the digital age** (see in depth Kilian & Kreutzer, 2022), further aspects must be considered. However, the **customer's brand experience** remains the focus (cf. Fig. 5.8). This brand experience is also shaped by the **interactions of third**

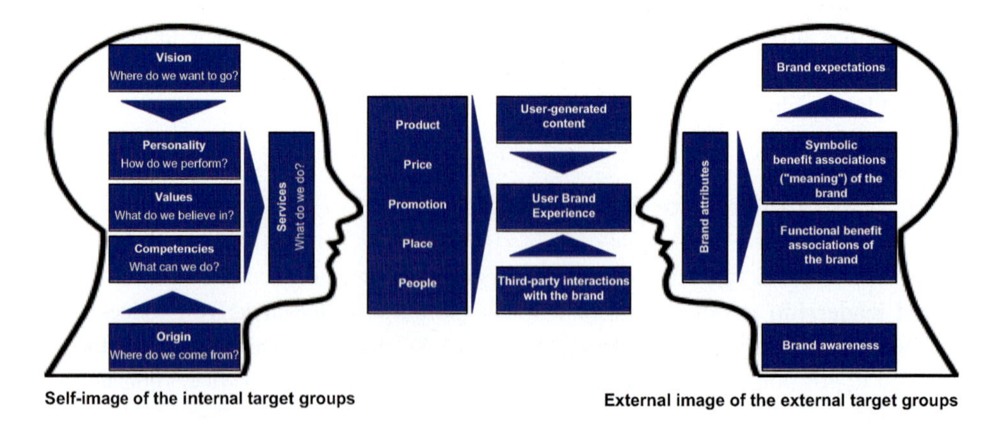

Fig. 5.8 Brand identityand brand image

parties with a brand. Other people can spread their experiences via analog, but especially via digital channels. These complement the customer's brand experience.

In addition, digital media offer stakeholders—and here especially customers—platforms for their own brand-related communication. Here, people—independent of the company—can develop and spread their own content. This **user-generated content** includes comments and reviews on the respective platforms (e.g., *Yelp, HolidayCheck, TripAdvisor*). This user-generated content also includes their own creations on *Instagram, Pinterest, TikTok* and *YouTube.* Activities in (customer-owned) blogs or communities also count. Figure 5.8 illustrates this development.

> **Note Box** In **(digital) brand management**, various influencing factors must be taken into account, which impact the **Brand Experience** and thus the **brand image.**

5.3.2 Building a Green Brand Identity

For the **construction of the brand identity**, one can orient oneself to the brand steering wheel presented in Fig. 5.9 **Brand steering wheel** (see Esch, 2019, p. 100). Based on the brand's own competence (Who am I?), which addresses the questions raised in Fig. 5.8, a concretization of the brand is now carried out with regard to

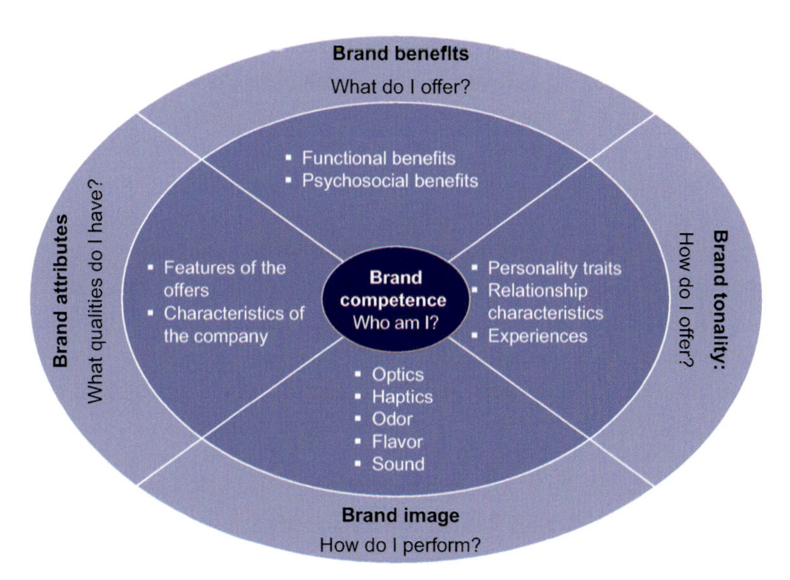

Fig. 5.9 Brand steering wheel. *Source* Adapted from Esch (2019, p. 100)

- Brand benefits,
- Brand tonality,
- Brand image and
- Brand attributes.

The **term "green"** is translated in this context as "sustainable" and addresses the topics that were particularly discussed in Chap. 4. Against this background, a **green brand** can be characterized as follows:

A **green brand** is understood to be a brand concept in which **ecological and social values** form an essential part of the brand core. To convey these values, functional and psychosocial benefit elements are highlighted as part of a **green brand management.** In the course of **functional brand positioning**, information is provided about the ecological or social advantages that the advertised brand has over competitive offers. Detailed data and relevant seals can be referred to in order to appeal to the **ratio of humans** (see also Errichiello & Zschiesche, 2021; Rieke & Schwingen, 2021).

The **emotio of humans** is addressed by conveying psychosocial benefit elements. This leads to an **emotional brand positioning.** Imagery contributes to this, showing, for example, "happy cows" for dairy products and "blooming landscapes" for cosmetic products. If positively connotated labels (e.g., *Demeter*) or symbols (such as a *frog*) are used, the emotional charge of a brand is further promoted. The "good feeling" conveyed by the brand can also contribute to this, giving customers the impression that they are acting in an environmentally and socially compatible way by choosing a green brand. If this positive action is visible to others, it can also yield prestige benefits.

In the course of brand management, the functional and psychosocial benefit elements are to be merged in order to achieve a **sustainable success of brand management.** Therefore, positioning as a sustainable, trustworthy, and green brand—especially for interested customers—should be supported by providing detailed background information. After all, the goal is to win over both the heart and the mind.

How do Green Marketing, Green Branding, and Corporate Social Responsibility (CSR) differentiate from each other? **Corporate Social Responsibility** describes a company's responsibility that goes beyond the actual purpose of the company and shows that companies are facing a larger responsibility. The focus here is often on the social impacts of corporate activity. Green Marketing and Green Branding or sustainable marketing and sustainable brand management define corporate responsibility much more broadly and particularly include the ecological effects of corporate activities.

5.3.3 Guidelines for Green Brand Management

Which **guidelines of a green brand management** are relevant is shown in Fig. 5.10. First of all, every responsible company should stipulate in its guidelines to refrain from any

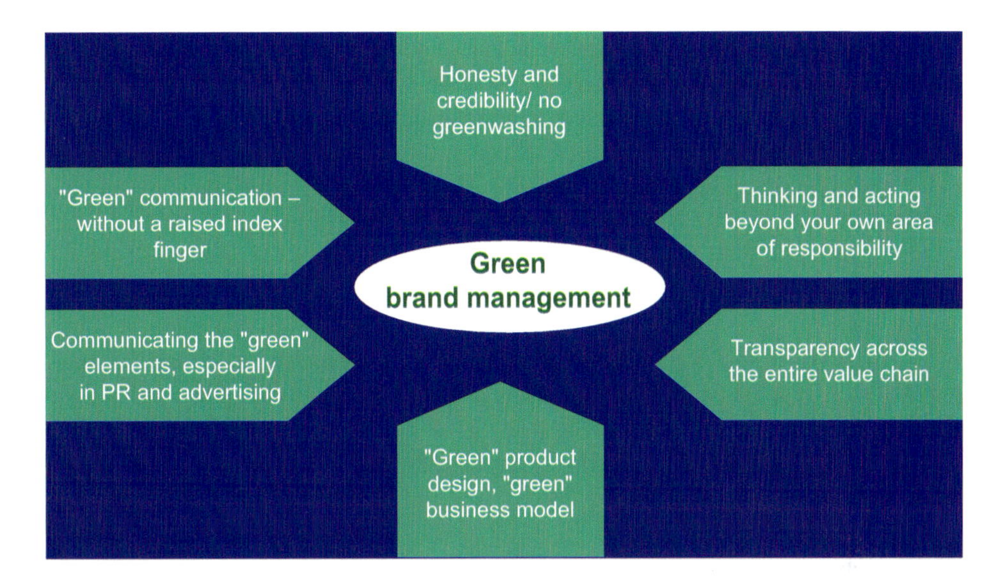

Fig. 5.10 Guidelines of green brand management

form of **greenwashing**. A sustainable anchoring as a green brand can only be achieved through a high degree of **honesty and credibility** (see in depth Chap. 7).

An important step towards sustainable and honest communication is to establish a **thinking and acting beyond one's own area of responsibility** within the company. After all, all companies are also integrated into the ecosystem "Earth" and therefore have to meet further requirements. Therefore, the question of sustainability should not end at departmental or area boundaries. Already in the design of a product or service, it should be checked to what extent aspects of the circular economy can be incorporated (see Sect. 4.4.1.1). The range here extends from the processing of recycled materials to resource-saving production and use to the possibility of recovering resources at the end of a product's usage phase.

This requires—for all participants—a high degree of **transparency about one's own value chain** and its effects on resource consumption and emissions. This transparency must be ensured regardless of legal obligations. The following questions, among others, need to be answered, which were already addressed in Sect. 4.4.1:

- What raw materials are used today?
- How sustainably were these raw materials obtained?
- Are recycled materials used for the products themselves and for packaging?
- Can the recycling rate be increased?
- Were the employees involved in the production process paid fairly?
- How reputable are your own suppliers?

- To what extent can resources already be saved today and harmful emissions in the manufacturing process be avoided?
- Where are there further optimization possibilities in production?
- Where can further resources be saved in logistics in the future?
- What possibilities exist to limit resource consumption and the generation of emissions in the usage phase?
- What does sustainable disposal of packaging or product look like after use?

Answers to these questions are not only relevant for managers in the areas of R&D, procurement and production, but also for employees in marketing, sales and logistics. For HR managers, it may be important to identify which qualification profiles need to be covered by the company's own resources in the future. The communication of such information is a core task of internal marketing or internal brand management. An important source of information for this is sustainability monitoring or sustainability controlling (see Chap. 6).

A core area of green brand management is the **"green" product design.** Here, the question for manufacturers is about the production program and for retailers about the offered assortment. Both can be more or less strongly oriented towards sustainability criteria (see Sects. 4.4.1.2 and 4.4.1.6). Going beyond brand management, the development of a **"green business model"** that is developed in addition to or as an alternative to the existing business model (see in depth on business model innovations Kreutzer, 2021).

How communication about a "green" offer design can succeed in companies that are on the path towards sustainability is demonstrated by the example of ***Aldi.*** Then it is important to report transparently and credibly about the **path to sustainability**. This is achieved with the following statement from Aldi (2022):

> "We have successfully committed ourselves to gradually abolishing chick culling by the end of 2021 and converting our entire shell egg range to organic, free-range and barn eggs. In the production of shell eggs, male chicks were usually killed because they cannot lay eggs and are not suitable for fattening. Unlike broilers, they hardly put on any meat."

Here, a self-commitment was openly communicated—the path and the goal became visible. And: The goal was achieved by the end of 2021—and communicated again. In addition, it was reported in 2022 that already 60% of the recipes of the cosmetic own brands are free of microplastics and more and more fruits and vegetables are being sold unpackaged (cf. Aldi, 2023).

L'Oréal has defined a **sustainability commitment** for its further business development until 2030 (cf. L'Oréal, 2023):

> "We at L'Oréal have committed ourselves to respecting the so-called 'Planetary Boundaries' in our activities, i.e., what the planet can withstand according to the definition of environmental science. In order to reconcile our needs with the preservation of a planet with limited resources, we are raising the bar and introducing a new internal transformation program.

> Unlike the first program, our new commitments will not only focus on our direct impacts. They will also deal with our indirect, extended impacts, which are, for example, related to the activities of our suppliers and the use of our products by consumers."

L'Oréal also describes the path to be taken. After all, this company cannot achieve its ambitious goals overnight either.

An important question is: Which **industries** should be dealing with green brand management today? The discussion on green brand management already covers all industries today. Initially, **food** was at the center of the sustainability discussion because food is directly consumed by people. Unhealthy food has a direct impact on consumers. Therefore, advocates for the production of "healthy" food were found early on to demand and promote it. In addition, agriculture, with its impact on the ecosystem, is one of the central drivers of climate change (cf. Sect. 4.4.1.2).

The **trade**—online and offline—plays a central role in this context. After all, trading companies can already pay attention to green brands when designing their product range. In addition, important purchasing impulses towards sustainability can be given by placing and promoting green brands (cf. Sect. 4.4.1.6).

The harmful effects of the fashion industry are being discussed particularly intensively today. The **fashion industry** is one of the world's largest polluters. The trend towards fast fashion contributes massively to this, as was exemplified by the example of *SheIn* (cf. Sect. 4.5).

All other **manufacturing companies** should also necessarily deal with the topic of green brand management. As described in Sect. 4.4, the entire value chain should be taken into account: from procurement to production to logistics. Only then can the goals of the circular economy be achieved. The regulations on the supply chain law and other legal requirements will increasingly oblige companies in the future to scrutinize their entire value chain with regard to sustainability (cf. Chap. 2).

The question of sustainability is also becoming increasingly urgent in the **service sector.** For example, in **tourism,** discussions are being held on how vacations can be designed without (long-term) destroying the respective destinations and/or unduly disturbing the people living there in their everyday life. In addition, the question is asked: Is it still permissible to fly, not only on short-haul routes but also on long-haul routes, when one considers the associated emissions? What forms of compensation are relevant here or how can CO_2 emissions be sustainably reduced?

In the **financial services sector**, the question arises as to what sustainable investments might look like. Should financial resources be invested in the extraction of fossil fuels? Are coal, gas, or nuclear power plants to be considered sustainable investments? What criteria should investment decisions that bear the label "green" or "sustainable" be based on? In these decisions, the so-called ESG criteria play a special role (see Sect. 2.2).

> ▶ **Note Box** The topic of green brand management is relevant for all industries—and simple solutions often do not meet the complex requirements.

5.3.4 Nudging and Signaling as Concepts of Green Communication

Most companies and brands still have a long way to go in achieving sustainability goals. This requires not only good ideas and financial resources but also time. This should be understandable and acceptable to most customers. However, only if the credible impression is conveyed that a company has already embarked on the journey towards sustainability. Therefore, a **green communication** is another important area of green brand management. Here, we deliberately speak not only of advertising but more broadly of communication. After all, aspects of sustainability are particularly often incorporated into public relations measures. PR activities, for example, also target investors and political decision-makers.

Of course, "green" aspects must also be reflected in the **advertising.** For this purpose, it is necessary, for example, to communicate "green" elements via **labels and seals.** After all, the sustainability message must become visible on the product and in the store. In addition, politics, society, and the economy in many areas wish that people show a certain behavior. However, coercion and regulations are often met with **reactance**—the opposite of acceptance. From our own storm and stress phase, we are also well acquainted with this behavior as **defiance reaction.**

What alternatives exist to guide the behavior of as many people as possible in one direction—for example, towards more sustainability? It is important that the **"green" communication** comes without a wagging finger. Who wants to be educated by brands that one chooses and then also pays for?

To be successful in **gentle behavioral guidance**, the so-called **Nudging** is recommended (see Thaler & Sunstein, 2010; Grunwald & Schwill, 2022, pp. 92–95). Literally translated, nudging means "nudging" or "pushing". Nudging aims to lead target persons to a certain behavior change without exerting force or issuing prohibitions. Nudging also refrains from economic incentives to achieve a certain behavior.

A **Nudge** is such a "nudge" or a "push"—usually in the sense of a thought-provoking impulse. Such thought-provoking impulses are important for increasing sustainable consumption and buying behavior if one does not want to work with sanctions. The goal is always to guide the behavior of the target persons in a certain direction through nudges. However, these persons still have the **freedom of choice,** because they do not have to respond to the nudges.

Note Box

Nudging attempts to change people's behavior through **thought-provoking impulses**. Here, the **freedom of choice** remains because nudges are easy to avoid. The use of prohibitions or commands as well as the use of economic incentives is therefore not part of nudging.

Nudges are therefore only supposed to "nudge" individuals so that they make a supposedly "better" decision. The use of nudges therefore changes the **decision-making environment** for people—by providing additional (decision-relevant) information.

When the state acts as a nudger and legally prescribes nudges (e.g., the horror images on cigarette packs), this is also referred to as **libertarian paternalism**. "Libertarian" here stands for "freedom". After all, people still have the freedom to choose one offer or another. "Paternalism" stands for a style of politics based on a guardianship relationship between the rulers and the ruled. Here, the rulers believe they have to encourage citizens to behave in a "better" way, however that may be defined.

Nudges are supposed to influence the decision-making process by intervening in the **conditions of decision-making.** For this purpose, certain information can be provided, warning notices can be attached, or social norms can be emphasized. The following **forms of Nudging** can be distinguished (see Rometsch, 2021):

- **Disclosure ofproduct components**
  To enable customers to make an informed decision, the respective ingredients can be declared on the products. However, it is necessary that the ingredients are presented in **understandable language.** This is often not the case. Long **lists of components** are regularly declared on products or packaging, the meaning of which is not apparent to the average consumer.
  This includes potentially problematic ingredients in cosmetic products such as methylparaben, ethylparaben, sodium laureth sulfate, alcohol denat., ethanol, paraffinum liquidum, and others. Often, the list of ingredients is also in **small font size** and often in **capital letters,** to make reading as difficult as possible. It becomes apparent that providers of traditional offers usually do not want to steer behavior towards healthier or more sustainable products.
  Nudging becomes visible when cosmetic products have labels like "No addition of alcohol, citric acid, and parabens", "Ingredients 100% of natural origin" or "Free from aluminum". Nudges are also the various **certificates,** that signal sustainably sourced raw materials or more sustainable production (see in-depth Sect. 5.3.5).
- **Disclosure of a product rating**
  A simplification compared to the indication of individual product components is given when these are incorporated into an overall evaluation of the product. A whole range of labels and seals has now been established for this purpose, which can or could promote nudging. An example of this is the traffic light scale for the **energy efficiency classes** of electrical appliances (see Fig. 5.11). This label is usually prominently displayed on the front of the devices and can provide important impulses in the purchasing process due to its transparency. This is not the case with many other evaluations (see Sect. 5.3.5).
  The interesting question is: To what extent does a low energy efficiency class affect actual purchasing behavior? How effective are these nudges in everyday life, especially when the price is of greater importance when shopping?
- **Use of symbols**
  The use of symbols can make it easier for the user to understand even complex processes and to show the desired behaviors. Whether this is already successful in the the

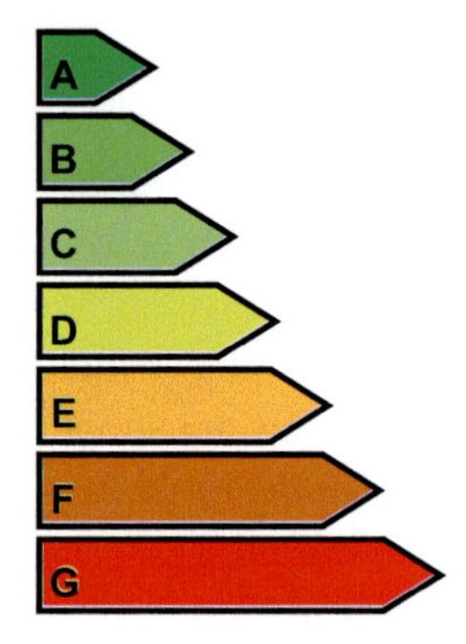

Fig. 5.11 Scale for energy efficiency classes

labeling of products and packaging is shown in Fig. 5.12. It is not to be assumed that all customers know the exact contents of these various symbols.

The 1st symbol stands for single-use (see Fig. 5.12). It belongs to the *German Deposit System GmbH* and labels cans or bottles as a **single-use product**. Products with this sign can be returned with a deposit (see DPG, 2023). However, no reuse takes place. Doesn't the symbol rather convey the idea of a cycle?

The 2nd symbol in Fig. 5.12 stands for the already over However, reusable products do not have a uniform sign. It is also up to each user to decide how to handle such a product. It is desired and expected that bottles marked as reusable will be returned to the dealers to be collected there. The customers then get their deposit back. The bottles and crates are usually returned to the manufacturers via the beverage wholesale trade. Ideally, the usage cycle then starts again (see Mehrweg, 2023).

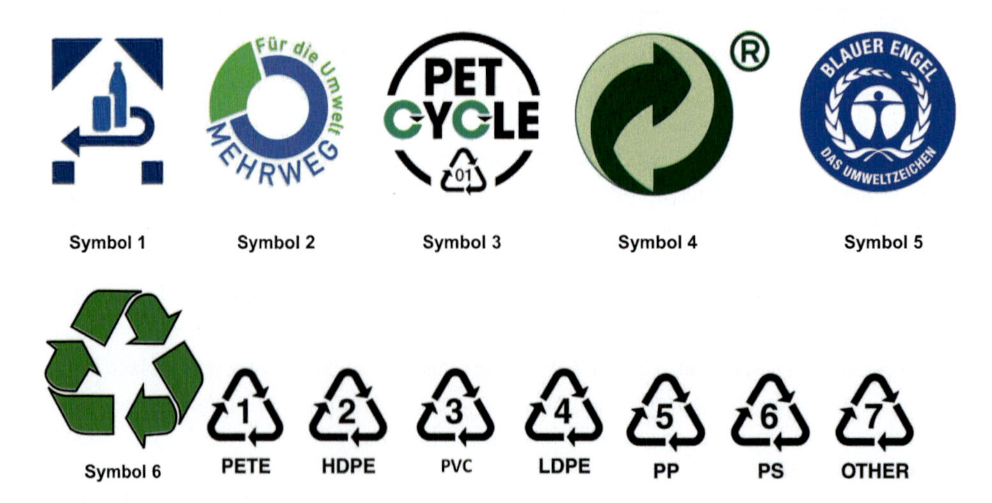

Fig. 5.12 Various symbols for labeling offers

The 3rd symbol in Fig. 5.12 comes from the company ***Petcycle.*** This combines a pool system with a **recycling cycle for PET beverage bottles**. For this purpose, reusable crate solutions are used, but with PET bottles that can only be filled once. The collected empties are processed into recyclate and used to manufacture new PET bottles. New *Petcycle* bottles consist of at least 75% recycled material (cf. Petcycle, 2023).
The 4th symbol from Fig. 5.12 is probably most familiar. It is the ***Green Dot.*** The *Green Dot* on a product or packaging means that it can be disposed of in the *Yellow Bin*. The company *The Green Dot—Dual System Germany* is supposed to ensure that the materials marked in this way are properly collected, sorted and recycled. There is no deposit on products with the *Green Dot*. What exactly happens to the materials remains unknown to the customer (cf. Green Dot, 2023).
The 5th symbol from Fig. 5.12 is also well known to many. It is the symbol ***Blue Angel*** with the addition: "Good for me. Good for the environment." This is a **government environmental label,** which has been in use for over 40 years. The *Blue Angel* represents an impartial and economically independent, voluntary product labeling. The *Federal Ministry of the Environment* is the owner of the label. The *Federal Environment Agency* is responsible for the scientific development of the award criteria. The *Jury Environmental Label* is the relevant decision-making body and the *RAL gGmbH* acts as an independent certification organization. The *Blue Angel* is a TYPE I—environmental label. This is based on the international standard DIN EN ISO 14024 (cf. Blue Angel, 2023).
The *Blue Angel* awards environmentally friendly **products in the non-food sector** and **services.** The *Blue Angel* has already been awarded to more than 20,000 products and services from over 1,600 companies. Products and services from the following categories can be awarded the *Blue Angel* (cf. Blue Angel, 2023):

- Household/Drugstore (such as detergent, toilet paper, clothing)
- Furnishing/Textiles (e.g. bathroom/kitchen furniture)
- Green-IT/Electrical appliances (such as shredders, printers, telephones)
- Building products (e.g. paints, insulation materials, floor coverings)
- Heating/Energy (insulation materials, wood pellets, wood-burning stoves, room air conditioners)
- Paper/Stationery (such as envelopes, labels, file covers)
- Vehicles/Mobility (e.g. cruise ships, car sharing)
- Trade/Municipalities (e.g. glass recycling containers, electric buses, street sweepers, garbage cans, de-icing agents)

The *Blue Angel* is intended to provide customers with reliable **guidance for environmentally conscious shopping.** This is intended to increase the **demand for environmentally friendly products** and promote **ecological product innovations** in order to reduce environmental impacts overall.

Food for Thought

With the *Blue Angel*, the emphasis is on **"environmentally friendly"**—not eco-friendly, not sustainable, not CO_2-free or similar. If a product or service is less harmful than others, that can already be enough for an award. This point is particularly emphasized with the *Blue Angel:*

The*Blue Angel*only awards the better products or services in a category.

That's why even the cruise ship *AIDAnova* can proudly carry the *Blue Angel,* which is critically discussed in Sect. 7.2. This relative perspective also leads to very loud garden shredders still being awarded the *Blue Angel* because they cause less noise pollution than others.

Therefore, with the *Blue Angel,* it only applies: **relatively environmentally friendly—only in comparison to other offers in the same category.**

The 6th symbol in Fig. 5.12 is the general sign for **Recycling**. It is similar to the *Green Dot,* but has a different meaning. The recycling symbol only indicates that the product or packaging can be recycled. If the *Green Dot* is not also printed, these materials do not belong in the *Yellow Bin.* The recycling symbol is printed in conjunction with a number or a code, as seen in Fig. 5.12. This explains what type of plastic it is (see plastic alternative, 2023). But who knows what's behind each code? The user is not informed about how to dispose of the material marked in this way. So it either ends up in the residual waste bin (tendency "burning"), unjustifiably in the *Yellow Bin* or is thrown onto the street or into nature.

Many other **eco-labels** and **eco-seals** are used to label ecologically produced materials and products. These can also—if perceived by the customer—act as nudges (see in depth Sect. 5.3.5).

▶ **Note Box**

This confusion of labels makes it clear:

A **nudge towards sustainability** is currently not being carried out in recycling.

- **Useof warnings**
  Nudges can also appear in the form of specific warnings. These include the **shock images on cigarette packs.** The successes achieved by these photos and notes can be observed by anyone when customers stand at the airport with several cartons of cigarettes or take a cigarette from packs adorned with horror images and consume it with pleasure. There are also **warnings on wine and other alcoholic beverages** that alcohol should be avoided during pregnancy. These are also forms of nudging. After all, people are still free to decide how they behave.
- **Reference to social norms**
  To motivate people for a desired—sustainable—behavior, it can be pointed out that many other people have already shown the desirable behavior. Such nudges are regularly used in hotels. Then it says, for example:
  - "9 out of 10 of our hotel guests use their towel multiple times" or
  - "Have you ever thought about how much water and detergent have to be used in hotels to provide fresh towels every day?"

Such statements—prominently placed in the bathroom—are intended to influence the behavior of guests towards multiple use.
Instead of considering making a **Veggie Day** mandatory in canteens, the benefits of a—at least partial—vegetarian diet can be highlighted through nudging. Example:
"Have you ever thought about the resources needed to produce 500g of meat compared to 500g of vegetables? Which is probably better for our environment?"

- **Presets**
  Presets—so-called default values—are a particularly effective form of nudging. The **default settings** offered to the user already point in the desired direction. This can, for example, occur in a **donation request** that is pre-filled with 10 € instead of 5 €.
  Also, in an online flight booking process, the **option to compensate for emissions** can already be preset. Here, the user must actively make a change if they do not want to follow the preset compensation.
- **Placement**
  In **retail**, nudging can succeed in many ways. Sustainable products are then positioned at eye level of customers and at high-traffic places. Also in online retail, sustainable products can be presented at the forefront in the search results or in the online shop to align the purchase with "sustainability". Thus, even those customers who are not interested in sustainable offers automatically stumble upon these products.
  Motto: **Visibility creates demand!**
  In **meeting rooms,** the popular cookies could be placed on the sideboard, while fresh fruit is offered directly on the conference table. In **canteens,** healthier foods could be presented at eye and reach level to draw customers' attention to them.
- **Immediate feedback**
  Nudges can also take the form of a **feedback,** whether a certain behavior was positive or negative. This form of nudge is used to inform a driver via a smiley or frowney (frown stands for "frown") whether the desired speed was maintained—or not.
- **Appeal to one's own possibilities and abilities**
  A nudge can also point people to their own potentials for "saving the world"—oriented towards the saying of *Confucius:*
  Often what you seek is already within you!
  A legendary nudge is also the sentence of *John F. Kennedy* from 1961:

> "And so, my fellow Americans: ask not what your country can do for you—ask what you can do for your country.
>
> My fellow citizens of the world: ask not what America will do for you, but what together we can do for the freedom of man." (https://1000-zitate.de/4734/Frage-nicht-was-Dein-Land-fuer.html)

- **Creating meaning**
  The high school of behavior control are nudges that convey a higher meaning to people. These include donation appeals to alleviate the concrete **need of people** or—in a broader sense—to make a **contribution to saving the world.** Sustainable buying

behavior can be linked to such a narrative. This is particularly convincing for the vision brands described in Sect. 5.4.
Through a **feedback mechanism**, an appeal can simultaneously be made to the **self-efficacy**. Then it becomes visible and comprehensible to the person acting accordingly that their commitment has actually changed something. An attitude like "I alone can't do anything anyway!" can be overcome in this way. Various organic seals can work in this sense (see Sect. 5.3.5).

Every company is called upon to try out which forms of nudging are better suited to achieving sustainable behavior. However, green brand management always faces the challenge of balancing the **long-term needs of the world** with the **short-term needs of customers** and the **short- and long-term corporate goals.** After all, green brand management cannot be successful if customers reject the "green offers" and/or the company cannot design its performance profitably. Here we have the triad in mind again: Planet—People—Profit.

Companies should not use a **re-educating, moralizing tonality** in their communication for green brand management. The art must succeed in charging "sustainability" not with renunciation and frustration, but with fun and joy of life—also for future generations. Sustainable, green consumption should be able to be integrated into everyday life without great mental and economic costs. In communication, but also in the offers themselves, valuable hints should be given that certain (self-defined) sustainability requirements were taken into account in the development and marketing of a product.

Ideally, it is possible to create a **sustainable feel-good factor:** Attractive, sustainable products that do not harm the environment and for the purchase of which one does not have to justify oneself in one's own circle of friends. Nudging can make an important contribution in this regard.

▶ **Note Box** Such **"green" communication** should ideally take place **without a wagging finger.** This is a critical undertaking when customers may demand more sustainability than the company itself already provides. Moreover, such admonitions are often not particularly effective—in education in general, but also in shopping behavior. This is why **nudging** is so central here.

▶ Green communication should ideally create a desire for sustainability.

For providers of "green" or "sustainable" offers, it is crucial that there is no phenomenon of an **adverse selection**. The term "adverse" means "disadvantageous" or "negative". Therefore, we also speak of a **negative selection**. Here, decisions are made and selection processes take place that are not to be welcomed from a higher perspective. What is the cause of this? The reason is an **information asymmetry** between the market participants involved before a transaction. This unequal knowledge can refer to hidden characteristics of an offer that the potential buyer does not recognize. Through adverse selection, providers can drop out of the market process whose offers are equipped with desirable characteristics—but were not recognized. The market then primarily remains those providers whose offers are to be rated as negative or less good. Since good quality is displaced from the market by negative selection, we speak of a **market failure** (cf. fundamentally Akerlof, 1970).

What significance does **adverse selection have for sustainable products**? Here it can happen that buyers cannot distinguish sustainable from less sustainable products. The providers themselves—especially the manufacturers, but also the dealers—however, know the different quality of their offers. Consequently, there is an information asymmetry. A normally informed customer will only expect an average quality for all offers. The providers of sustainable products will have to demand higher prices to cover their costs than the providers of less sustainable products. If customers cannot distinguish between sustainable and non-sustainable products, they will not want to pay the higher prices for sustainable products. Due to the **information deficit of the buyers,** their willingness to pay remains at a low level. As a result, sustainable offers can be displaced from the market. In the extreme case, no sustainable products are offered anymore, although there would be enough buyers for them. What is the consequence? There is a a **market failure,** because "good" offers disappear from the market due to lack of demand (cf. Erlei & Szczutkowski, 2022).

How can the underlying information problem be overcome? It is crucial that the identified **information asymmetry** is overcome. This can be achieved through **signaling**. In this **signaling method**, the better-informed market partner sends signals to overcome the information asymmetry. The goal is to increase the perceived value of one's own offer by providing additional information. However, this should be "verifiable" data to be trustworthy. For sustainable products, these could be certificates, for example, that make essential product characteristics visible to the buyer. The labels and seals presented in Sect. 5.3.5 can make an important contribution to this. However, the question arises as to how high the trust is today that is placed in brands and which dimensions are important for this.

> **Note-Box**
> In **Signaling**, it is essentially an informational obligation of the providers so that customers can make "good" decisions. By emphasizing the benefits of sustainable offers, customers' attention is drawn to the advantages of sustainable products. This can contribute to customers being willing to invest more money in sustainable products.
>
> Also, **Social Proof** can contribute to signaling. By using customer reviews or recommendations from other customers who have already bought sustainable products, more customers can be motivated to invest in more sustainable solutions as well.

In summary, by placing and highlighting sustainable offers prominently, providers significantly contribute to customers discovering them more easily and thus being more willing to invest in these offers.

Consequently, it is interesting to look at the currently important **performance dimensions for brand trust** in Germany (see Fig. 5.13; see Statista, 2021a). What importance is—in the eyes of the customers – **sustainability issues** already attributed? In a survey, 4000 people (18 years and older) were asked the following question: "Which of these performance dimensions of a brand are most important to you, so that it earns your trust and you would recommend it to friends or family?"

The dimension **"Environmental Protection"** only ranks first for 7% of the respondents, and second for another 9%. A **focus on sustainability in consumption** cannot be inferred from this. Perhaps for some customers, environmental protection is already part of the quality that ranks first here. However, based on these study results, it cannot be determined whether such an interpretation is permissible.

5.3.5 Use of Green Labels and Seals

If we are honest, one thing must be stated: Hardly any customer today has the time and desire to thoroughly inform themselves about how "green" the respective offer is every time they purchase a product or select a service. Not only is the information too inconsistent for this—often there is simply a lack of expertise to understand this flood of data. Therefore, a challenge for green brand management is to communicate "green" elements of one's own offer in a simple form. For this purpose, many eco **eco- and sustainability labels** have already been developed in the past to act as **nudges** or as **signals** (see Sect. 5.3.4). The following illustrates the wide range of green labels and seals used.

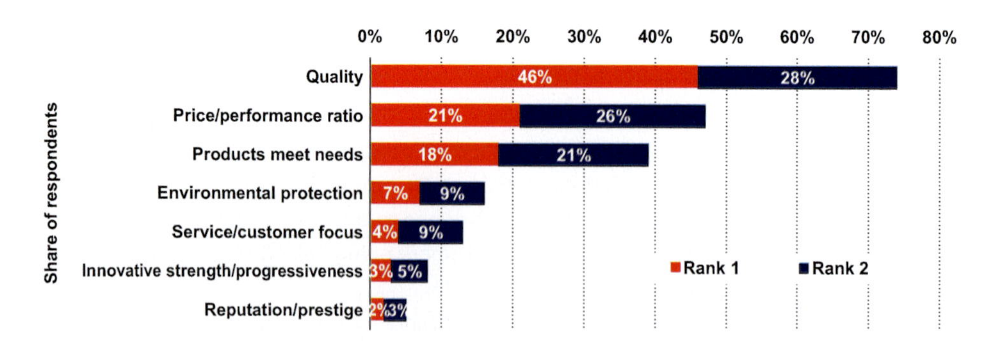

Fig. 5.13 Important performance dimensions for brand trust in Germany—2021. (Data source: Statista, 2021a)

- On products, you can find the official ***EU organic seal*** as well as the ***German organic seal according to EC organic regulation***. Producers also communicate the labels ***Bioland, Naturland,Demeter*** and ***Fairtrade.*** Others communicate ***EcoVin,*****"without genetic engineering"** and **"from certified sustainable fisheries"*****(MSC).***
- Poultry meat from France is marketed under the labels ***Nature & Respect*** and ***Label Rouge***. Elsewhere it is called ***"for more animal welfare".*** The Society of Agricultural and Food Industry created the ***QS mark***.
- In addition, the already mentioned environmental label of the Federal Government ***Blue Angel - Good for me. Good for the environment.*** is supposed to provide orientation. With the *Blue Angel*, environmentally friendly products and services in the non-food sector are marked. These include, for example, paints, furniture, detergents, and recycled paper (see in depth Sect. 5.3.4).
- Retail companies have launched ***Aldi-Bio***, ***DM-Bio,Edeka-Bio,Lidl-Bio*** and ***Rewe-Bio***. For branded products, from ***Lavera Natural Cosmetics,Alterra Natural Cosmetics,Alverde Natural Cosmetics,*** from ***Klar EcoSensitive***, from ***Edding Eco-Line*** and from ***Biobaula—for a better world*** is spoken.

As shown in Fig. 5.14 (see Statista, 2022c), many companies are already using an organic **label**. The number of products with organic labels in Germany has been steadily increasing in recent years. In 2022, over 99,000 products in Germany already had an organic label. The most important **product groups with organic labels** are hot beverages, herbs/spices, and bread/bakery products.

A particular challenge is to provide **food buyers** with **guidance** that goes beyond the labels and seals presented, which are sometimes difficult to understand. In Germany, however, **food manufacturers** have long resisted installing a food **traffic light**. Only with the entry into force of the *Regulation on the Nutri-Score* on November 6, 2020,

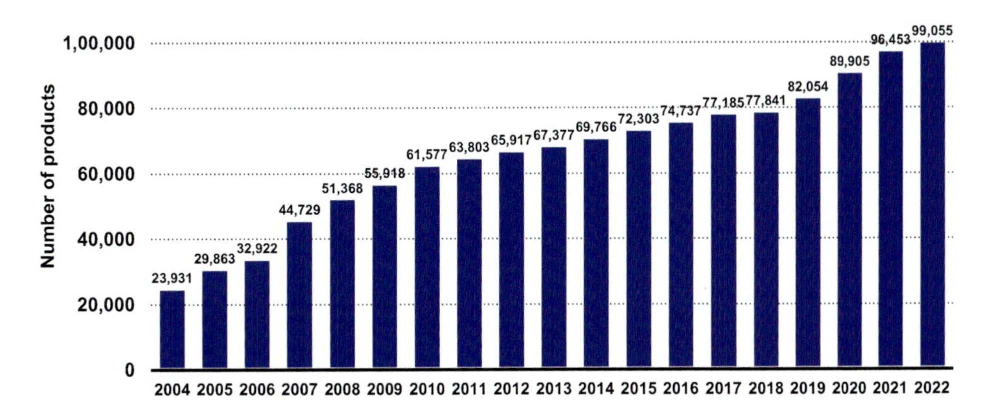

Fig. 5.14 Products with organic labels in Germany up to 2022. (Data source: Statista, 2022c)

has the use of the cross-provider food label **Nutri-Score** in Germany become legally secure. To determine the Nutri-Score, good nutrients are offset against bad ones. The scale shown in Fig. 5.15 is used as a result. The Nutri-Score is calculated on a 100-gram basis. Here, **positively rated nutrients and ingredients** (such as fiber, protein, vegetables, nuts, fruit) receive negative points. Why positive substances get negative points is not necessary to understand. In contrast, **negatively rated nutrients** (including saturated fatty acids, salt, and sugar) receive positive points. Both point values are offset against each other. The lower the total score achieved, the better the overall rating (see BMEL, 2022).

The five-level **traffic light of the Nutri-Score** shows buyers on the packaging whether a product is recommended or rather to be avoided. This is a **form of nudging,** because every customer can still choose a product from group D or E (see Sect. 5.3.4). By the way: The Nutri-Score does not say whether a food is healthy or unhealthy. After all, only foods that are safe for health can be sold! Here, only favorable and unfavorable nutrients and ingredients are evaluated. The Nutri-Score therefore only helps to compare the same foods, e.g., yogurt A and yogurt B, to find the better alternative here. A comparison between the product categories yogurt, buttermilk, and ready-made pudding is not possible. In addition, the Nutri-Score does not say anything about the eco-balance of the product marked in this way.

The **use of the Nutri-Score** is **voluntary** and **free of charge** for food companies. However, registration and agreement to the terms of use with the French trademark owner *Santé publique France* are necessary, an agency in the business area of the French Ministry of Health (see BMEL, 2022). ***Nestlé*** points out in its **Sustainability Report 2021** that 750 products have already been labeled with the Nutri-Score (see Nestlé,

Nutri-Score	Scale for dishes	Scale for drinks
Group A	-15 to -1 point(s)	Water only
Group B	0 to 2 points	-15 to 1 point(s)
Group C	3 to 10 points	2 to 5 points
Group D	11 to 18 points	6 to 9 points
Group E	19 to 40 points	10 to 40 points

Fig. 5.15 Calculation of the Nutri-Score

2022). By the end of 2022, about 650 food companies had registered over 1000 brands for the use of the Nutri-Score. If we assume about 170,000 food brands in Germany (see Lebensmittelmagazin, 2022), then as of the end of 2022, 0.6% of these brands have a Nutri-Score.

Note Box

The Nutri-Score does not provide any indications that would be helpful for more sustainable consumption. Moreover, it is still up to the companies to place the corresponding information on the front of a food product.

And: It takes entrepreneurial courage to label one's own products with a Nutri-Score D or E.

In France, where the Nutri-Score was introduced in 2017, further labels have been developed to take into account the **environmental effects of food**. Since 2021, an **Eco-Score** developed by ten French companies has been used in supermarkets. In addition, the **Planet-Score**—also in France—has already been developed as an alternative to the Eco-Score to further increase transparency. What is behind these labels, which are now also being tested in German retail?

The **Eco-Score** and the **Planet-Score** of a food product are displayed as a total value on a five-level color scale from A to E, similar to the Nutri-Score. The basis for the evaluation is the ***Agribalyse* database** of the French ***Agency for Environmental Protection and Energy***. This database provides the reference data on the environmental impacts of agricultural and food products. Based on this data, the **environmental footprint of products** is determined—the so-called **Product Environmental Footprint** (PEF). The following **16 impact categories** have been defined for its determination, whose individual values are combined into a total value (see Organic Farming, 2023):

- Climate change
- Ozone depletion
- Photochemical ozone formation
- Terrestrial eutrophication (eutrophication describes the enrichment of nutrients in originally nutrient-poor areas caused by human activities)
- Aquatic eutrophication of the seas
- Aquatic eutrophication of freshwater
- Toxic to humans: carcinogenic
- Toxic to humans: non-carcinogenic
- Particulate matter
- Ionizing radiation (this involves "knocking out" electrons from the shell of atoms or molecules)
- Ecotoxicity (recording the effects of substances on the living environment)
- Land use
- Water use

- Acidification
- Use of fossil fuels
- Resource consumption: minerals and metals

In addition to the **environmental impacts** covered in the Product Environmental Footprint, additional criteria are included in the calculation of the Eco-Score through a bonus-malus system. These include the following factors (see Organic Farming, 2023):

- Sustainability labels such as organic, *Fairtrade, MSC* (certified sustainable fishing), *Rainforest Alliance*
- Origin of the product or ingredients (calculated based on transport distance)
- Environmental practices in the production countries (based on the *Yale Environmental Performance Index*)
- Type of packaging (regarding the recyclability of a packaging)
- Impact on biodiversity (such as overfishing, deforestation)

Depending on the extent of these factors, additional plus or minus points can be awarded for the food item being evaluated. Like the Nutri-Score, the Eco-Score is also awarded as an overall score on a colored scale (see Fig. 5.16, left).

It is noteworthy that in Germany, the discount store *Lidl* was the first retail company to test the Eco-Score labeling in 2022. *Lidl* wants to enable its customers to make a conscious purchasing decision. Therefore, the Eco-Score is used for selected products. This involves discussions with representatives from society, associations, and politics. In addition, it is examined in Berlin branches whether customers notice the labeling and how they react to it. For this purpose, *Lidl* uses price tags with the Eco-Score for a total of around 140 coffee, tea, and milk products from its own brands. This is intended to determine whether a **uniform and easily understandable sustainability label for food** can be established in Germany (see Lidl, 2022).

The result showed that customers welcome the **introduction of the Eco-Score**. However, there is a risk of confusion with the Nutri-Score. Since the labeling with the Eco-Score only takes place on the price tags on the shelf, the environmental label is hardly noticed. Consumers expect such labeling directly on the product packaging. Overall, the Eco-Score has so far only slightly influenced the purchasing decision. This could also be

Fig. 5.16 Eco-Score and Planet-Score. *Sources* abzonline (2023)

due to the fact that the label was only recently introduced. In summary, consumers favor an **easy, understandable, transparent sustainability label** (see Ökolandbau, 2023).

The **Planet-Score** also shown in Fig. 5.16 (right) displays the subcategories pesticides, biodiversity, and climate in addition to the colored, five-level overall score from A to E. This provides a more differentiated picture of a product's environmental impact. The Planet-Score was initiated by French consumer protection and environmental associations and developed with the participation of the French *Research Institute for Organic Agriculture and Food.*

The Planet-Score evaluates food based on an **extended life cycle analysis**. For this purpose, the **Product Environmental Footprint** is supplemented by the following indicators (see Ökolandbau, 2023):

- Impacts of synthetic nitrogen fertilizers and pesticides on biodiversity
- Human health (for example, due to the impact of pesticides)
- Ensuring animal welfare
- Influence on the climate
- Decline in soil fertility
- Consideration of planetary boundaries
- Effects of pollutants on the ecosystem (here in terms of marine and terrestrial toxicity)
- Systemic approach to agriculture

By additionally **displaying the criteria of pesticides, biodiversity, and climate**, the Planet-Score enables a quick comparison between products of the same category in terms of their environmental impacts. This addresses a criticism of the Nutri-Score. Especially the providers of organic products had criticized the Nutri-Score for not sufficiently taking into account the use of pesticides and the influence of production conditions on biodiversity.

The ***Bio Planète Ölmühle Moog*** in Germany was one of the first companies to participate in a test run with the Planet-Score and advocates for the introduction of the environmental label in Germany. Selected oils are prominently labeled directly on the bottle with the Planet-Score. Parallel to its use, customers are asked about their acceptance of the Planet-Score (see Bio Planète Ölmühle Moog, 2022). Other companies are also testing the Planet-Score. The *Federal Association of Organic Food and Natural Products* promotes the Planet-Score for environmental labeling through a position paper. According to this federal association, currently only the Planet-Score is capable of providing a **comprehensive and transparent evaluation of food** (see Federal Association of Organic Food and Natural Products, 2023).

The EU Commission plans to propose a **legislative initiative for a sustainable food system** as part of the **Farm-to-Fork Strategy** by the end of 2023. This is intended to make the European food system more sustainable and healthier across the entire value chain. This project will definitely lead to an **environmental label for food** in the next few years as part of the ***Green Deal*** of the *European Commission*. However, it is still open what this will be.

> **Food for Thought** As long as the criteria used to assess the environmental friendliness of a product are not determined by a state-appointed body independent of the food industry, any provider can develop labels and seals at their own discretion. However, this does not help the consumer—and often the environment as well.

There are also a multitude of seals in the **textile sector** that are intended to act as nudges. These include:

- ***Green Button***
  This state seal for sustainable textiles signals that products have been manufactured taking into account social and ecological criteria. The standards are continuously adapted to reflect the entire supply chain in perspective (see BMZ, 2023).
- ***Oeko-Tex Standard 100***
  This is a globally uniform, independent testing and certification system for textiles at various stages of processing. The goal is to reduce pollutants and signal that products have been produced in an environmentally and socially compatible manner. When awarded the *Oeko-Tex Standard 100* label, all components of this article (including all threads, buttons and other accessories) have been tested for harmful substances. This test is carried out by independent *OEKO-TEX® institutes* based on an extensive catalog of criteria (see Oeko-Tex, 2023a).
- ***Made in Green by Oeko-Tex***
  Made in Green by Oeko-Tex is a traceable product label. It identifies all types of textiles and leather goods that have been produced in environmentally friendly companies and at safe, socially acceptable workplaces. The label also signals that this textile or leather product is made from materials tested for harmful substances. For a product marked in this way, the production process can be traced using a unique product ID or by scanning the QR code. This makes it visible in which production facilities a textile or leather item was produced and in which countries the manufacturing took place (see Oeko-Tex, 2023b).
- ***Global Organic Textile Standard (GOTS)***
  The *Global Organic Textile Standard* is based on a comprehensive catalog of criteria. This standard is therefore more meaningful than terms like "sustainable fashion", "ethical clothing" or "fair production". *GOTS* is a leading global textile processing standard for organic fibers, which includes ecological and social criteria in the production of textiles. The entire textile supply chain is subject to an independent and transparent certification process. End products, fiber products, yarns, home textiles, clothing, personal care products, mattresses, fabrics, and even textiles with food contact (such as cheese cloths) can receive a *GOTS* certification.

This standard applies exclusively to natural textiles and identifies products that come from strictly controlled organic production. Only products that consist of at least 70% organically produced natural fibers may carry the seal. Processors and manufacturers of textile products can export their products with a globally accepted organic certification if they meet the requirements (see Global Standard, 2023).

- ***Cradle to Cradle***

  The *Cradle to Cradle Products Innovation Institute* is dedicated to the **promotion of the circular economy** through products and systems. The label ***Cradle to Cradle Certified***® is a **global standard for products** that are safe, recyclable, and responsibly manufactured. Brands, retailers, manufacturers, and designers throughout the value chain use the ***Cradle to Cradle Certified Product Standard,*** to ensure that the impact of products on people and the environment is positive. For more than a decade, *Cradle to Cradle Certified* has been helping companies optimize materials and products according to globally advanced and scientifically based standards (see Cradle to Cradle Products Innovation Institute, 2023).

  Cradle to Cradle Certified evaluates the safety, recyclability, and responsibility of materials and products based on **five categories of sustainability** (see Cradle to Cradle Products Innovation Institute, 2023):
  - **Health compatibility of the material used:** Ensuring the safety of materials for humans and the environment
  - **Recyclability of products:** Enabling a circular economy through regenerative products and process design
  - **Clean air and climate protection:** Protecting the air, promoting renewable energies, and reducing harmful emissions
  - **Responsibility for water and soil:** Protecting clean water and healthy soils
  - **Social fairness:** Respecting human rights and contributing to a fair and just society

It is not always transparently communicated what requirements underlie the various seals. Often it also remains unclear how intensively their compliance is checked. Usually, this knowledge is only gained through extensive research by the customer. Therefore, this multitude of labels and certificates contributes more to **confusion** than to **informed guidance** in the purchasing process. After all, who wants to take the time to check the underlying criteria in depth—with every single purchase? As long as customers—even unchecked—trust such signals, nudging can still succeed.

The confusion of customers by labels and seals is sometimes even systematic. Hardly any customer knows that terms like **"organic cosmetics"“** or **"natural cosmetics“"** are not protected. Here, there are no binding legal definitions for cosmetic products, which range from decorative cosmetics (such as lipsticks or make-up) to hygiene and care products (e.g., soaps and skin creams). This means that each company can decide for itself whether it wants to advertise its cosmetic products with the label "organic" or "natural". The only requirement is that the cosmetics contain—also—organically produced raw materials.

To reinforce a supposed "organic message", images of green landscapes as well as plants and fruits are often shown on the packaging as nudges, which, however, should not be misunderstood as an **indicator for organic or natural cosmetics**. Only reliable seals or certificates provide transparency here to identify the corresponding offers.

Only **certified organic and natural cosmetics** avoid—checked—problematic ingredients. These include microplastics and petroleum-based substances. The plant ingredients of natural cosmetics do not necessarily have to come from controlled organic farming. The requirements here vary from seal to seal.

The company **Ecocert** (2023) supports companies through **certifications,** to make environmentally friendly and socially responsible practices visible through appropriate seals. In total, over 150 programs are certified. Selected certificates are presented below (see Fig. 5.17).

Based on a ***COSMOS* certification**, products can be marketed worldwide as **organic or natural cosmetics** if the following requirements are met (see Ecocert, 2023):

- All ingredients are of natural origin, with the exception of a limited list of tested ingredients (such as preservatives) in small quantities.
- A beauty care product is only certified as *COSMOS ORGANIC* if at least 95% of the plant ingredients are "organic" and at least 20% organic ingredients are included in the entire formulation.
- For rinse-off products (products are rinsed off, such as shower gel or shampoo) and for powders, the required proportion is 10%.
- It is important that water and minerals are not considered "organic" as they are not agricultural products.
- Avoidance of petrochemical ingredients (exception: approved preservatives): parabens, phenoxyethanol, synthetic fragrances and dyes
- Responsible handling of natural resources
- Promotion of biodiversity
- Avoidance of genetically modified organisms
- Use of environmentally friendly and health-safe manufacturing processes
- Free from animal testing
- Recyclable or biodegradable packaging

Fig. 5.17 Seals awarded by the company Ecocert (excerpt). © with kind permission of Ecocert Germany GmbH

Other seals for natural cosmetics are awarded, for example, by the company **Natrue** (2023). Also **Demeter** (2023) certifies cosmetics and body care products.

> **Food for Thought**
> In total, it becomes clear here to what extent companies are striving to make sustainable processes and products visible to the customer. However, the flood of different labels and seals with their different requirements makes it almost impossible to get a meaningful overview.
> The question arises:
> Is such confusion sought by (some) companies—or the inevitable chaos before a consolidation of labels and seals?

5.3.6 Sustainability of Advertising Activities

5.3.6.1 Measurement of Emissions from Advertising Activities

Companies are beginning to also examine advertising activities in terms of their ecological footprint. This is intended to design communication measures according to sustainability criteria.

In order to make its own advertising activities more sustainable, initiated ***Nestlé*** a test concept to first measure the emissions associated with advertising. For Switzerland, the **effects** of a media campaign were determined. The entire value chain was recorded—from production to broadcasting, including the activities of the employees involved. The CO_2 emissions were estimated at five to 70 t CO_2 (see Beuchler, 2022, p. 58 f.). If these values are transferred to Germany, which is about ten times larger, you end up with 50 to 700 t CO_2—for a single campaign.

To support companies in capturing the impact of advertising campaigns and materials on the environment, ***A.L.I.C.E.*** was developed. This acronym stands for *Advertising Limiting Impacts & Carbon Emissions*. Behind it is a **CO_2-calculator** to determine the carbon footprint of advertising. The platform available through *A.L.I.C.E.* allows advertising teams to simulate the CO_2 emissions of various advertising campaigns. For this purpose, the variables format, production process, and energy consumption can be varied (see CGLR, 2023).

The results of *A.L.I.C.E.* are aligned with the ***GHG*-protocol** . This refers to the ***Greenhouse Gas Protocol***, which defines the accounting of greenhouse gas emissions (keyword "Carbon Accounting"). The methodology and calculation of CO_2 emissions are monitored by an independent third party. More than 80,000 employees of the *Publicis Groupe* worldwide now have access to this platform. This is intended to raise awareness among advertising clients about the CO_2 emissions of advertising campaigns. According to analyses by *A.L.I.C.E.*, an average of up to 80% of a campaign's emissions can be saved (see Beuchler, 2022, p. 60).

> **Note Box** Compared to the production sector, advertising campaigns—in absolute terms—offer only a small CO_2 saving potential. Relatively speaking, the picture looks different if up to 80% of CO_2 emissions can be avoided.

Ad Net Zero describes the advertising industry's endeavor to reduce the CO_2 emissions associated with the development, production, and deployment of advertising to zero. To this end, a **five-point action plan** by ***Ad Net Zero*** was developed. This serves as a guide for the advertising industry to transition to emission-free operations (see Ad Net Zero, 2023):

- **Action Area 1: Get Your House in Order**
  All agencies and marketing service providers are called upon to join this *Ad Net Zero* initiative. This comes with the obligation to reduce operational CO_2 emissions to achieve a net-zero balance. To do this, it is first necessary to measure one's own CO_2 footprint and systematically reduce it. This can be achieved, for example, by reducing travel and using renewable energy. Unavoidable emissions must be offset (see Sect. 4.4.2).
  In addition, all participating companies are asked to internally inform about the goals of *Ad Net Zero*. Furthermore, employees are encouraged to become aware of their own CO_2 footprint and to increasingly adopt sustainable behaviors.
- **Action Area 2: Reduction of Emissions from Advertising Production**
  The participating agencies and production companies are encouraged to commit to waste-free and carbon-free production. *AdGreen,* an initiative for sustainability in advertising production, provides information and tools to support the industry in this transition.
- **Action Area 3: Reduction of Emissions from Media Planning and Media Procurement**
  Advertising partners are asked to consider CO_2 emissions in media planning and media buying. This goal is to become an essential part of the *Ad Net Zero* plan, in agreement with all advertisers.
- **Action Area 4: Reduction of Advertising Emissions through Awards and Events**
  The *Ad Net Zero* initiative aims to ensure that the sustainability and climate impact of campaigns are taken into account in the evaluation—and not primarily the price determines the implementation of a campaign. In addition, event and conference organizers are encouraged to increasingly consider sustainability criteria in planning and execution.
- **Action Area 5: Use the Power of Advertising to Support a Change in Consumer Behavior**
  Ad Net Zero calls on the advertising industry to put climate protection at the center of their work through their commitment, thereby changing consumer behavior towards sustainability.

This ***Ad Net Zero* initiative** was launched in the UK in 2020 so that the advertising industry can contribute to sustainability. Global agency holdings, advertising associations, and not least advertisers such as *Google, Meta, Reckitt, Sky* and *Unilever* have decided to roll out the program internationally. An exciting concept is in place. Now it's up to the involved partners to implement it with concrete measures.

5.3.6.2 Abandoning Printed Advertising Media

Even in the **design of advertising activities** itself, the question arises to what extent more sustainable solutions can be used here. More and more companies have abandoned printed communication in recent years. As early as 2019, the last large ***OTTOcatalog*** was published. In 2020, *IKEA*—also with a view to customer behavior—discontinued the printing of the ***IKEAcatalog*** after 70 years. The *IKEA* catalog was published worldwide in an edition of 200 million copies—printed in up to 32 languages. In Germany alone, the catalog was sometimes published in an edition of up to 23 million copies. The last *IKEA* catalog had a circulation of about 8.5 million.

By now, other large and individual retailers have stopped the **printing of advertising material** or are planning to do so. In addition to changed customer behavior, the exit from printed advertising is justified by the rising costs of paper and the pursuit of higher sustainability. Each year, more than 28 billion advertising brochures end up in German mailboxes.

Rewe plans to stop producing flyers and advertising brochures from 2023. So far, the trading company has distributed about 25 million pieces every week. According to the company, the abolition of these paper materials would save more than 73,000 tons of paper, 70,000 tons of CO_2 and 380 million kilowatt hours of energy per year. ***Obi*** has already stopped printed advertising since 2022 (see Diemand, 2022, p. 23). Customers are now encouraged to download and use the *heyOBI* app instead of waiting for print media. A video chat is also offered. The slogan is: "Less paper. More benefits. The heyOBI app."

Whether the calculation will work out to reach customers in the same way via digital channels as previously paper-based will have to be seen—and not just among older customers. After all, an app—at least without permission for push notifications—is a **pull medium**. Here, the user has to become active on their own to become aware of new offers. Print ads and brochures, on the other hand, are classic **push media**, which fight for the attention of the recipients by intruding into the recipients' field of vision.

However, the **ecological balance of online advertising** also needs to be critically examined. After all, data centers as service providers and customers as users with their online activities cause significant emissions. The extent to which online advertising performs better than print advertising—from an ecological point of view—depends on the specific online behavior of consumers. After all, energy is also consumed with online use.

> **Food for Thought**
>
> He who stops advertising to save money can just as well stop his clock to save time.
>
> *Henry Ford*

5.3.6.3 Sustainable Promotional Items

Promotional items are of great importance even in a time when many people already have almost everything. In Germany, €3.65 billion was spent on promotional items in 2021. Medium-sized and larger companies in particular expect further increasing expenditures for promotional items. The main argument for promotional items is the chance of a **more sustainable advertising effect** compared to many other online/offline advertising measures. Tangible advertising media are inherently sustainable, especially since 90% of promotional items are actually used. 40% of promotional items are even owned by the recipient for more than two years. These are at least the results of a study by the *German Association of the Promotional Products Industry,* which is why the results should be viewed critically (cf. GWW, 2022).

To achieve a **sustainable advertising effect**, two factors must be met. On the one hand, the promotional item must be liked so that it is not immediately discarded unused. On the other hand, the buyers must be open to it. This is already the case. Today, two thirds of promotional item buyers already pay attention to sustainability. Companies are even willing to spend an average of up to 10% more for sustainable promotional items.

The company ***Greengiving***, founded in 2009, sees itself as a specialist for **sustainable, environmentally friendly promotional gifts.** For this purpose, the company has defined the following *Green Rules* as the basis of its business model (Greengiving, 2023):

- "All our products are free from child labor.
- All our products are reusable or made from reusable materials.
- All our products are shipped to our customers in an environmentally conscious manner.
- All our products are BPA-free.
- All our products are green, sustainable and ecologically responsible."

"BPA-free" means that the products do not contain the plasticizer Bisphenol A. Such an offer allows the giving companies to demonstrate their own sustainable awareness through the selection of promotional items used.

▶ Motto: Give and be gifted—but with a clear conscience.

The range of *Greengiving* includes, for example, energy-saving gifts such as solar-powered products, items made from environmentally friendly raw materials (such as Fairtrade cotton bags) or ecological promotional gifts such as seed paper or flower bulbs (cf. Greengiving, 2023).

The company shows: **Sustainable promotional gifts are possible!**

5.3.6.4 Sustainable Trade Fair Economy

The **trade fair industry** is also striving to reduce the CO_2 emissions of its activities. To this end, a ***Net Zero Carbon Pledge for the Events Industry*** was made in 2021. This is a **net zero emissions pledge** by the events industry. In this way, this industry wants to contribute to achieving the goals of the *Paris Climate Agreement*. Therefore, the companies involved commit to supporting the goal of net zero greenhouse gas emissions by 2050 and reducing these emissions by 50% by 2030.

Why is the events industry particularly challenged to reduce emissions? After trade fairs, which often last only a few days or weeks, heavy equipment is often used to literally flatten the expensive trade fair constructions. Large piles of waste are then the remainder of expensive trade fair installations—a "solution" that is increasingly less accepted today.

To achieve ***Net Zero Carbon Events***, the possible **measures for emission reduction** must first be identified and prioritized. The relevant areas of action include energy, water and material management, reducing food and beverage waste, and sustainable procurement. To offset the non-reducible emissions, trustworthy **compensation certificates** should be purchased. In parallel, budgets, time and/or expertise should be provided to develop a **collective milestones plan** for achieving a CO_2-free events industry. This requires cooperation with partners, suppliers and customers to achieve changes throughout the value chain (see Net Carbon Events, 2023).

Those who join this *Net Zero Carbon Pledge* should publicly announce this promise and drive the goal of Net Zero across the entire events industry. In addition, a **due diligence review of the procurement process** must be carried out. In addition, a **measurement and tracking of Scope 1, 2 and 3 greenhouse gas emissions** is required. This poses a particular challenge for Scope 3. Often, over 90% of the emissions are attributable to Scope 3—in the upstream and downstream supply chain of the trade fair organizers. Therefore, appropriate **systems for measuring CO_2 emissions** must also be installed at the involved partners. In addition, the organizers must report on the progress made at least every two years and support an **exchange of best practices and gained experiences** (see Net Carbon Events, 2023).

If as many trade fair companies as possible join the initiative, the trade fair industry can make a significant contribution to reducing emissions. In this way, one cogwheel engages with another to achieve the overarching goal of Net Zero.

Note Box

Marketers should generally adhere to the O_5C_2 rule (see Nitsche, 2022):

- **C**alculate Impact (an early focus on the achievable effects of an advertising stimulus is indispensable)
- **O**ptimize Strategy (all possibilities to optimize advertising campaigns based on previous insights should be consistently used)
- **O**ptimize Circulation (by focusing on the relevant target groups, by a high quality of the addresses used, by avoiding unwanted multiple approaches through duplicates, the circulation can be optimized)
- **O**ptimize Creatives (by using smaller font sizes, less text, reducing black areas and using fewer photos, emissions can be avoided)

- **O**ptimize Production (by choosing environmentally friendly suppliers and local production of advertising materials, emissions can be further reduced)
- **O**ptimize Distribution (distribution partners who reduce or compensate their own emissions through intelligent concepts contribute additionally to sustainability)
- **C**ompensate Impact (through own compensation services, further influences on the environment can be balanced)

5.3.7 Sustainability of Sales Channels

Sustainable management also includes the question of how sustainably the integrated sales channels operate. This topic was already addressed in the context of **sustainable logistics processes** in Sect. 4.4.1.5.

With regard to customers, the question arises as to how they assess the **sustainability of the sales channels**. In a study, 2084 people in Germany were asked: "Is it more sustainable to shop online or in a store?" (see Fig. 5.18; cf. Statista, 2021c). This was not about facts, but about the individual perception of supposed sustainability. The majority considers stationary stores to be more sustainable (44%), only 12% the online shops. Equally, 32% see both forms of distribution. 11% do not know. However, most respondents probably just guessed.

This also raises the question of the **perceived sustainability of online deliveries.** To this end, 2084 people in Germany were asked: "Is it more sustainable to shop online or in a store?" (see Fig. 5.19; cf. Statista, 2021d). 50% consider online shopping to be less sustainable—20% consider it more sustainable. Just under a third simply do not know.

To compare the sustainability of the two forms of distribution, it is worth looking at a phenomenon that only exists in online shopping: the **return of shipments.** For this,

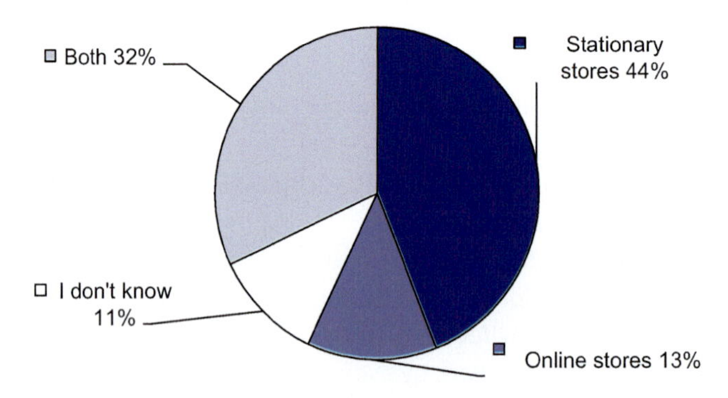

Fig. 5.18 Perceived sustainability of stationary stores and online shops. (Data source: Statista, 2021c)

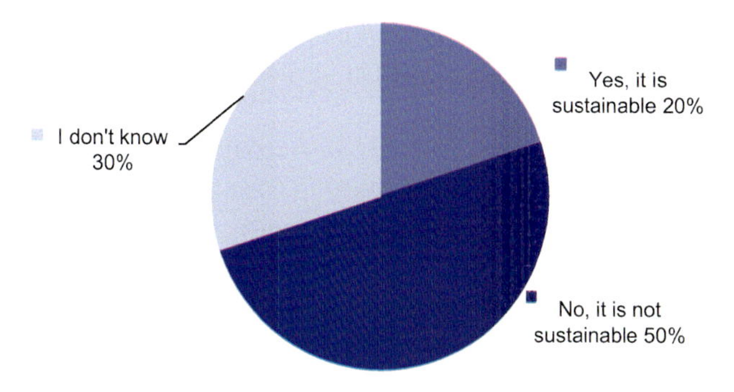

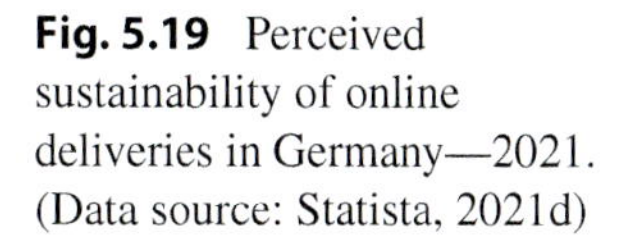
Fig. 5.19 Perceived sustainability of online deliveries in Germany—2021. (Data source: Statista, 2021d)

a **study on returns management,** the ***European Return-o-Meter*** of the Returns Management Research Group at the *Otto-Friedrich University of Bamberg,* provides exciting results. This is based on around 500 questionnaires filled out by European online retailers. The sample includes 411 online retailers. These account for an e-commerce turnover of almost 60 billion €. At the same time, they were responsible for the shipping and return of 1.25 billion packages in 2020. The following insights were gained from this study (cf. Otto-Friedrich University of Bamberg, 2022):

- In 2021, people in Germany ordered almost **100 billion € worth of goods** on the internet and at the same time triggered **530 million return shipments**.
- The return rate is particularly high in the fashion segment: 83% of all shipments and 91% of ordered items were returned.
- Germany has the highest return rate in Europe—**Germany is the infamous return champion of Europe.**
- In Germany, almost **every fourth package is returned**—exactly 24.2%. This corresponds to about 530 million returns with around 1.3 billion items.
- The **reasons for the high return rates in Germany** are most likely due to three factors—across all product clusters:
  - German customers more often order **goods on account** (28.8% compared to the EU average of 9.5%).
  - Online retailers in Germany allow **longer return periods** (51.7 days compared to 28.1 days in the rest of the European Union).
  - In Germany, the **return** is usually **free of charge** (88.7% of the surveyed online retailers compared to 52.4% in the rest of the European Union).

Additionally, the question is exciting: What **environmental impacts** are associated with this **return behavior**? The surveyed companies estimate the ecological footprint at approximately 1500 grams of CO_2 equivalents per returned item. The returns alone in Germany in 2021 would then have caused approximately 795,000 tons of CO_2 equivalents. The same amount is produced when an average car travels 5.3 billion kilometers.

How online retailers deal with the **ecological footprint of returns** is shown by a look at the following results (see Otto-Friedrich-University Bamberg, 2022).

- Over 80% of the surveyed online retailers do not record this ecological footprint.
- 15% were unable or unwilling to make any statements on this.
- Less than 5% measure the CO_2 footprint of the returns.

From a sustainability perspective, the question also arises as to what **costs the returns** cause and how the **reuse of the returned goods** turns out (see Otto-Friedrich-University Bamberg, 2022)?

- A **returned item** on average causes €2.85 in **transportation and processing costs.**
- For a **return shipment** with several items, the average **transportation and processing costs** are €6.95.
- 1.3% of the returned items are directly disposed of by the German online retailers. In Germany, an estimated 17 million returned items were disposed of in 2021. This proportion is lower in Germany than in other European countries.
- Disposal by resellers and disposal by customers in the case of a refund without return is not recorded.
- 93.2% of the returned items were sold as **like new.**

These figures thus only represent part of the problem. In addition, it can be asked whether the data could have been influenced by effects of social desirability (see Sect. 1.3.2).

> **Food for Thought** Online shopping is a very convenient process. Fast delivery and easy return—usually free of charge and even after longer consideration—are possible. The question arises whether buyers who return goods are aware of the ecological footprint they create—not only when returning, but already at the time of delivery itself. A look at the retailers shows: The majority does not record this—significant—ecological footprint.

In this context, the question also arises about the **importance of sustainable e-commerce packaging**—from the perspective of buyers. For this purpose, 1047 people aged 18 to 90 years in Germany were surveyed representatively. The results are shown in Fig. 5.20; see Statista, 2021e).

Online retailers need to check whether the high importance of suitable, reusable and/or plastic-free packaging is also reflected in individual purchasing behavior. After all, a stated "importance" does not say whether it will also become effective in behavior. Moreover, so far only a few companies share information about how environmentally friendly the packaging is during the online purchasing process—and only a few consumers so far seek such information.

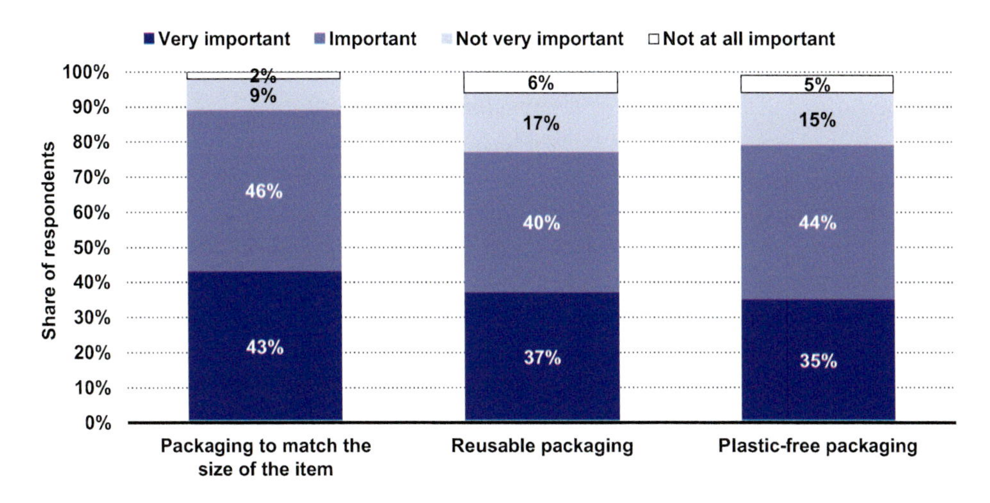

Fig. 5.20 Importance of sustainable e-commerce packaging for buyers in Germany—2021. (Data source: Statista, 2021e)

> **Food for Thought** Just for comparison: Those who want a **packaging** in **stationary retail** mostly have to pay for it these days.

Companies are increasingly striving to make packaging more sustainable in general. An example of this is the project ***HolyGrail 2.0.*** This is an initiative of the *AIM—European Brands Association,* which is supported by the *Alliance to End Plastic Waste.* So far, more than 160 companies and organizations from the entire value chain of the packaging industry have joined this initiative.

What is the *HolyGrail 2.0?* This project advocates for the **use of Digital Watermarks**. By applying postage stamp-sized **digital watermarks** to packaging, the different types of waste in waste processing plants are to be identified and directed to the appropriate recovery process. Ideally, this would be the reuse of a package or at least the recycling of the raw materials used. Today, such processes fail because materials are not cleanly identified—or the packaging materials are not recyclable at all (cf. Digital Watermarks, 2023).

Procter & Gamble is supporting the semi-industrial **trial run of digital watermarks** with more than 100 products in Europe. For this purpose, the product packaging was provided with digital watermarks to improve the processes in the sorting of packaging and achieve higher recycling rates (cf. Procter & Gamble, 2023).

> **Food for Thought**
> If one dreams, an **ideal packaging world** would look like this. First of all, all unnecessary and multiple packaging is avoided. This is the most important step (guiding principle **"Refuse"**). All indispensable packaging is—worldwide—fed

into high-performance collection systems. A digital watermark is found on all packaging. This is read out in the sorting plants. In this way, all packaging could be directed to the ideal further use (guiding principles **"Refuse"** and **"Recycle"**).

If packaging that cannot be reused or recycled were charged at higher prices, economic incentives would lead to such packaging disappearing from the market—worldwide!

"Brave new world" or "realistic goal"? It is definitely worth taking this important step towards a circular economy in order to achieve higher recycling rates for packaging—and ideally not only in the EU.

In this context, the question also needs to be answered why people rely on **Click & Collect**. A representative survey of the population aged 18 to 90 years in Austria (n = 1054), Germany (n = 1047) and Switzerland (n = 1054) provides exciting results. Multiple responses were possible here (cf. Fig. 5.21; cf. Statista, 2022b).

The most important reason for Click & Collect in all countries is the **saving of shipping costs.** It is interesting that in Germany, with 48%, the **support of local businesses** is already closely followed in second place. In Switzerland, where *Amazon* is much less widespread, this value is only 28%. **Flexibility** and **speed** are other reasons. The **environmental friendliness** ranks in Germany with 36% and in Switzerland with 23% in 5th place. This criterion lands in Austria with 24% only in 6th place, even after convenience/less stress.

In another **study on e-commerce**, additional need for action becomes apparent. Here, 1048 online shoppers (16 years and older) were asked: "Which of the following statements do you agree or disagree with?" The results of "fully agree" or "rather agree" are shown in Fig. 5.22 (see Statista, 2021b). It becomes clear that customers have high demands on the retailers. At the same time, these results could also signal that customers simply know nothing about how online retailers actually behave—or do not want to know.

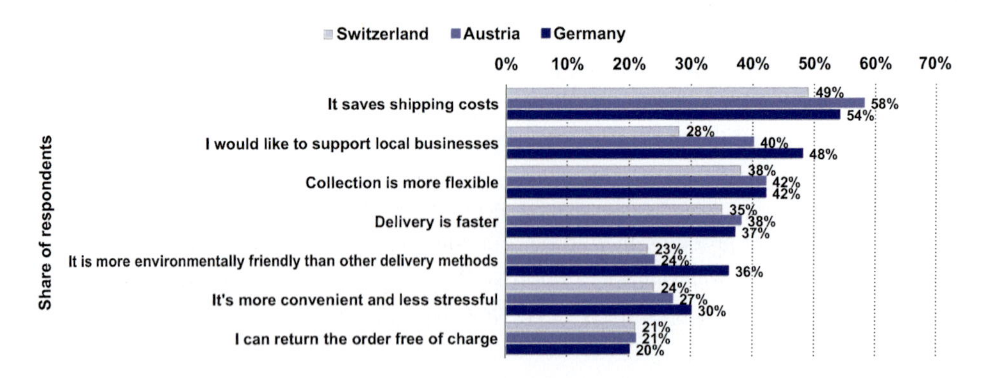

Fig. 5.21 Reasons for Click & Collect in the German-speaking area—2021. (Data source: Statista, 2022b)

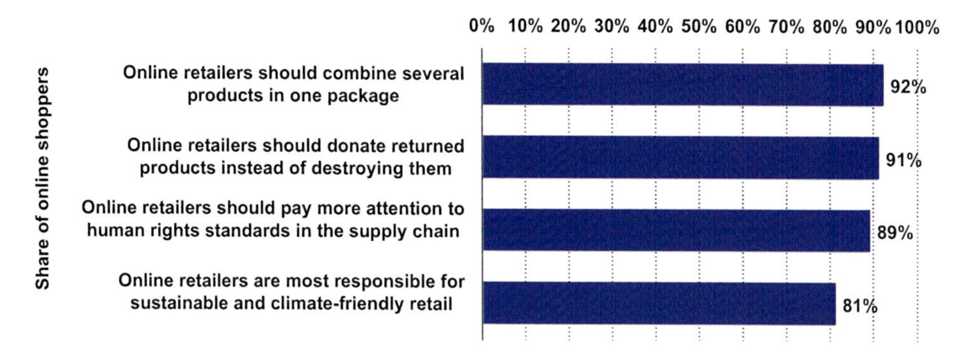

Fig. 5.22 Perceived responsibility of online retailers in combating climate change in Germany—2021. (Data source: Statista, 2021b)

From these responses, one gets the impression that online retailers are responsible for everything and should also handle all corresponding challenges on their own. This is what a simple delegation of responsibility looks like when customers are presented with a wish list in a study.

> **Note Box** In sum, based on the various study results, it can be stated that sustainability has so far not been given much importance in online purchasing.

The possibilities to **compensate** for the **CO_2 emissions** caused by shipping were already discussed in Sect. 4.4.2. Such emission-compensating behavior is made visible to the customer through labels. Here, each sender individually needs to check what importance a **CO_2-neutral shipping** has for the customers. If a climate-neutral shipping takes place, this can contribute to the customer's **feel-good factor** (see Sect. 1.3.4)—and motivate further online purchases.

5.3.8 Claims to Visualize One's Own Sustainability

How can, how may, and how should generally be communicated on the subject of sustainability and a so-called "climate neutrality"? Many companies already use a **neutrality claim** and describe themselves as **climate-neutral company.** However, this term is not used uniformly. For this neutrality claim to have a positive effect, it needs to be specified what is meant by **climate neutrality** exactly . For this purpose, the ***Intergovernmental Panel on Climate Change (IPCC)*** has defined the term.

The *IPCC*—also called the ***World Climate Council***—is an intergovernmental institution to inform political decision-makers about the state of scientific research on climate change and to create the conditions for science-based decisions, without giving recommendations for action itself. This ***IPCC* definition of climate neutrality** reads (see IPCC, 2023, p. 545):

Climate neutrality is a state in which human activities have no net impact on the climate system. To achieve such a state, the remaining emissions would have to be balanced with the reduction of emissions (carbon dioxide). In addition, the regional or local biogeophysical effects of human activities would have to be taken into account. This also includes impacts on the local climate and the albedo.

The term **Albedo** is used to measure the **reflectivity** of diffusely reflecting surfaces or objects that do not emit light themselves. Albedo is therefore the proportion of solar radiation that is reflected by a surface or an object. This reflected portion of the sun's rays is expressed in percent. While snow-covered surfaces have a high albedo because they reflect a lot of light, the albedo is low for very dark soils. Surfaces with vegetation and oceans also have a low albedo.

The *IPCC* definition of climate neutrality makes it clear that the goal is not just to achieve **net-zero emissions**. Rather, it is about avoiding any impact of human activity on the climate. This includes not only refraining from deforestation but also the subsequent effects triggered by human and corporate actions—such as the sealing of surfaces. Also, the melting of the poles or the increase in climate disasters should be considered (see atmosfair, 2023).

▶ **Note Box** The goals of climate neutrality—according to the *IPCC* definition—are hardly or not achievable by a single company. Therefore, companies must handle the term "climate neutrality" very carefully.

Aldi Süd also had to experience this in 2021. The company had advertised with the term "first climate-neutral food retailer". This brought the company a **cease and desist lawsuit from the Competition Centre**. *Aldi Süd* was accused of having used the term "climate neutral" misleadingly. This creates the impression that climate neutrality is achieved through emission-avoiding measures. Instead, *Aldi Süd* has "only" purchased CO_2 certificates to offset its own emissions. However, this is not the content of "climate neutrality". After all, the activities of *Aldi Süd* had a negative impact on the climate, which led to the need for compensation.

▶ **Food for Thought**

It is undisputed that humans on earth do not live climate neutrally. However, animals and plants do not do this either, because their life also affects the climate—positively and negatively. Even inanimate objects like the oceans, mountains, and deserts have an impact on the climate. Neither a human, a tree, a cow, and certainly not a company can ever truly be "climate neutral". Everything has an impact on the ecosystem—positively and negatively!

So the question arises: **Can there be real climate neutrality at all?**

Against this background, the claim by companies to be **climate neutral** through the purchase of emission credits is repeatedly criticized by environmental NGOs and referred to as **greenwashing**. After all, compensation for CO_2 emissions does not say anything about the extent to which a company is striving to reduce its own pollution. And one thing a company does not become through the payment of compensations: climate neutral. Through compensation measures, pollutions caused by companies are merely offset elsewhere. How well this can actually succeed was discussed in Sect. 4.4.2.

> **Food for Thought**
> Can companies claim **CO_2-neutrality** for themselves if they offset their emissions in full? The answer is: No!
> By offsetting emissions, no **CO_2-neutrality** and certainly no **climate neutrality** is achieved. Companies can only work towards achieving global **CO_2-neutrality** by supporting **projects for emission reduction**.

Against this background, the question arises: How should companies communicate that cannot (yet) completely avoid emissions and therefore rely on **compensation**? The idea of compensation is that unavoidable emissions must be removed from the atmosphere in equal amounts. Compensation is intended to ensure a **neutralization of harmful entries**. Compensation is primarily to be used in the transition phase to possible CO_2 neutrality to offset the emissions that cannot yet be avoided. However, it is also known that not all companies will be able to completely prevent CO_2 emissions. Therefore, the possibility of compensation was created.

> **Note Box** The **purchase of emission credits** does not neutralize the company's own emissions. The money collected through the emission certificates only finances a balance of the global emission balance.

Companies can highlight a **CO_2-neutrality** or **Net-Zero-Emissions** ("net zero") if this goal was achieved without compensation through emission avoidance. If a company is still on the way to "Net Zero", the claim can be:

> "CO_2-neutral through compensation."

The preservation of forests and the replanting of trees contribute to this. In this argument, the claims of logistics service providers are not precisely enough defined as long as these companies still rely on compensation.

If projects of other companies or countries are supported to reduce their emissions, this is very honorable. After all, the companies are taking responsibility beyond their own performance area. However, their own emissions remain uncompensated. This is the case, for example, when building wind turbines or distributing low-emission stoves, if these activities are independent of the company's business operations. In these examples, nothing is compensated, but the additional pollutant output is reduced elsewhere through alternative technologies. The slogan "CO_2-neutral through compensation" may not be used because the company's own emissions are not compensated. Correct claims would be:

- "CO_2-friendly through climate investment in XY" or
- "climate-friendly through technology investment in country XY".

When developing **sustainability claims** for such activities, it is important to avoid the impression that a company would offset emissions in its own value chain with the emission reductions from another project. After all, this is not the case. The company's own emissions remain uncompensated. In addition, additional emissions from third parties are reduced (see atmosfair, 2023).

In summary, the EU Commission, the German Bundestag, and also courts are increasingly critical of **environment-related advertising claims**. Courts have already uncovered many cases of **greenwashing** and banned **false advertising claims**. Every company is called upon to closely follow further developments. It is quite possible that in the future statements such as "environmentally friendly", "climate-friendly", "environmentally friendly", "ecological", "CO_2-neutral", "energy-efficient" or "biodegradable" will no longer be permissible without further checks.

It could also be decided that no product can be labeled "made with recycled material" if only the packaging is made from recyclate. Also, further **sustainability labels** could be banned if they do not base on a **certification process** or were not set by government agencies. In addition, **legal minimum requirements for certification** can be defined (see Stegmann, 2022, p. 16).

> **Note-Box**
> In legal practice, **advertising with environmental reference** is now subject to similarly strict standards as **advertising with health reference**. Also, **environment-related statements** are increasingly classified by courts as a **significant feature**. Therefore, every company is well advised to be able to substantiate its advertising statements concretely—and otherwise to refrain from them.
>
> The general rule is: **Act first, then communicate!**

Due to the now ubiquitous **War for Talents** at all qualification levels, green brand management also affects the **Employer Brand**. The purpose, vision, mission, and values of a company are becoming increasingly important in choosing a workplace. For more and more—and not just young—people, the consideration of ethical and ecological aspects as an expression of responsible corporate management is becoming increasingly important in choosing an employer. Consequently, when selecting claims to visualize a company's sustainability, possible effects on (potential) employees should also be considered.

> **Note-Box** Sustainable corporate management for employee retention and recruitment!

5.4 Examples of Green Brand Management

5.4.1 On the Way to the "Green" Brand

An indispensable prerequisite for green brand management to succeed is the already mentioned **honest and credible communication** (see Fig. 5.10). Companies are well advised to show that they are "first" on the way to a green brand. No serious stakeholder—except perhaps *Last Generation* and *Extinction Rebellion*—will expect a company to become "green" overnight.

For the necessary "**Greening**" processes need to be developed or adapted within the companies. New suppliers must be identified, more sustainable packaging and product presentations developed, and more environmentally friendly production methods worked out and tested. Activities that no longer meet the company's self-defined standards may also have to be abandoned. To secure these decisions, a wealth of information needs to be gathered and evaluated. This was made clear in Sect. 4.4.

At the same time, **business partners** must be committed to **new requirements**, who then also have to further develop their products, services, and processes. In addition, new partners often need to be involved if the previous value-adding partners are unwilling or unable to meet the further requirements in terms of sustainability.

> **Note-Box** It is a **sign of honesty and credibility** when a company openly communicates about the necessary **adjustment steps of its own "greening"**. On the way to green brand management, facts and verifiable evidence count—not mere promises!

Which companies have dealt with green brand management early on? Perhaps the starting signal for some companies came at a time when hardly anyone was dealing with green brand management. What drove these companies back then, how did they proceed, and where are they today? These questions are answered below.

5.4.2 The Example of Frosch

The brand Frosch (2023b) positioned itself as "green" as early as 1986, when hardly anyone was talking about green brand management. Under the brand *Frosch*, the company *Werner & Mertz* positioned itself as an ecological and sustainable manufacturer of various household products. The cleaning products marked with ***Frosch*** were the first phosphate-free offers in Germany. The name *"Frosch"* with green and red color in the logo still visually transports this green positioning today.

Since 2012, the company has been focusing on the ***Frosch Initiative.*** This initiative shows that a **circular economy for plastic** is sensible, possible, and necessary. The company pursues the following vision (Frosch, 2023b):

> "We use plastic over and over again, instead of burning it, shipping it, throwing it into the sea, or hoarding it on gigantic landfills somewhere in the world."

The packaging developed according to the **Design for Recycling** is aimed at future recycling. The basis for this is the **concept of the circular economy.** How can a customer recognize such packaging? Four features are at the forefront:

- Light colors
- Mono-material
- Removable labels
- Separable components

This consistently implements the **Cradle-to-Cradle concept** described in Sect. 1.2. In addition, the brand ***Frosch*** strives to develop powerful cleaning agents whose ingredients are as environmentally friendly as possible. Moreover, these products are based on European plant oils to reduce the transport of raw materials. These messages are clearly and understandably conveyed in the online and offline communication of the brand across various customer touchpoints.

The brand *Frosch* pursues a comprehensive **circular principle**. For this purpose, the packaging is designed in such a way that it can be kept in a **closed material cycle**. As early as 2012, the company *Werner & Mertz* developed an innovative process with industrial partners to obtain high-quality recycling materials (recyclate or old plastic) from plastic waste. Today, the *Frosch* bottles consist of 100% old plastic. For this purpose, the company processes plastic from the *Yellow Bag* or the *Yellow Bin* (cf. Frosch, 2023a).

The award-winning **Recyclate Initiative** is designed as an **Open Innovation solution**. This allows other partners to get involved as well. This seems necessary considering that about 9 million tons of plastic waste end up in the sea every year. According to a study by the *Ellen MacArthur Foundation*, it is feared that by 2050 there will be more plastic than fish in the sea. However, the necessary solution is not to be found in the sea, but on land. The more plastic materials are kept in the cycle, the less plastic waste is burned or ends up in landfills or in the sea (cf. Frosch, 2023a).

The Recyclate Initiative therefore wants to prevent a large part of the plastic waste from the *Yellow Bag* from being down-cycled, i.e., recycled in a low-quality way, exported to other countries, or burned. The challenge is called **Upcycling**, which is only used to a small extent today. The potential for this is high, as plastic can be recycled almost infinitely. As early as 2019, the *Öko-Institut* had evaluated the **climate friendliness**

of various recycling processes using the example of a 0.5-liter bottle. Even when considering the entire energy expenditure of all recycling processes, the reuse of plastic through recycling is still significantly more climate-friendly than the new production of the same material. Further information on the study can be found at *initiative-frosch.de.*

The commitment of *Frosch* has paid off. In 2009, *Frosch* was awarded the **prize** in the category "Most Sustainable Brand". In 2021, *Frosch* was awarded the *German Sustainability Design Award* for circular packaging in the "Design" category for visionary, sustainable packaging concepts (cf. Frosch, 2023a).

What successes has the company *Werner & Mertz* been able to achieve with *Frosch* through this strategy? This question can be convincingly answered by the results of the market research institute *GfK*. The needs of buyers primarily include sustainability, demand, and security. It is predicted that brands that can meet these needs will continue to grow most dynamically in the future.

The ***GfK*** distinguishes between four categories for the **manufacturer brands** today (cf. Kecskes, 2021; cf. Fig. 5.23):

- **Functional brands**
  These are usually familiar, large and well-known umbrella brands that make a promise of **function** based on **tradition** and **proven reliability**. Consumers receive proven, high-quality quality here. However, little or no additional social benefit is provided. Most manufacturer brands fall into this category. Their share of sales amounts to 65%.

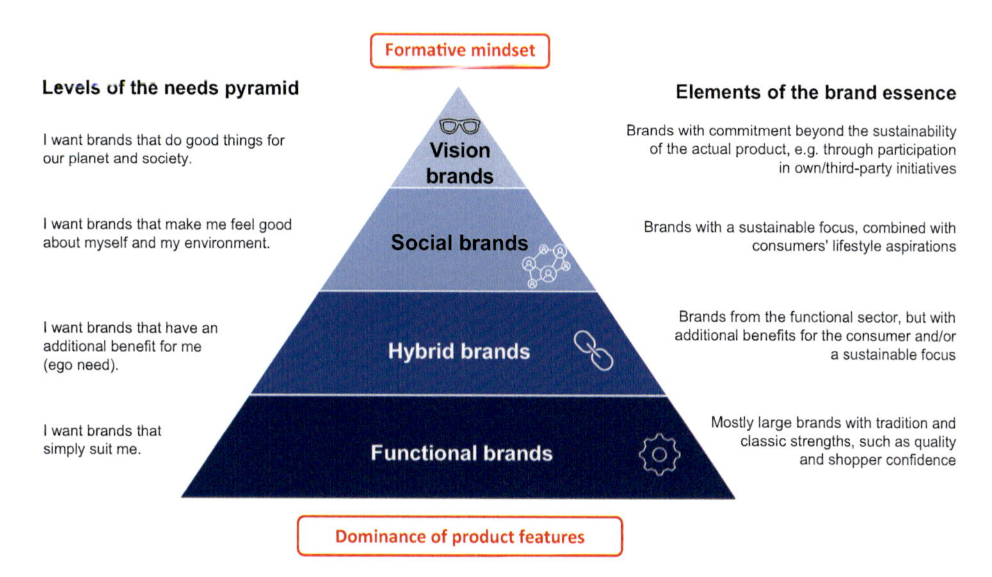

Fig. 5.23 The GfK brand concept—from product features to mindset. *Source* Adapted from Kecskes (2021)

The former slogans of the detergent *Persil* fit these functional brands: *Persil remains Persil* or *Persil: You know what you're getting.* The former slogan of *Volkswagen* also falls into this category: *Volkswagen—The Car.* You can hardly position yourself more confidently.

- **Hybrid brands**
  Hybrid brands include the functional brands that highlight a **specific additional benefit**. Hybrid brands have a market share of 28%.
  Such an additional benefit can, for example, emphasize **regionality**. Motto: *"The beer from your region"*. Specific aspects of **sustainability** can also highlight such an additional benefit. For example, a body lotion from *Weleda* states: *Ingredients 100% of natural origin.*
- **Social brands**
  Social brands have a much more comprehensive **sustainable orientation,** which is visible and emotionally tangible for the consumer and has a high relevance. The market share of social brands amounts to 3%.
  An example of social brands are the natural cosmetics products from *Dr. Hauschka.* Accordingly, the slogan here is: *Dr. Hauschka Cosmetics is 100% certified natural cosmetics.*
- **Vision brands**
  Vision brands go far beyond the brand messages conveyed in the previous sections. These umbrella brands embody—in content and communication—a **purpose,** a **vision,** based on a visible **value compass.** These brands often promise to be not only good for the users, but also good for the entire planet. Motto: These brands are good and do good! However, the market share of vision brands is currently only 4%.
  To this end, vision brands adopt a **holistic perspective** that does not stop at the ingredients or packaging, but also includes, for example, participation in relevant initiatives. With a balanced price-performance ratio and high availability, these brands contribute to an **eco-democratization**. *Frosch, Patagonia, Veja* and *Chameleon* (Sect. 5.4.5) are convincing examples of this positioning as a vision brand.

The positioning of the brands presented here is based on the **needs pyramid of *Maslow*.** **Functional brands** primarily cover basic needs. **Hybrid brands** and even more so **Social brands** also cover needs for social relevance. At the top of the pyramid are the **Vision brands,** which as "world-improving products" also satisfy needs far beyond the basic use of a product or service. *Frosch* is positioned in the pyramid as a vision brand (see Kecskes, 2021). The products of this brand are readily available in food retail and drugstores. In addition, the company behind *Frosch*—as already described—is characterized by various initiatives for a more sustainable economy. The sharply positioned vision brands have recorded the highest growth rates in recent years. Market share winners here are primarily ecological brands in the segment of washing, cleaning and cleaning products.

According to the *GfK*, it is primarily the younger generations—increasingly also baby boomers—who increasingly distinguish social and vision brands from pure functional brands.

5.4.3 The Example of Patagonia

The company ***Patagonia*** – provider of outdoor clothing and equipment for climbing, surfing, skiing, snowboarding, fly fishing, trail running etc.—has its **Purpose** completely focused on sustainability (Patagonia, 2023a):

> "Patagonia is in business to save our home planet."

To achieve this ambitious goal, the company must reduce its impact on the environment—regardless of how many products it sells. How can this be achieved? *Patagonia* has taken many steps over the years to permanently position itself as a **green brand**. It has been using **organic cotton** for its products for many years. To change even more, *Patagonia,* tried to encourage other clothing companies to use organic cotton. Here, the company itself states that it has failed in this goal (see Patagonia, 2023a).

To become more sustainable, *Patagonia* uses, for example, **more environmentally friendly dyeing processes** and **recycled synthetic fibers**. As early as 1993, the company sold its first clothing items made from recycled polyester. The **processing of old plastic bottles into clothing fabric** was a groundbreaking innovation. This reduced the dependence on fossil fuels and thus also the CO_2 footprint. It should be noted that the term **"CO_2 footprint"** has only been part of the general vocabulary since the mid-1990s.

Since 2005, a new technology has enabled the **chemical recycling of polyester**. This made it possible to recycle old clothes and use them to produce new clothes. This led to the entry into the circular economy: Worn products from *Patagonia* could be reintegrated into the production cycle. However, this process did not allow **scaling to industrial scale**. In 2016, it finally succeeded in developing a high-performance insulation with a 55% share of recycled polyester. This enabled *Patagonia* to prevent more than two million plastic bottles from ending up in the landfill in the first year alone (see Patagonia, 2023a).

Ryan Gellert, CEO of *Patagonia,* explains that **growth** at *Patagonia* is not a dominant goal—they want to grow, but in a reasonable extent. This is also supported by **growth-critical advertisements** that read something like this:

Don't buy this jacket!

The company promotes durable products on the one hand, but on the other hand points out that the consumer should only buy what he actually needs. Instead of replacing worn products with new ones, *Patagonia* consistently focuses on repair, reuse and recycling. As part of a so-called *Worn-Wear-Tour*, a workshop van from *Patagonia* travels through tourist places to repair damaged outdoor clothing on site. In addition, the company donates one percent of its sales to environmental organizations every year (see Gellert, 2021).

Patagonia does not shy away from entering the political arena with **provocative slogans**. In 2020, it launched a campaign against US politicians who denied the climate crisis. In *Patagonia* shorts, labels with the slogan "Vote the assholes out" were sewn in. The response: All such equipped pants were sold out in no time. Here, the **commitment of a visionary brand** becomes clear: *Patagonia* acts as a **Political Citizen**—as a political citizen who actively participates in ongoing discussions. Is the commitment worth it? The answer: *Patagonia*'s turnover in 2021 was one billion € (cf. Gellert, 2021).

In 2022, the founder and owner of *Patagonia, Yvon Chouinard,* transferred his company to **non-profit foundations**. Through this, the owner makes his wealth entirely available for environmental protection. The **company value** of the company was around three billion € at this time. The company's profits that are not reinvested in the company are to be used in the future through special foundations for the fight against global warming and for nature conservation. Profits in the order of one hundred million € are generated each year. A trigger for this step was—according to *Chouinard*—the mention of his name on the *Forbes* list of super-rich. There, his wealth was stated as 1.2 billion US$ (cf. Lindner, 2022, p. 20).

Here is a quote from the 2022 published **statement by *Yvon Chouinard*** (Patagonia, 2023b):

> **"The Earth is now our only shareholder. [â€¦] The resources of the Earth are not infinite, and it is clear that we have already exceeded its limits. But it is also resilient. We are convinced: We can save our planet if we commit to it."**

It is important to note that *Patagonia* still sees itself as a profit-oriented company and competes with convincing products. This is made clear by the words of the Chairman of the Board *Charles Conn* (Conn, 2022, p. 20):

> "We are supposed to make the best raincoats, and we want to beat other companies in competition. We believe in capitalism as a driver of innovation. But we believe it can be practiced in a way that is better for the environment and society."

5.4.4 The Example Veja

The brand *Veja* (pronounced: wehscha) positions itself as a provider of **eco-sneakers.** Since 2005, this company has been combining the production and marketing of sneakers with **social projects** and the **pursuit of economic justice** using **ecological materials.** For example, Brazilian and Peruvian **organic cotton** is used for the production of the sneakers. Other parts of the sneaker production use materials made from **recycled plastic bottles** and **recycled polyester**. Here, the already introduced concept of **upcycling** is used (see Sect. 1.2). Thus, plastic bottles, recycled cotton from textile industry waste, and recycled polyester are used for shoe production (see Veja, 2023a).

In the **value chain**, the company focuses on **fairness** and **transparency**. Since 2004, *Veja* has been setting the prices for the purchase of organic and agroecological cotton in consultation with producer associations from Brazil and Peru in advance. This decouples the price from market fluctuations. This provides the producers with greater financial security. In this context, it can happen—as in 2017—that *Veja* purchased its cotton at a price that was on average twice as high as the market price (see Veja, 2023a).

The use of fair trade and organic raw materials, as well as fair payment of the involved value chain partners (including factories with high social standards), result in ***Veja* sneakers costing three to seven times more in production** than sneakers from major brands. To determine this, a Chinese factory was asked to provide a cost estimate for the production of a specific *Veja* model. Manufacturing in the *Veja* factories in Brazil costs €18.21. In the Chinese factory, only costs of €5.30 would be incurred. However, the shoes would then not be made with socially and environmentally friendly materials and processes (see Veja, 2023b).

How can it be then, that *Veja* sneakers can compete in terms of price? Here, *Veja* refers to a simple observation: 70% of the costs of a normal large sneaker brand are spent on advertising. *Veja* consistently refrains from advertising. This way, the *Veja*-sneakers can be sold at the same price as the major brands of the competition. This allows a brand to be green and still compete in price (see Veja, 2023a).

The following figures document the successes that the **visionary brand** *Veja* has achieved with a triple-bottom-line approach (see Veja, 2023b):

- Since 2004, 8.1 million pairs of shoes have been sold.
- *Veja* shoes are sold through more than 3000 retailers in more than 60 countries.
- From 2004 to the end of 2021, more than 1052 tons of organic cotton and more than 1953 tons of rubber from the Amazon region were purchased.
- Four of the top five earners in the company are women.
- A significant portion of the energy that *Veja* consumes comes from renewable sources.

> ▶ Food for Thought The measures implemented by *Veja* as well as the results achieved may motivate many companies to initiate similar steps.

5.4.5 The Example of Chameleon

Guest contribution by Ingo Lies, founder and owner of Chameleon

For us as a **provider of worldwide group tours, sustainability is not a trend.** It has played a central role since our founding in 1996, when I organized the first trips from my living room and accompanied them as a tour guide. From the beginning, I wanted to share my fascination with the world and make it a little better with each trip. This was not a sophisticated marketing strategy, but a love for a very specific type of travel.

In this way, we invented a completely **new style of travel**. At the center is the fact that we travel with **no more than twelve guests**. This creates a win-win situation: Travelers experience a country up close. And on site, especially smaller local companies and their families benefit on a social, economic, and ecological level. Interestingly, this type of travel hardly needed to be advertised in the early years. Word spread naturally and people came back because they always wanted to travel this way. Therefore, for us, **social or ecological commitment and economic success have never been opposites**. On the contrary.

Our efforts are repeatedly recognized by both our guests and official bodies. Most recently, I was elected as the ***Travel Industry Manager 2022*** on behalf of the entire company. In the justification, our role as a **pioneer in the industry for sustainability** is highlighted. This applies to both the **compensation of the CO_2 footprint** caused by travel, as well as the **fair treatment of customers, travel agencies, and business partners** worldwide—especially during the Corona crisis.

Sustainability by Plan

Sustainable corporate management is not a matter of course. It requires courage. You need commitment, perseverance, and not least financial investments. You need employees who feel responsible in the company and always consider sustainability issues—employees who actively search for innovations and present them in the company. At *Chameleon*, we have therefore created a separate **department for sustainability issues**. Here, ideas for sustainable tourism can be sought, even aside from purely business considerations or other constraints.

Of course, this does not change the fact that the search for these ideas and their implementation ultimately have to be financed by the **company's profit**. In the final analysis, this is then reflected in the travel price, which we openly deal with. **We are not cheap.** But we see this as a strength of our strategy. As long as we manage to communicate this process transparently, our guests are willing to explicitly support these efforts.

Traveling with a clear conscience in times of climate change—is it possible?

A **long-distance trip**, even with *Chameleon,* inevitably causes **CO_2 emissions.** For us, **transparency** is the magic word here. Even before booking, our guests are informed about the exact **CO_2 footprint** for travel, local transport, and accommodation. The **compensation measures** are already **included in the travel price.** It is not left to the decision of the customers what and how emissions are offset. We, as the organizer, take over the **compensation.** In the Amazon, we have already purchased over 18 million square meters of rainforest and placed it under nature protection. Each guest receives a symbolic certificate of a personal piece of forest, the size of which is determined by the

Fig. 5.24 Symbolic certificate of a personal piece of forest from *Chameleon*

scientifically calculated emissions of the respective trip (see Fig. 5.24). The CO_2 emissions of 2.21 t associated with a 19-day vacation in Cuba are offset here by an investment in 152 square meters of rainforest.

Since 2014, every guest traveling with *Chameleon* receives a **durable water bottle** (see Fig. 5.25). This saves the environment 650,000 plastic bottles or 22 tons of plastic waste every year. In addition, there are of course many other measures, such as the **planting of a tree** in Mexico by the initiative *"Plant for the planet"* for each traveler, as well as the **preference for direct flights,** the **train transfer to the airport** included in the travel price, and the **support of specific environmental projects** worldwide.

Fig. 5.25 Nature Bottle for every Chameleon guest—Ingo Lies visiting a project supported by Chameleon in Peru. *Source* © Chameleon

Socially Fair Tourism on an Equal Footing

For us, sustainability is not limited to purely ecological aspects. In addition to climate and species protection, the **respectful encounter on an equal footing** with the people in the travel countries is part of the ***Chameleon*-DNA**. We have known many of our worldwide partners for many years. With many, we maintain friendships today that go far beyond pure business relationships. These **partnership networks** are also lived sustainability for us. And they enable the kind of travel we desire, because it's not about being served, but about encounters on an equal footing.

We place great value on ensuring that the **majority of the travel price** reaches local companies and families. In the travel country, we focus on **ecological and socially compatible accommodations** that are mostly owner-managed, small, and certified. To make these factors visible to our guests, we calculate for each trip the **"local earnings".** This is the portion of the travel price that stays in the visited countries and contributes to economic and social development there. For the previously mentioned Cuba trip, the local earnings are 54%. With a trip to Africa, Latin America, or Asia, each guest secures on average five jobs for a month.

In addition, the **company's own foundation** supports more than 50 **educational projects for children and adults, initiatives for environmental and animal protection** and for the **preservation of tradition and culture**. Since its foundation in 2012, these **foundation projects** have been additionally supported with **€1,523,720**.

In **Peru**, these funds are used to support a village located in the Andes. Visits to the village are an integral part of our Peru trips. The villagers welcome the guests with traditional food and offer their handicrafts (see Fig. 5.25). This creates a **secure and calculable source of income** for the residents, which enables the preservation of the village community with all its traditions.

How closely commitments like this bind our customers to *Chameleon* is shown by the fact that the **foundation** received **donations over €200,000** even in the almost travel-free year 2020.

Interest-Free Loans in Crisis for Business Partners
With the close business connections, of course, comes a **special responsibility**. During the Corona crisis, part of the **global *Chameleon*-network** threatened to collapse. To ensure that the families in the lodges, the partner agencies or even the drivers in the travel countries still had a secured income, we set up a large **support fund of over €1 million** at the end of 2020.

Smaller monetary contributions, which served as a bridge, were **given away** to lodges, social projects or partner agencies in the travel countries. Larger amounts we have given as **interest-free loans**, such as to the partner agency in Sri Lanka, which we supported with €40,000 in 2021. Half given away, the other half as an interest-free loan—indefinitely. In the difficult Corona year 2021, with this money, all 80 employees and tour guides in Sri Lanka could continue to be paid—and thus retained. An advantage that the company now clearly feels as tourism is picking up again. Because while qualified personnel worldwide have migrated from tourism to other sectors, *Chameleon* can still rely on experienced employees.

With courage, know-how, and new technologies into the future
Through *Chameleon*, over 30 **accommodations** have already been created. We continue to assist in the creation, planning, and financing of such projects. **Training** creates an **awareness of ecological issues.** In implementing **climate-friendly technologies** such as solar panels, greywater toilets, or plastic-free gastronomy, we support our local partners with advice and action. Our goal is for all accommodations used by us to be emission-neutral by 2030.

In the popular *Aqua Beach Resort & Spa* in Tanzania, for example, the drinking water is treated on-site, plastic has been completely banned, and some of the electricity comes from the in-house solar system (see Fig. 5.26). Shopping is done at the neighboring fisherman's and seasoned with herbs from the own garden.

And also in the **company headquarters**—the ***Chameleon House*** in Berlin—**sustainability** has always been lived. The electricity comes from the roof, a heat pump system

Fig. 5.26 Solar technology of the accommodations ensures emission neutrality—Chameleon House in Berlin. *Source* © Chameleon

provides heating and cooling in the summer (see Fig. 5.26). All employees also receive free tickets for public transport. The company covers the CO_2 offset of business and private trips. The mostly vegetarian lunch in the in-house company canteen is prepared from organic, regional ingredients and is free for all employees. Resource-saving work is emphasized in everyday office life.

Being sustainable is not a static state, but means always finding new ways. **Change is the keyword here.** Travel a lot yourself, talk to the people on site and the experts, listen carefully and then at some point stop talking so much and start doing. Only in this way can something be created. We are currently working hard to further reduce the **use of plastic** on our trips so that we can offer all trips completely plastic-free as soon as possible. By the international tourism fair in March 2023, *Chameleon* is expected to be the first tour operator to be certified by *Travelife,* an initiative for implementing sustainable principles in the tourism industry.

The journey continues for *Chameleon* itself â€¦

5.4.6 The Example of Debatin

An interesting example of a **B2B company,** that strives for ecological, economic, and social sustainability, is ***Debatin.*** The company produces **logistics and film packaging**—an area that is particularly scrutinized in terms of sustainability. The corporate action is based on the following **mission,** which is derived from the company's guiding principles, values, and vision (Debatin, 2023b, p. 4):

> "Develop the best packaging solutions for all areas of life while considering social, ecological, and economic aspects."

A central point of reference for the **sustainability strategy of *Debatin*** is the *Agenda 2030 for Sustainable Development* adopted by the *United Nations* in 2015 with the *Sustainable Development Goals* (see Fig. 1.3). The company's products are already climate-neutral. At the same time, the range of **climate-friendly products made from recycled materials** is being further developed. Many products have received the **"Made for Recycling" certificate**. This certifies these products with high recyclability. In parallel, there is a continuous further development of a **climate-friendly production** and a company-owned **recycling cycle**. This is intended to implement the **Cradle to Cradle** principle—so that raw material for new films is obtained through a material cycle. *Debatin* provides customers with a **recycling kit** on request to further advance this principle.

The sustainability-oriented activities also include marketing. A sustainable trade fair stand was developed for the external appearance. Through **investments in climate protection projects**, a **compensation** for the emissions not yet avoided is achieved. This involves the purchase of climate protection certificates. These promote hydroelectric projects in India and various social and environmental projects in Togo. Through these compensation measures, trade fair appearances can also be designed to be "climate neutral" or more precisely "climate compensated" (see Debatin, 2023a, b, p. 7 f.).

By orienting itself to the **Cradle-to-Gate approach**, the company not only offsets the emissions of its own production, but also includes the emissions that occur during the production of raw materials. These include, among others, foils, paper, and adhesives. The perspective extends from the cradle (**Cradle:** raw material extraction) to the gate (**Gate**) of the finished products. This achieves comprehensive compensation up to the handover to the customers.

In parallel, the company exclusively uses papers from exemplary managed, *FSC*-certified forests and other controlled sources. The **environmental seal *FSC*** *(Forest Stewardship Council)* stands for forest and wood certification. By using *FSC*-certified wood, a contribution can be made to the preservation of forests. In 2017, the company began converting document pockets to a ***Blue Angel*-quality.** As a result, customers can purchase "climate-neutral" products certified with the *Blue Angel*. In addition, the company uses its own seals with ***Deriba Greenline.*** Products with this seal guarantee a qualitative and sustainable packaging. This consists of at least 80% PCR material in *Blue Angel*-quality and is 100% recyclable, achieving a particularly high purity level of the recyclate in the process (see Debatin, 2023a).

To also take the customers on the journey towards sustainability, *Debatin* developed a **CO_2-calculator**. This allows customers to calculate online the **CO_2-savings** when **using the sustainable product lines** compared to using products made from new, conventional foil. This calculation is based on data from the *Federal Environment Agency*. This makes customers aware of the consequences of their purchasing decision. This is an interesting example of **signaling** to avoid negative selection (see Sect. 5.3.4).

A **social sustainability** starts within the company itself. That's why *Debatin* relies on solid training and offers a wide range of further education opportunities. Flat hierarchies, high family-friendliness, comprehensive health management, and team spirit are intended to create an appealing working atmosphere. The company is a partner in the *"Alliance for Family Bruchsal"* and is also involved in the corporate network *"Success Factor Family"*. Since 2015, the company has held the predicate *"Family-friendly in the Bruchsal economic region"*. In 2022, *Debatin* is among the 1% of the best employers in Germany at ***Leading Employer*** (see Debatin, 2023a).

5.4.7 The Example of Continental

The example of *Continental* is intended to show what an **organizational anchoring of "sustainability"** can look like in order to achieve ambitious goals. At the same time, it becomes clear that a **commitment to sustainability** must also be reflected in the **compensation structures of the company management**. In addition, the importance of **communicating sustainability initiatives internally and externally** becomes apparent.

The company ***Continental*** has anchored its **sustainability strategy** in the corporate strategy in the cornerstone *"Consistent use of opportunities"*. This makes it clear: *Continental* sees **sustainability as an important driver of innovation** in the dynamic change of mobility and industries. *Continental* is therefore an example of a "creative approach" to the challenges of sustainability (see Fig. 4.1).

In the **Group Sustainability Ambition**, the company describes how the transition towards sustainability should take place in the relevant subject areas. This primarily involves the core topics of clean mobility, climate protection, circular economy, and sustainable supply chains. This includes sustainably operating production facilities, good working conditions, safe mobility, social commitment, and of course, sustainable profitability.

It is worth noting here that the **ultimate responsibility for sustainability** is located at the board level—in the *Group Human Relations* and *Group Sustainability* departments. The **organizational anchoring of sustainability management** is supplemented by sustainability functions in the various business areas and by coordinators in several business sectors and countries. The *Group Sustainability Steering Committee*—also *Group Sustainability Committee*—is responsible for assessing the overarching issues. This is also where the weighing of opportunities and risks takes place to prepare board decisions (see Continental, 2022).

A **Group Sustainability Scorecard** compiles the **performance indicators for sustainability** (see also Sect. 6.2.6). It is based on defined quality criteria and is continuously developed further. The scores shown here provide the basis for integrating **sustainability into the company's processes**. This scorecard is approved annually by the board and thus defines the goals of sustainability management. Since the fiscal year 2021, the topic of **sustainability has become an integral part of corporate strategy development**. This leads to important investments being routinely checked and evaluated for their contribution to sustainability during the approval process. In the business areas Automotive, Tires, and *ContiTech*, concrete **roadmaps for the implementation of sustainability goals** have been gradually developed (see Continental, 2022).

Both the board and globally active executives are also evaluated based on the **achievement of sustainability goals**. For this purpose, the long-term compensation components have been linked to sustainability goals. In order to promote a **cultural change towards sustainability** within the company, the **internal communication on the topic of sustainability** has been intensified. A large number of sustainability events have taken place. Above all, the topic has also been integrated into the internal communication formats for executives. Dialogue with employees takes place through town hall meetings, employee surveys, webcasts, and employee representations.

In addition, the **external stakeholders** were informed about the sustainability ambition of *Continental* through various channels (see Continental, 2022). The **customers** are primarily informed about this through sales or key account management, through cooperations, and at trade fairs. **Investors and shareholders** receive the important information through the annual general meeting as well as accompanying webcasts and roadshows. The dialogue with the **general public** takes place, among other things, through surveys, trade fairs, engagement projects, and open house days. Only through such comprehensive exchange with the relevant stakeholders can the different perspectives be captured and taken into account—as fuel for the further development of the sustainability strategy and the reporting on it (see Continental, 2022).

Continental is a convincing example of how "sustainability" must be anchored in the organizational structure and processes, in the compensation structures, and in the internal and external communication. In this way, a **"sustainability of sustainability"** is ensured procedurally and structurally.

5.5 Overall Status Quo of Green Brand Management

If one analyzes the **search behavior for "green brand management"** with *Google Trends,* then it says there in August 2022 with a focus on Germany:

"Your search does not contain enough data!"

The latest studies, which were published by *Interbrand* among others, on the topic of **Global Green Brands** are already several years old. In the last published *Interbrand* study on **Best Global Green Brands** from 2015, *Ford* was in first place, followed by *Toyota* and *Honda* in second and third place. These companies would probably not occupy a podium position today! The internet address used at the time, www.bestglobalgreenbrands.com now only leads to the *Interbrand* homepage—and not to current results. *Interbrand* seems to have lost interest in this topic.

The company ***Green Brands*** therefore wants to provide some guidance in this area. According to its own statements, it is an international, independent and self-sufficient **brand evaluation organization** with its headquarters in Germany. This organization, together with other independent institutions and societies in the environmental/climate protection and sustainability sector, awards **ecologically sustainable brands**. Partners of Green Brands include (see Green Brands, 2023):

- *Sustainable Europe Research Institute,* Vienna (http://www.seri.at)
- Focus: scientific supervision and development of the criteria catalogs for the *Green Brands Index*
- *Ipos* (https://www.ipsos.com/de-de)
- Focus: representative market research studies in the relevant countries
- *Institute for Sustainability of the University of Economics and Law, Berlin* (https://www.ina.hwr-berlin.de)
- Focus: collaboration in the further development of the approach and the criteria for validation
- *Allplan*, Vienna (https://allplan.at)
- Focus: development of the criteria catalogs and execution/evaluation of the validation
- *Federal German Working Group for Environmentally Conscious Management* (https://www.baumev.de/)

The company *Green Brands* awards a ***Green Brands Seal of Approval.*** This is a registered **EU guarantee mark for ecological sustainability**. The brands of companies, products or services that demonstrably make a special effort to protect and preserve natural resources are awarded (see Green Brands, 2023). The current results are listed on the website https://green-brands.org/ausgezeichnete-marken/germany/ for various countries.

The brands *Green Brands* **awarded** from 2020 to 2022 include, among others:

- **Service providers**
  Aldi Süd, Bio Company, dm Drogeriemarkt, Norma, SuperBiomarkt
- **Products**
  Alterra, Alverde, edding Ecoline, Fischer Greenline, Klar EcoSensitive, Lavera, Uhu
- **Food**
  dm Bio, Green Petfood, Vio Bio Limo

> **Food for Thought** Many big players are still missing among the brands awarded by *Green Brands*. There is still a lot of room for improvement!

This is also evident in a **twelve-country study.** In 2021, 350 decision-makers in consumer goods and retail companies were surveyed, who achieved an annual online turnover of at least 500,000 US$. The **sustainability initiatives of brands** mentioned in Fig. 5.27 were the focus (see Statista, 2022a).

When asked about **sustainability initiatives** in 2021, almost half of the companies stated that they will invest in simple solutions for the **recycling** of their products in the

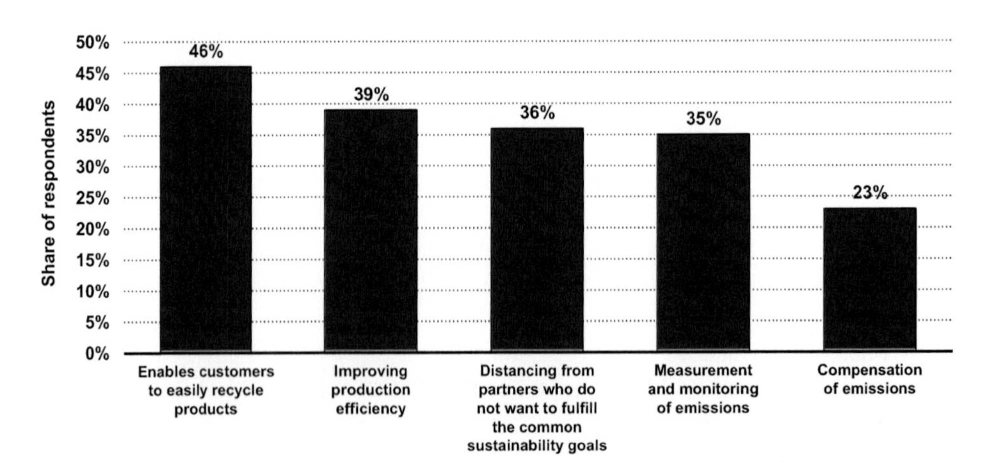

Fig. 5.27 Most common sustainability initiatives of brands worldwide—2021–2022. (Data source: Statista, 2022a)

next year. However, it was already clear that recycling in the circular economy is not a top priority and should only be considered after refuse, reduce, and reuse.

More than a third of the surveyed companies also plan to **distance themselves from partners** who cannot meet common sustainability goals. Only just over a third of companies aim to **measure and monitor emissions,** and only just under a quarter of brands worldwide have **offsetting emissions** on their agenda. There is still a great need for action here.

Questions you should ask yourself

- How comprehensive is our marketing research focused on sustainability developments in the micro and macro environment?
- Which of our customer groups are particularly interested in sustainable solutions?
- Which stakeholders—besides customers—are bringing demands for more sustainable solutions to us?
- What exactly do these demands refer to?
- How consistently are we working on "green" marketing concepts and a "green" brand management that builds on concrete measures of our company?
- Where do we risk slipping into greenwashing?
- What mistakes have we and our competitors already made in this area?
- How consistently do we consider the relevant guidelines for green brand management?
- Are we already using nudging ideas for behavior control or should we?
- How convincingly are we already using signaling to highlight our sustainable offers?
- Has our company already dealt with green labels and seals?
- Which sustainability certificates would be particularly exciting for our company?
- Who deals with labels, seals, and certificates in our company?
- How sustainably are our advertising activities already aligned?
- Are we working together with relevant initiatives in this area?
- In which fields can we further increase sustainability in advertising?
- What sustainability potentials are there in our sales activities?
- Are we already using claims to communicate our sustainability activities?
- How legally secure are these claims?
- What other ideas can we gain from the examples of green brand management for ourselves?
- What initiatives for sustainable marketing are there and in which should we get involved?
- Where do the threads of these questions come together?
- Is there support from top management (personnel and budget) to tackle these tasks?
- Are our incentive systems aligned to promote sustainable corporate management?

References

Abzonline. (2023). "Eco-Score" trifft auf "Planet Score". https://www.abzonline.de/nachrichten/aktuell/kennzeichnung-eco-score-trifft-auf-planet-score-151314?crefresh=1. Accessed 3 Jan 2023.

Ad Net Zero. (2023). Advertising's response to the climate crisis. https://adnetzero.com/. Accessed 4 Jan 2023.

Akerlof, G. A. (1970). The market for "Lemons": Quality uncertainty and the market mechanism. *The Quarterly Journal of Economics, 84*(3), 488–500. https://doi.org/10.2307/1879431.

Aldi. (2022). Wir schaffen das Kükentöten ab. https://www.aldi-sued.de/de/nachhaltigkeit/lieferkette/produktionsstandards/tierwohl/ohne-kuekentoeten.html. Accessed 23 March 2022.

Aldi. (2023). Bewusste Ernährung leicht gemacht. https://www.aldi-sued.de/de/nachhaltigkeit/nachhaltige-produkte/bewusste-ernaehrung.html?utm_campaign=uf_cr_heute%20fuer%20morgen_22_na_na&utm_content=shorturl. Accessed 2 Jan 2023.

Atmosfair. (2023). Atmosfair 2.0. https://www.atmosfair.de/de/. Accessed 2 Jan 2023.

Bauer, M. J., & Sobolewski, S. (2022). *Grüne Marketing-Kommunikation. Green Communication im Marketing-Mix nachhaltigkeitsorientierter Unternehmen.* Springer Gabler.

Beuchler, T. (2022). *Grüne Farbe bekennen. Interview. Horizont, 36–37,* 58–60.

Bio Planète Ölmühle Moog. (2022). Bio Planète wirbt für Planet Score. https://www.bioplanete.de/aktuelles/bio-planete-wirbt-fuer-planet-score. Accessed 9 Sept 2022.

Blauer Engel. (2023). Das deutsche Umweltzeichen. https://www.blauer-engel.de/de. Accessed 1 Jan 2023.

BMEL. (2022). Nutri-Score. https://www.bmel.de/DE/themen/ernaehrung/lebensmittel-kennzeichnung/freiwillige-angaben-und-label/nutri-score/nutri-score_node.html. Accessed 9 Sept 2022.

BMZ. (2023). Der Grüne Knopf. https://www.gruener-knopf.de/. Accessed 2 Jan 2023.

Bundesverband Naturkost Naturwaren. (2023). Planet-Score: Echte Nachhaltigkeitskennzeichnung statt Greenwashing. https://n-bnn.de/planet-score. Accessed 1 Jan 2023.

Burmann, C., Halaszovich, T., Schade, M., & Piehler, R. (2018). *Identitätsbasierte Marketingführung. Grundlagen—Strategie—Umsetzung—Controlling* (3rd. edn.). Springer Gabler.

CGLR. (2023). Meet A.L.I.C.E.; Publicis Groupe's Carbon Calculator. https://councilgreatlakesregion.org/meet-a-l-i-c-e-publicis-groupes-carbon-calculator/. Accessed 1 Jan 2023.

Concept M. (2022). *Nachhaltigkeit und Marketing*. Concept M.

Conn, C. (26. September 2022). In unserer idealen Welt wären Gewinne niedriger. *Frankfurter Allgemeine Zeitung,* p. 20.

Continental. (2022). Nachhaltigkeitsbericht 2021. https://www.continental.com/de/nachhaltigkeit/nachhaltige-unternehmensfuehrung/nachhaltigkeitsberichte/. Accessed 15 Sept 2022.

Cradle to Cradle Products Innovation Institute. (2023). Made for Tomorrow. https://www.c2ccertified.org/. Accessed 2 Jan 2023.

Debatin. (2023a). Unsere Offensive für Nachhaltigkeit. https://www.debatin.de/nachhaltigkeits-offensive-fuer-dokumententaschen/. Accessed 2 Jan 2023.

Debatin. (2023b). Verpackungslösungen für die Zukunft. Sicher, nachhaltig und innovativ. https://www.debatin.de/unternehmen/#historie. Accessed 2 Jan 2023.

Demeter. (2023). Demeter—konsequentes Bio seit 1924. https://www.demeter.de/. Accessed 1 Jan 2023.

Diemand, S. (29. July 2022). Den Werbebroschüren droht das Aus. *Frankfurter Allgemeine Zeitung*, p. 23.

Digital Watermarks. (2023). Pioneering Digital Watermarks for Smart Packaging Recycling in the EU. https://www.digitalwatermarks.eu/. Accessed 3 Jan 2023.

DPG. (2023). Das DPG Einwegpfandsystem—Ein wertvoller Beitrag zur Kreislaufwirtschaft. https://dpg-pfandsystem.de/index.php/de/. Accessed 2 Jan 2023.

Ecocert. (2023). Handeln für eine nachhaltige Welt. https://www.ecocert.com/de-DE/home. Accessed 1 Jan 2023.

Erlei, M., & Szczutkowski, A. (2022). Adverse Selektion. https://wirtschaftslexikon.gabler.de/definition/adverse-selection-26952/version-121089. Accessed 12 Sept 2022.

Errichiello, O., & Zschiesche, A. (2021). *Grüne Markenführung: Grundlagen, Erfolgsfaktoren und Instrumente für ein nachhaltiges Brand- und Innovationsmanagement* (2nd. edn.). Springer Gabler.

Esch, F.-R. (2019). Identität der Corporate Brand entwickeln und schärfen. In F.-R. Esch, T. Tomczak, J. Kernstock, T. Langner, J. & Redler (Hrsg.), *Corporate Brand Management, Marken als Anker strategischer Führung von Unternehmen* (4. edn., pp. 81–105). Springer Gabler.

Frosch. (2023a). Unsere Recyclat-Initiative. https://frosch.de/Nachhaltigkeit/Saubere-Meere.html. Accessed 2 Jan 2023.

Frosch. (2023b). Willkommen bei der Initiative Frosch. https://initiative-frosch.de/, Accessed 2 Jan 2023.

Gellert, R. (2021). Die Outdoor-Marke Patagonia will nicht mehr wachsen. https://magazin.nzz.ch/wirtschaft/patagonia-will-nicht-mehr-wachsen-sagt-der-ceo-der-outdoor-marke-ld.1596786?reduced=true. Accessed 8 Sept 2022.

Global Standard. (2023). Global Organic Textile Standard. https://global-standard.org/de. Accessed 2 Jan 2023.

Green Brands. (2023). Die Organisation. https://green-brands.org/ueber-uns/die-organisation/. Accessed 1 Jan 2023.

Greengiving. (2023). Alles über Greengiving. https://www.greengiving.de/uber-greengiving. Accessed 2 Jan 2023.

Grunwald, G., & Schwill, J. (2022). *Nachhaltigkeitsmarketing*. Schäffer-Poeschel.

Grüner Punkt. (2023). Kreisläufe schließen—gemeinsam für unsere Umwelt! https://www.gruener-punkt.de/de/. Accessed 2 Jan 2023.

GWW. (2022). GWW legt Ergebnisse des aktuellen Werbeartikel-Monitors vor. Werbeartikelumsatz klettert erstmals über 3,6-Milliarden-Marke. https://gww.de/2020/02/gww-legt-ergebnisse-des-aktuellen-werbeartikel-monitors-vor-2/. Accessed 19 Sept 2022.

IPCC. (2023). Glossary. https://www.ipcc.ch/sr15/chapter/glossary/. Accessed 2 Jan 2023.

Kecskes, R. (2021). Purpose—von Produkt und Bedeutung. https://www.horizont.net/planung-analyse/nachrichten/purpose--von-produkt-und-bedeutung-was-starke-marken-der-zukunft-auszeichnen-wird-193080. Accessed 8 Sept 2022.

Kilian, K., & Kreutzer, R. T. (2022). *Digitale Markenführung*. Springer Gabler.

Kreutzer, R. T. (2021). *Toolbox für Digital Business. Leadership, Geschäftsmodelle, Technologien und Change-Management für das digitale Zeitalter*. Springer Gabler.

L'Oréal. (2023). L'Oréal for the Future, our sustainability commitments for 2030. https://www.loreal.com/en/commitments-and-responsibilities/for-the-planet/. Accessed 1 Jan 2023.

Lebensmittelmagazin. (2022). Tag der Lebensmittelvielfalt: 170.000 Lebensmittelprodukte auf dem deutschen Markt. https://www.lebensmittelmagazin.de/wirtschaft/20200731-tag-der-lebensmittelvielfalt-170000-lebensmittel-produkte-auf-dem-deutschen-markt. Accessed 30 Dec 2022.

Lidl. (2022). Lidl testet als erster deutscher Händler die Eco-Score Kennzeichnung. https://unternehmen.lidl.de/verantwortung/fokusthema-ernaehrung/eco-score. Accessed 9 Sept 2022.

Lindner, R. (16. September 2022). Abschied vom Milliardärsdasein. *Frankfurter Allgemeine Zeitung*, p. 20.

Mehrweg. (2023). Mehrwegsystem. https://www.mehrweg.org/mehrwegsystem/. Accessed 2 Jan 2023.

Natrue. (2023). Unsere Mission ist, Natur- und Biokosmetik für das Wohl der Verbraucher weltweit zu fördern und zu schützen. https://www.natrue.org/de/. Accessed 1 Jan 2023.

Nestlé. (2022). Nestlé Deutschland in der Gesellschaft. https://www.nestle.de/verantwortung/nachhaltigkeitsbericht. Accessed 16 Sept 2022.

Net Carbon Events. (2023). Net Zero Carbon Events. https://netzerocarbonevents.org/the-pledge/. Accessed 1 Jan 2023.

Nitsche, M. (2022). *Grünes Marketing*. DDV.

Oeko-Tex. (2023a). STANDARD 100 by OEKO-TEX®. https://www.oeko-tex.com/de/unsere-standards/standard-100-by-oeko-tex. Accessed 2 Jan 2023.

Oeko-Tex. (2023b). MADE IN GREEN im Überblick. https://www.oeko-tex.com/de/unsere-standards/made-in-green-by-oeko-tex. Accessed 2 Jan 2023.

Ökolandbau. (2023). Nachhaltigkeitslabel für den Handel—Eco-Score und Planet-Score unter der Lupe. https://www.oekolandbau.de/handel/unternehmensfuehrung/nachhaltig-wirtschaften/nachhaltigkeitslabel-fuer-den-handel-eco-score-und-planet-score-unter-der-lupe/. Accessed 2 Jan 2023.

Otto-Friedrich-Universität Bamberg. (2022). Deutschland ist Retouren-Europameister. https://www.marketing-boerse.de/news/details/2236-deutschland-ist-retouren-europameister/187164. Accessed 7 Sept 2022.

Patagonia. (2023a). Warum Fair Trade. https://eu.patagonia.com/de/de/why-fair-trade-card.html. Accessed 1 Jan 2023.

Patagonia. (2023b). Die Erde ist ab sofort unsere einzige Anteilseignerin. https://eu.patagonia.com/de/de/home/. Accessed 1 Jan 2023.

Petcycle. (2023). Petcycle—das effiziente und nachhaltige System für Getränkeverpackungen. https://www.petcycle.de/. Accessed 2 Febr 2023.

Peterson, M. (2021). *Sustainable Marketing* 2nd edn. Sage

Plastikalternative. (2023). Kleines Zeichenlexikon. https://www.plastikalternative.de/zeichenlexikon/. Accessed 3 Jan 2023.

Procter & Gamble. (2023). Pressemitteilungen. https://pgnewsroom.de/pressemeldungen/pressemitteilung-details/2021/Projekt-HolyGrail-2.0-erreicht-die-nchste-Stufe-Digitale-Wasserzeichen-starten-europaweit-in-die-semi-industrielle-Testphase/default.aspx. Accessed 4 Jan 2023.

Rieke, N., & Schwingen, H.-C. (2021). *Wie Werte Marken stark machen: Mit dem Leitsystem für werteorientierte Markenführung mehr gesellschaftlichen Impact erzielen*. Haufe.

Rometsch, K. (2021). 10 Arten von Nudges aus dem Alltag. https://www.die-debatte.org/nudging-listicle/. Accessed 6 Sept 2022.

Statista. (2021a). Welche dieser Leistungsdimensionen einer Marke sind Ihnen am wichtigsten, damit diese Ihr Vertrauen verdient und Sie diese Freunden oder der Familie weiter empfehlen würden? https://de.statista.com/statistik/daten/studie/1125480/umfrage/wichtigste-leistungsdimensionen-markenvertrauen-in-deutschland/. Accessed 30 May 2022.

Statista. (2021b). Perceived responsibilities of online retailers in fighting climate change in Germany in 2021. https://www.statista.com/statistics/1175192/online-retailers-responsibility-to-fight-climate-change-germany/. Accessed 1 June 2022.

Statista. (2021c). Opinion on which type of stores, brick-and-mortar or online stores, are more sustainable in Germany in 2021. https://www.statista.com/statistics/1284874/perceived-sustainability-of-stores-ecommerce-germany/#:~:Text=A%20survey%20carried%20out%20in%20Germany%20in%202021c,considered%20physical%20stores%20as%20sustainable%20as%20online%20stores. Accessed 1 June 2022.

Statista. (2021d). Elektroschrott. https://de-statista-com.ezproxy.hwr-berlin.de/statistik/studie/id/101889/dokument/elektroschrott/. Accessed 27 July 2022.

Statista. (2021e). Importance of sustainable e-commerce packaging among shoppers in Germany in 2021. https://www.statista.com/statistics/1288788/sustainable-e-commerce-packaging-germany/. Accessed 30 May 2022.

Statista. (2021f). LOHAS in Deutschland 2021. https://de-statista-com.ezproxy.hwr-berlin.de/statistik/studie/id/61700/dokument/lohas-in-deutschland/. Accessed 24 March 2022.

Statista. (2022a). Leading sustainability initiatives brands are investing in worldwide in 2021 and 2022. https://www.statista.com/statistics/1305916/main-sustainability-initiatives-by-brands-worldwide/. Accessed 1 June 2022.

Statista. (2022b). Leading reasons to opt for click & collect in German-speaking countries in 2021. https://www.statista.com/topics/7830/click-and-collect/#dossierKeyfigures. Accessed 1 June 2022.

Statista. (2022c). Anzahl der Unternehmen, die das Bio-Siegel nutzen, in Deutschland in den Jahren 2004 bis 2022. https://de-statista-com.ezproxy.hwr-berlin.de/statistik/daten/studie/421382/umfrage/produkte-mit-bio-siegel-in-deutschland/. Accessed 1 Aug 2022.

Stegmann, O. (27. Juli 2022). Beim "Greenwashing" sieht die EU-Kommission rot. *Frankfurter Allgemeine Zeitung*, p. 16.

Thaler, R. H., & Sunstein, C. R. (2010). *Nudge: Wie man kluge Entscheidungen anstößt*. https://www.amazon.de/Nudge-Wie-kluge-Entscheidungen-anst%C3%B6%C3%9Ft/dp/3548373666/ref=sr_1_1?__mk_de_DE=%C3%85M%C3%85%C5%BD%C3%95%C3%91&crid=HSCMPF1XUWFF&keywords=Nudge%3A+Wie+man+kluge+Entscheidungen+anst%C3%B6%C3%9Ft&qid=1662465067&sprefix=nudge+wie+man+kluge+entscheidungen+anst%C3%B6%C3%9Ft%2Caps%2C68&sr=8-1. Ullstein.

Veja. (2023a). Fair Trade. https://project.veja-store.com/en/single/fairtrade. Accessed 2 Jan 2023.

Veja. (2023b). Transparency. https://project.veja-store.com/en/single/transparency. Accessed 2 Jan 2023.

Weigand, H. (2020). Green Marketing—nachhaltig erfolgreich. In M. Stumpf (ed.), *Die 10 wichtigsten Zukunftsthemen im Marketing* (2. edn., pp. 47–69). Haufe.

Zalando. (2021). *It takes two*. Zalando.

6 Monitoring and Controlling to Ensure Sustainable Corporate Governance

Scheitern ist nicht das Gegenteil von Erfolg. Es ist ein Teil davon.

Lebensweisheit

Abstract

A sustainable orientation towards sustainable corporate management cannot be achieved without sustainability monitoring and sustainability controlling. Here, the concepts of eco-audit, life cycle assessment, sustainability analyses, and approaches for determining the corporate and product carbon footprints are presented. In addition, a dashboard for monitoring the Green Journey and a Balanced Scorecard with a Sustainability module are presented. The establishment of such concepts helps to align companies sustainably.

6.1 Basics of Sustainability Monitoring and Sustainability Controlling

The action levels of Planet, People, and Profit outlined in Chap. 1 must also be reflected in sustainability monitoring and sustainability controlling. The **sustainability monitoring** (also **Green Monitoring**) focuses on the monitoring of ongoing processes. This involves the systematic recording of processes to document their effects on the sustainability of the company. Various technologies are used for this purpose.

© The Author(s), under exclusive license to Springer Fachmedien Wiesbaden GmbH, part of Springer Nature 2024

R. T. Kreutzer, *The Path to Sustainable Corporate Management*,
https://doi.org/10.1007/978-3-658-43974-3_6

> **Note Box** Through the **sustainability monitoring**, it should be determined whether the ongoing processes are taking the desired course. For this purpose, defined threshold values must be adhered to. If this is not the case, timely intervention can be made.

The **sustainability controlling** (also **Green Controlling**) is intended to prepare decision-relevant information on all areas of sustainable corporate management. In sustainability controlling, the classic economic control variables are supplemented by social and ecological indicators. This is intended to ensure transparent and responsible handling of all relevant impacts of corporate activities. In this context, the **controllers** as **business partners** should support and advise all areas of the company so that they can achieve the defined sustainability goals (cf. Schaltegger, 2022; Colsmann, 2016).

> **Note Box** The **sustainability controlling** is an indispensable performance component to permanently align the transformation of the value creation logic and thus the entire business model towards sustainability.

The following **Key Performance Indicators of sustainability monitoring and sustainability controlling** may be relevant for companies—depending on the respective business model (cf. in depth Sect. 4.4.1):

- **Material use** (oil, gas, electricity, water, land area, etc.)
- **Emissions** (CO_2, polluted water, contaminated soils, noise)
- **Waste quantities** and achieved **rates in refurbishing, remanufacturing, and recycling**
- **Quantity and quality of labor input in the value chain**
- **Type and extent of compensations made**
- **Equality** and **diversity**
- ...

Further criteria can be derived company-specifically from the **legal requirements** (cf. Chap. 2) as well as from the different **fields of action of the circular economy** (cf. Chap. 4). Based on such **sustainability indicators**, controlling can evaluate the processes of performance creation and their results.

> **Note Box** In **sustainability controlling,** it is essential that the economic criteria as well as the criteria for social and ecological results are analyzed and evaluated on the same level of relevance. Social and ecological impacts should not be seen as a "side condition" of entrepreneurial action, but should stand equally alongside the economic criteria. This is the core of the Triple-Bottom-Line.

6.2 Concepts of Sustainability Controlling

Sustainability indicators can be integrated into various concepts of **sustainability controlling** (see also Balderjahn, 2021, pp. 184–192). The most important approaches are presented below. They can also be combined to illuminate all relevant facets of sustainability.

6.2.1 Eco-Audit

An **eco-audit** ("audit" stands for "review" or "accounting check") is a process in which a company analyzes its own behavior with regard to sustainability goals. Based on the results obtained, the company's activities can be optimized in various areas. Such an audit can be more strategic in nature and scrutinize the company's strategies for their sustainability. But it can also be operationally oriented to determine the extent of sustainability already achieved in daily operations.

In the course of an eco-audit, the following questions can be examined (see Baumgarth & Binckebanck, 2018, pp. 295–297):

- Does an **anchoring gap** of the propagated sustainability values exist?
  Here, it is necessary to check whether the proclaimed sustainability values are fully lived in all areas of the company.
- Is an **implementation gap** noticeable?
  This involves determining whether there is consistent implementation of sustainability aspects in all relevant fields of action.
- Is there an **experience gap**?
  Here, it is necessary to analyze whether the defined sustainability values also become apparent in the customer experience. It should be checked whether the measures aimed at sustainability become visible and tangible for the customers (see for avoiding adverse selection Sect. 5.3.4).
- Is there a credibility gap?
  This involves investigating whether the company's propagated sustainability values are believed by the relevant stakeholders. In addition, it should be determined whether the messages are perceived as authentic.

The need for the establishment of a risk management audit was already described in the discussion of the implementation of the supply chain law (see Sect. 4.4.1.3).

6.2.2 Life Cycle Assessment

The **Life Cycle Assessment** is a tool that can be used to systematically record, analyze, and evaluate environmental impacts. For Life Cycle Assessments, the term **Life Cycle**

Assessment (LCA) is also used. Impacts can be caused by individual products or services, by processes, or by the company as a whole. To document and evaluate these impacts, the sustainability criteria already presented are used.

The Life Cycle Assessment covers the entire cycle: from the production of raw materials and other purchased products and services, through the production and packaging process, distribution and consumption or use, to disposal. In the Life Cycle Assessment, the ecological and social effects are recorded for each stage of this process. The results shown by a Life Cycle Assessment can not only be used for **optimizing processes** or for **developing sustainable production**. These results also provide a **decision-making aid for the award of the *Blue Angel*** (see Sect. 5.3.5).

In the **development of Life Cycle Assessments**, two important **principles** must be observed (see Federal Environment Agency, 2022):

- **Cross-resource consideration**
  All relevant potential harmful effects on the environmental resources soil, air, and water must be taken into account in the recording.
- **Material flow-integrated consideration**
  In addition, all material flows that are associated with the system under investigation must be taken into account. This includes the use of raw materials, emissions from supply and disposal processes, from energy generation, from transport, and from other processes.

The **standards for Life Cycle Assessment** were established internationally in the **ISO standards** 14040:2006 and 14044:2006. Through DIN EN ISO 14040 and DIN EN ISO 14044 these were transferred into the **German standardization system.** It is defined here that a Life Cycle Assessment has the following four areas (see Federal Environment Agency, 2022):

- Definition of the objectives and the respective investigation framework
- Creation of a material balance
- Estimation of the impacts
- Evaluation of the data collected

The ***Federal Environment Agency*** actively participates in the development and updating of these ISO standards and the methods used. Which **analysis fields a Life Cycle Assessment** covers is shown in Fig. 6.1.

▶ **Note Box** A **Life Cycle Assessment** differs from other concepts in that all environmental impacts are included. In the case of the **CO_2 footprint (Carbon Footprint)** as well as the **Water Footprint (Water Footprint)**, only one environmental dimension is considered. However, the calculation methods used for creating a Life Cycle Assessment are similar.

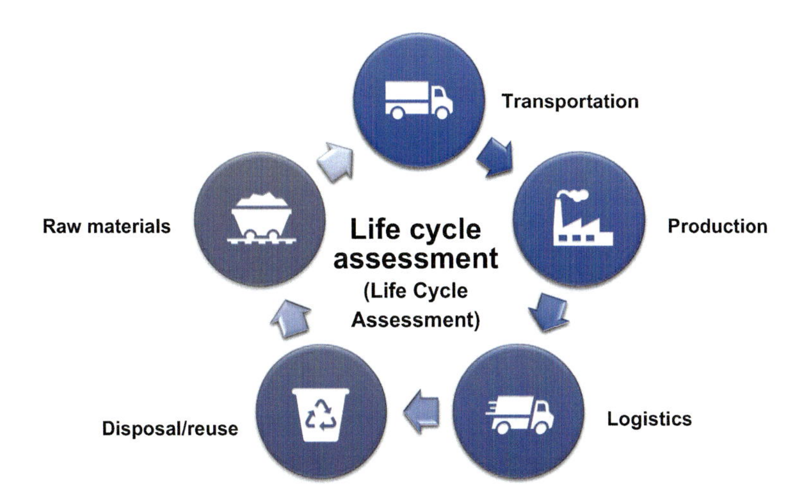

Fig. 6.1 Analysis fields of a Life Cycle Assessment

6.2.3 Sustainability Analysis of Products and Product Lines

At the center of a sustainability analysis are individual products or product lines (see Colsmann, 2016, p. 68). This involves capturing and evaluating the ecological, societal, and economic impacts of products or product lines over their entire life cycle. The criteria catalogs already introduced can be used for this purpose (see Sect. 4.4).

Criteria	Characteristics				
	--	-	+-	+	++
Ecological criteria - - -					
Social criteria - - -					
Economic criteria - - -					

Fig. 6.2 Structure of a sustainability analysis of products and product lines

As a result of a sustainability analysis, serious **weaknesses** may be revealed that need to be immediately eliminated. These include, for example, unacceptable working conditions in raw material extraction or in own or external production processes. It can also be determined that recycling systems do not function as they should (keyword "PET bottles"). In addition, **potential for improvement** can be identified, for example, to make packaging more easily recyclable or to largely do without it. The basic structure of a sustainability analysis is shown in Fig. 6.2.

The concept of sustainability analysis is designed to be flexible and can be aligned with the respective company situation. Based on this analysis, it can be determined year after year where new or additional fields of action arise.

6.2.4 Determination of Corporate and Product Carbon Footprints

For companies, it is a major challenge to calculate the **CO_2 footprint of their own company**. However, many companies are obliged to do this in order to meet their reporting obligations. How do companies proceed in this regard? To answer this, we will once again refer to the representative study by *Bitkom* already cited in Sect. 1.4. As part of this study, 506 companies with 20 or more employees in Germany were surveyed. The companies proceed as follows in recording and offsetting their own ecological footprint (see Bitkom, 2022):

- 28% of all companies carry out a digital measurement of their CO_2 emissions.
- A further 30% plan a digital measurement of their CO_2 emissions.
- 35% of all respondents already offset their CO_2 emissions.
- An additional 34% plan to offset their CO_2 emissions.

What measures can achieve the reduction of a company's own CO_2 footprint? The **management of the CO_2 footprint** is achieved through the following steps:

- Calculation of the Corporate Carbon Footprint
  To create a **CO_2 balance for the own company**—the **Corporate Carbon Footprint**—it is first determined what should be included in the evaluation. This decision is based on the company's objectives, the respective industry, and legal requirements. A **CO_2 footprint** is divided into "Scopes" for application or scope of validity according to the ***Greenhouse Gas Protocol,*** which were introduced in Sect. 4.4.1.4.
- Determination of the Product Carbon Footprint
  With the **Product Carbon Footprint**, the CO_2 emissions for each product are calculated. Here, the CO_2 emissions of a product or service along the entire value chain must be determined—from the raw materials used to production to delivery to customers. In addition, the usage and possibly also a disposal phase can be included. The aim is an approach that covers the complete life cycle of the product or service. Then

the approach **Cradle to Customer plus End of Life** is used. This includes raw material production, production, logistics, use, and disposal.

The determination of the Corporate or Product Carbon Footprint provides a comprehensive picture of the extent to which a company and in which phases and due to which components a product or service emits particularly much CO_2. At the same time, it can be determined which measures can have the greatest effect on reducing the CO_2 footprint of the company or individual offers. It also becomes clear in which phases avoidance is not possible and measures for compensation should be initiated.

Most companies rely on **external support** when **creating** a **CO_2 balance**. Various service providers help companies to merge the necessary data in cloud-based software systems. The tools used there enable companies to automatically convert activity data into **CO_2 equivalents**. For this purpose, various service providers have built up **databases with emission factors** that facilitate such conversion. The activity data to be converted can refer to electricity consumption, customer visits by sales, and commuting distances of employees. In the latter case, the kilometers driven or flown are converted into CO_2 equivalents. Based on the insights gained here, initial recommendations for reducing emissions can be derived.

Salesforce supports companies worldwide on their way to net-zero emissions through the **capture and analysis of CO_2 emissions.** The use of **precise data analyses** on a central platform helps in this. For **Carbon Accounting,** *Salesforce* provides data set templates from environmental protection organizations, the *IPCC,* and other organizations to correctly determine the CO_2 balance. Integrated instructions and user workflows help optimize data collection and develop a climate action plan. This is supported by analytics functions that assist in calculating energy consumption and developing optimization measures. A dashboard displays the relevant **sustainability KPIs** to better assess the environmental impact of the company. For this, the ***Salesforce Net Zero Cloud*** was developed (see Salesforce, 2023).

SAP offers with the ***SAP Cloud for Sustainable Enterprises*** a solution that follows the triple-bottom-line approach. This allows companies to holistically measure, manage, and optimize their own sustainability performance. The necessary data and processes are integrated for this purpose. These support management in developing sustainable products, services, and business models. Through appropriate measures, the CO_2 footprint can be reduced and the concept of the circular economy can be implemented. The transparency created by the *SAP Cloud for Sustainable Enterprises* helps to comply with global and local regulations. The single-cloud offering can be tailored to individual requirements and expanded with other cloud solutions. Integrated and automated sustainability and ESG reports support sustainable corporate management (see SAP, 2023a).

Through the ***SAP Sustainability Control Tower Device,*** companies can increase their **ESG transparency**. Holistic control is achieved by setting sustainability goals. With the help of this device, progress can be monitored and valuable insights for further optimization can be gained. For this purpose, sustainability-related data are merged to determine the central sustainability KPIs based on this. The setting of goals, the monitoring of achieved progress, and the derivation of actionable insights are supported by automatically generated reports. For this purpose, the necessary data from various applications are transferred to a central **Sustainability Data Warehouse**. This enables the calculation of sustainability metrics for the relevant areas of the value chain—oriented towards established ESG reporting standards. The spectrum of covered areas ranges from purchasing, production, finance to the HR area—divided by country, business unit, department, etc. The data are provided at the detail level required for financial, operational, and sustainable corporate performance. The following metrics are supported (see SAP, 2023a):

- **Stakeholder Capitalism Metrics** of the *World Economic Forum*
- Metrics of the **Global Reporting Initiative** (GRI) with a focus on diversity and corruption prevention
- Metrics of the **Science Based Targets initiative** (SBTi) related to climate change
- Metrics of the **Sustainability Accounting Standards Board** (SASB) on water and climate
- Qualitative metrics of the **Task Force on Climate Related Financial Disclosures** (TCFD) on governance, strategy, and risk management

The procurement and integration of these sustainability and ESG data are supported by the offered technology platform. The key metrics in a **Management Dashboard** give the company management an overview of the holistic sustainability performance.

SAP (2023b) provides further interesting **solutions for the path to sustainable corporate management:**

- ***SAP Product Footprint Management***
  Through this module, companies gain insight into the environmental impacts of their own products over their entire lifecycle. This knowledge enables them to optimize the products and the processes used. Within the *SAP* systems, the company's own business data can be supplemented with environmental factors. This allows the **Product Footprint** to be calculated. The cloud-based solution supports its company-wide scaling.
- ***SAP Product Footprint Management for Clean Operations***
  This solution is primarily aimed at the needs of small and medium-sized production and product-oriented companies. It supports these companies in measuring and managing the ecological footprint of individual products "from cradle to grave" across the entire value chain.
- ***SAP Responsible Design and Production***
  This concept supports the introduction and implementation of the circular economy. To this end, the solution combines the company's own data with data from third-party providers. In addition, global regulations are used to ensure compliant action. Furthermore, any fees and taxes can be determined in order to reduce such regulatory costs

through appropriate measures. In this way, a **circular-oriented product portfolio** can be developed step by step and zero-waste commitments can be met.

- ***SAP Document and Reporting Compliance***
  This module supports the **execution and management of legal and voluntary declarations and regulations.** Standardization and automation of processes ensure consistency between real-time documents and assist in the creation of legally required reports. For this purpose, e-documents and legally required reports can be managed globally. This module ensures ongoing adaptation to changing legal regulations. In addition, a smooth transition from regular, legally required reporting to continuous transaction control with an integrated solution is achieved.

Companies often rely on such solution concepts to meet the complexity of analysis and reporting obligations.

6.2.5 Dashboard for Analyzing Green Brand Management

For the **analysis of green brand management**, a **dashboard with key performance indicators** should be set up. These KPIs reflect the measures presented in Chap. 4. On an aggregated level, the following criteria can be analyzed and commented on, among others:

- **Sustainable revenue**
  Percentage of revenue that a company generates with sustainable products
- **Sustainable product range**
  Percentage of the product range that consists of sustainable products (measured by respective revenue)
- **Sustainableadvertising**
  Percentage of the advertising budget that a company uses for advertising sustainable products
- **Sustainable design**
- Percentage of products/services based on a sustainable design
- **Reduction of emissions and waste**
  Percentage of the reduction of emissions and waste, relative to the respective total amounts (by categories)
- **Procurement of raw materials from certified sustainable sources**
  Percentage of the total raw materials (by classes) that come from certified sources
- **Investments in sustainable offerings**
  Percentage of the total investment budget for the development of new offerings that is allocated to sustainable variants
- **Investments in sustainable processes**

Percentage of the total investment budget for process development that is aimed at sustainable processes

- **Number of employees working in "green areas"work**
  Percentage of the total workforce engaged in sustainable projects and programs
- **Proportion of variable compensation attributable to the achievement of sustainability goals**
  Percentage of the total variable compensation at target values (potential) and actual values (potential realization) that is attributable to sustainability goals

Using such a **marketing-oriented dashboard**, one's own progress can be determined and credibly communicated internally and externally.

> **Reminder Box** The insights of a critical status quo analysis provide the basis for green brand management—not the wishful thinking of (ir)responsible managers.

6.2.6 Balanced Scorecard with Sustainability Module

The Balanced Scorecard with Sustainability Module is an advancement of the classic **Balanced Scorecard** (see Kaplan & Norton, 1997). The Balanced Scorecard is the amalgamation of various perspectives and layers of corporate and/or departmental goals. This leads to a **multidimensional goal framework**—a kind of **control cockpit of the company**. In addition to financial goals, process, customer, and employee-related goals are also defined at the same hierarchical level in the classic Balanced Scorecard. The goals defined there need to be achieved in parallel. This eliminates a hierarchy of these goals.

The use of such a Balanced Scorecard is intended to ensure that multiple strategic perspectives are simultaneously considered in corporate management, which are relevant for the performance evaluation of a company. The attribute "balanced" expresses that a company strives for a **"balanced" goal achievement** in all performance areas. The Balanced Scorecard transforms a classic goal pyramid into a **goal cockpit.** This takes into account the concept of stakeholder management. According to the stakeholder concept, a company only achieves its overall goal when a balanced goal achievement is ensured across all fields defined in the Scorecard.

The concept of a Balanced Scorecard is intended to prevent the optimization of certain areas of the company (e.g., financial indicators) at the expense of other areas (such as the personnel sector or customers). The starting point in the **development of a Balanced Scorecard** is the purpose, vision, or mission of the company (see Chap. 3). From this, goals, indicators, or specifications are derived in the classic Balanced Scorecard—oriented towards the following questions:

- **Financial perspective**
  How should financial successes be documented to the shareholders? Here, financial sizes such as costs, sales, profit, and risks are depicted.
- **Customer perspective**
  How should it be measured to what extent customers could be convinced in the implementation of purpose, vision, and mission? Criteria such as price, acquisition costs for customer acquisition, customer loyalty, recommendation criteria, and customer reviews are used for this purpose.
- **Process perspective**
  How should it be determined in which processes are efficient and effective? All processes of service creation are examined here. The range extends from purchasing to research & development and production to sales and marketing.
- **Employee perspective**
  How should it be determined whether employees are convinced by the company's purpose, vision, and mission? How is it determined whether employees actively and successfully participate in their implementation? The learning perspective is also included here. It is analyzed what the level of know-how is and how high the satisfaction and motivation of the employees are.

A **Balanced Scorecard with a Sustainability module** adds the following module to these perspectives:

- **Sustainability perspective**
- How should it be determined what impact the company's demand for raw materials, production and packaging processes, distribution of offers, their use and consumption, and disposal have on people and the environment? Here, the entirety of the ecological and social effects of the company's actions must be recorded.

This question systematically includes the **ecological and social sustainability aspects** in the Balanced Scorecard. This brings together the three pillars of the sustainability concept Planet, People, and Profit in one tool. The placement of sustainability goals on the same level as all other goals is intended to promote the **company's awareness of sustainability** (see Fig. 6.3).

Alternatively to a fifth perspective in the Balanced Scorecard, **indicators for determining sustainability** can also be integrated into the classic four perspectives. However, greater transparency is achieved when a genuine sustainability perspective is included, as can be seen in Fig. 6.3.

> **Note Box** Every company is called upon to develop a **Balanced Scorecard with a Sustainability module** and use it for overall control of the company. This makes the Balanced Scorecard a **KPI Dashboard,** which also supports the controlling of sustainability processes.

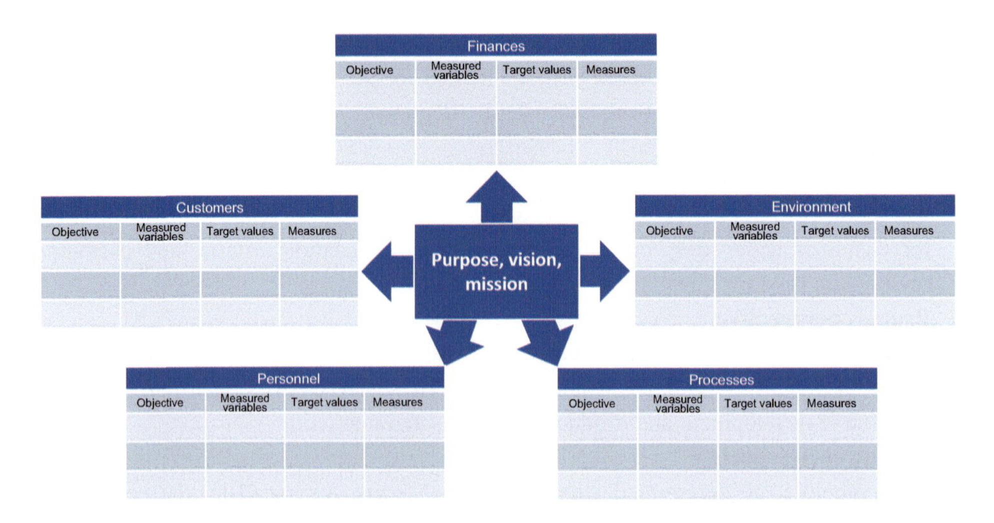

Fig. 6.3 Balanced Scorecard with Sustainability module

In cooperation with the compliance department (often located in the legal area), the **Sustainability compliance** (also **Green Compliance**) must be monitored. This primarily involves compliance with legal obligations. In some cases, companies also define obligations in the form of codes of conduct (see Sect. 3.2), which go beyond legal requirements. Their compliance is also to be checked by a sustainability compliance (see Sect. 8.4).

Questions you should ask yourself

- Is a sustainability monitoring (Green Monitoring) implemented in our company?
- Do we have a sustainability controlling (Green Controlling) to align the impacts of our actions with ecological and social sustainability?
- Do we understand the controllers as business partners who advise and support all company areas to achieve the defined sustainability goals?
- Do we use an eco-audit?
- Have we already developed an eco-balance?
- Do we regularly conduct sustainability analyses of products and product lines?
- What concepts are used to determine the corporate or product carbon footprint?
- Which service providers can support us in this process?
- Do we use a balanced scorecard with a sustainability module?
- Is there a training agenda on the topic of sustainability monitoring and sustainability controlling?
- In whose hands is the topic of sustainability monitoring and sustainability controlling?

References

Balderjahn, I. (2021). *Nachhaltiges Management und Konsumverhalten* (2nd edn.). UVK.

Baumgarth, C., & Binckebanck, L. (2018). CSR-Markenführung im B-to-B-Umfeld—Modell und Fallbeispiele. In C. Baumgarth, C. (Hrsg.), *B-to-B-Markenführung. Grundlagen—Konzepte—Best Practice* (2nd edn., pp. 289–302). Springer Gabler.

Bitkom. (2022). 9 von 10 Unternehmen setzen ihre Klimaziele mit digitalen Technologien um. https://www.bitkom.org/Presse/Presseinformation/Digitalisierung-und-Klimaschutz-in-Wirtschaft-2022. Accessed 27 July 2022.

Colsmann, B. (2016). *Nachhaltigkeitscontrolling. Strategien, Ziele, Umsetzung* (2nd edn.). Springer Gabler.

Kaplan, R. S., & Norton, D. P. (1997). *Balanced Scorecard, Strategien erfolgreich umsetzen.* Schäffer-Poeschel.

Salesforce. (2023). Reduzieren Sie Ihre CO2-Emissionen und fördern Sie die Nachhaltigkeit, indem Sie Umweltdaten nachverfolgen, analysieren und Bericht darüber erstatten. https://www.salesforce.com/de/products/net-zero-cloud/overview/. Accessed 2 Jan 2023.

SAP. (2023a). SAP Cloud for Sustainable Enterprises. https://www.sap.com/germany/products/cloud-for-sustainable-enterprises.html. Accessed 2 Jan 2023.

SAP. (2023b). SAP Product Footprint Management. https://www.sap.com/products/scm/product-footprint-management.html. Accessed 2 Jan 2023.

Schaltegger, S. (2022). Nachhaltigkeitscontrolling. https://www.controlling-wiki.com/de/index.php/Nachhaltigkeitscontrolling. Accessed 1 Aug 2022.

Umweltbundesamt. (2022). Ökobilanz. https://www.umweltbundesamt.de/themen/wirtschaft-konsum/produkte/oekobilanz. Accessed 15 Sept 2022.

7 Dos & Don'ts of Sustainable Corporate Governance

Ich kann, weil ich will, was ich muss.

Immanuel Kant

Abstract

Trust and credibility are central foundations of long-term business success. Therefore, companies should refrain from merely draping their own activities in a "green cloak" without aligning the core business activities towards sustainability. This often not only violates legal requirements, but also—legitimate—customer expectations. Instead, companies should consistently pursue various paths to build trust. Helpful rules for honest communication are presented for this purpose.

7.1 Trust and Credibility of Relevant Game Changers—the Edelman Trust Barometer

Before which challenges companies face today in building trust and credibility, underline the results of the ***Edelman Trust Barometer*** 2022. This determines annually the **state of trust and credibility.** The basis for this are 30-minute online interviews. More than 36,000 people participated in these. In the 28 included countries, at least 1150 people from the population were surveyed. Particularly important findings for sustainable corporate governance are summarized here (cf. Edelman, 2022).

- **Economy is once again the most trusted institution**
  With 61%, the economy is the most trusted institution—even before non-governmental organizations (NGOs) with 59%. Governments follow at a considerable distance

© The Author(s), under exclusive license to Springer Fachmedien Wiesbaden GmbH, part of Springer Nature 2024

R. Kreutzer, *The Path to Sustainable Corporate Management*,
https://doi.org/10.1007/978-3-658-43974-3_7

(52%). The media are only considered trustworthy by half of the respondents (50%). It is interesting that 77% of respondents trust their employer.

- **News sources cannot solve their trust problem**
  None of the major information sources is trusted as a source for general news and information. Trust in search engines is at 59%. This result can be irritating because search engines do not produce their own content. Traditional media follow in trust with 57% before company-owned media with 43%. Social media are trusted—far behind—by only 37%.
- **Concern about fake news is as high as never before**
  The concern about fake news or false information used as a weapon has reached a new high of 76%.
- **Societal fears are increasing**
  Without the trust that the existing institutions provide solutions or societal leadership, fears in society are growing. 85% are concerned about job loss and 75% are worried about climate change.
- **Economy must pay more attention to societal problems**
  The economy is ahead of governments in terms of attributed competence to solve societal problems with 53 percentage points and in terms of ethics with 26 percentage points. At the same time, respondents believe that the economy is not doing enough to address societal problems. These include climate change (52%), economic inequality (49%), retraining of workers (46%), and the provision of trustworthy information (42%).
- **Societal leadership is now a core function**
  60% of employees want their CEO to comment on controversial issues that are close to their hearts. 80% of the population want CEOs to appear personally when they discuss public policy or their company's work for the benefit of society with external stakeholders. In particular, CEOs are expected to influence discussions and policies on jobs and the economy (76%), wage inequality (73%), technology and automation (74%), and global warming and climate change (68%).
- **Economy must take the lead to break the cycle of mistrust**
  On every single issue, people want more commitment from the economy—not less. On climate change, 52% say the economy is not doing enough. Only 9% think that representatives of the economy are exaggerating here. The role and expectations of the economy have never been as clear as they are today, and companies must recognize that their societal role will be permanent.

The high trust that respondents place in business and companies is both recognition and obligation. Companies and their leaders are called upon to use this great potential for trust for the benefit of society and nature.

7.2 Ways to Build Trust

The results of the *Edelman Trust Barometer* underline several aspects. The economy and thus its representatives are highly trusted to find **solutions for the societal and ecological challenges** of our time. At the same time, however, it is expected that companies will also do this.

It has already been worked out how important **trust** is also with regard to sustainable corporate governance. But how can trust be characterized? Trust is the **subjective conviction** and/or the **subjective feeling** of the **honesty of another person or an institution** (e.g., a government, a company). This corresponds to a trust that this counterpart can be helpful and will certainly not cause harm. When trusting statements, this conviction includes that neither deception nor untruth is to be expected in the words. One's own actions can therefore be aligned with it without harm.

Those who trust in this sense believe that the statements and especially the actions of the other are true and sincere. Without such trust, neither a society nor an economy can function.

> **Note-Box** Trust is the glue that enables resilient relationships—with other people, but also with companies and brands.

How can trust be systematically built by companies? The following factors contribute to the **growth of trust** (see Fig. 7.1; for theoretical foundations Morgan & Hunt, 1994):

- **Consistency of words and actions ("Walk the Talk")**
  We all know the so-called **announcement champions** in both private and professional environments: They promise a lot and deliver nothing! Through announcements, a company can attract attention. However, trust is only established when delivery is also made. Consequently, consistency of communication and performance is indispensable for building trust.
  The motto here is: Underpromise—overdeliver!
  Only through a **fact-based communication about the climate protection strategy** of a company can it or its brands gain credibility and trust. The focus should be on the goals and measures that a company is striving for or implementing, even beyond legal obligations. For this purpose, scientific foundations should also be made available (e.g., via the corporate website or a company blog) for particularly interested stakeholders.
- **Protection**
  Indispensable for the emergence of trust is the **assumption of a protective function** by the company and its offerings. Customers will only trust a company if they have the certainty or at least the feeling that a company only offers products and services that do not endanger the health and well-being of the buyers. This includes, for exam-

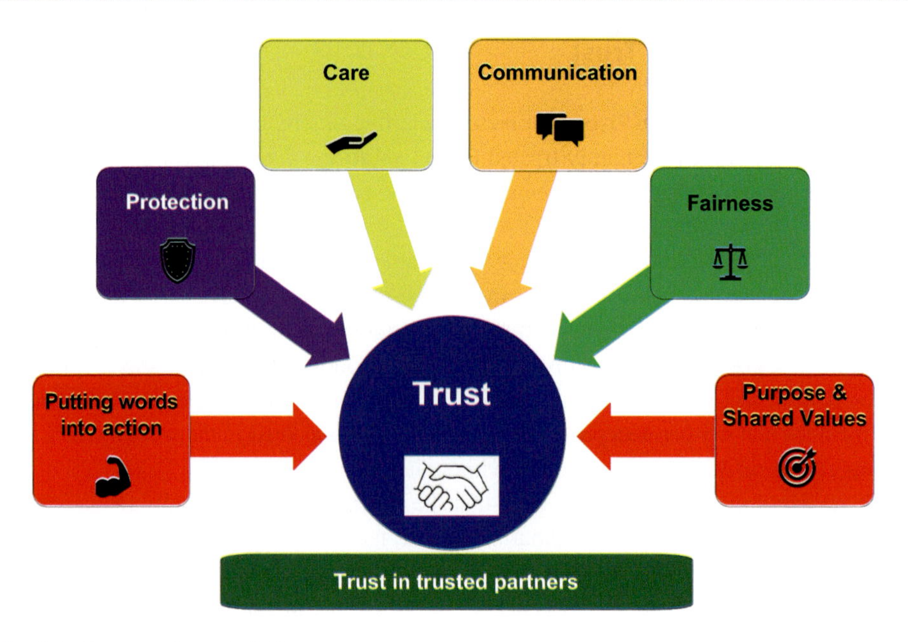

Fig. 7.1 Factors of trust building

ple, that neither food nor drugs nor leisure activities are offered that are highly dangerous for the buyers.

- **Care**
  Whether a company acts with care is often only seen in **crisis situations** or when **errors in products or services** occur. Is the company still approachable and are reliable information and solutions offered?
  Those who now spontaneously think of **airlines** realize why there is increasingly a lack of trust here. For example, in responses of the ***Lufthansa*** to requests for refund of the costs of cancelled flights: "We ask for understanding that the response is currently delayed due to the extraordinarily high volume of incoming mail and that processing will extend into the summer." This response was sent in January 2022! It can also happen here that you have to wait 1.5 hours in the telephone queue (starting at position 18) before you can speak to an employee.
  The company ***Adidas*** also failed to fulfill its duty of care at the beginning of the Corona pandemic. The first **Corona lockdown** was decided on 16.03.2020 and came into effect on 22.03.2020. As early as 28.03.2020, *Adidas* announced that it would stop paying rent for April 2020 due to the closure of the stores. Other chains followed this—bad—example. Just a few days later, the sports equipment manufacturer apologized after a real shitstorm for its step and announced that it would pay the rents as agreed. How caring the company ***Chameleon*** acted during the pandemic was made clear in Sect. 5.4.5.

Those companies that do not wait for legal requirements to align their activities towards higher sustainability also act with care. Those actors who proactively align their value chains towards higher environmental friendliness act with care.

- **Understandable Communication**
  Also indispensable for building trust is **understandable communication.** Anyone who hides their actions behind incomprehensible cascades of words will not gain trust. Then the recipient does not feel valued and may even feel misled.
  What, for example, is a consumer supposed to do with the following sentence on the **packaging of *Bahlsen* cookies**: "Baked with wheat flour from controlled contract cultivation." The term wheat flour is still understandable. But what is being controlled here? The cultivation method? If organic cultivation were present, it would certainly be indicated. And what exactly is "contract cultivation"?
  And what is the meaning of the statement on the packaging of *Griesson* cookies: "For EU applies: Chocolate from EU." Are cocoa beans now being grown in Europe?
- **Fairness**
  Fair treatment—with everyone—also contributes to building trust. It is important that business partners treat each other with dignity and generally accept each other. In this context, fair treatment also refers to suppliers and their working conditions. A "fair" treatment of natural resources also plays a central role in the topic of sustainability.
- **Consistent Sense, Convincing Orientation (Keyword "Purpose")**
  Above all, a consistent sense and a convincingly presented and lived orientation promote the building of trust. If sustainable corporate management is already defined in the purpose and consistently declined throughout the entire value chain based on defined values, solid trust can grow. Especially shared values lead to engagement and trust among all stakeholders. This can then also lead to the development of vision brands (see Sect. 5.4.2).
- **Trust in the "Trust Partners"**
  A company can also build trust by "trusting" its customers to make an important contribution themselves. This can be facilitated by impulses that appeal to the **self-efficacy of the customers**. Spaces must be created for this that enable initial successes. This could be, for example, a simple hint on how to dispose of the components of a package. This is the case with the already cited packaging of the **cookies from *Bahlsen***. It simply says: "This packaging is 100% recyclable. Help out and dispose of the outer box in the waste paper and the foil and the cookie tray in the yellow bag or yellow bin if possible."

Food for Thought

We do not need the all-caring state or the all-caring company. People need institutions that trust people and demand something from them.
As said ***Martin Luther*** convincingly and simply at the same time:
"Efforts make healthy and strong."

In order not to squander the *Edelman Trust Report* documented **trust in the economy** and the associated attributed **competence**, companies must absolutely refrain from a **Purposewashing**. In Purposewashing, the term "Purpose" would be misused for commercial purposes. After all, it is clear to everyone:

> Just with a nice declaration of intent that the company will only do "good" in the world, nothing changes at first.

After all, the intentions must be translated into concrete actions. Anyone who only defines a **sustainable purpose** but changes nothing else, betrays customers and all other stakeholders. At the same time, the trust that is still placed in companies today would be lost in the long run.

Also, a **Greenwashing**—the **pretense of a "green" orientation** of companies and brand—should be avoided. In Greenwashing, companies try to simply put on a "green coat"—without following up their words with corresponding actions. You can also put it bluntly:

> In Greenwashing, a company lies to its stakeholders.

The following **manifestations of Greenwashing** can be distinguished (cf. Grimm & Malschinger, 2021, p. 193 f.):

- **Green Partial Services**
  An offer is positioned as "green" even though only individual characteristics are environmentally friendly. Due to these "green" partial services, the entire offer or even the whole company is supposed to "go green".
- **Sustainability is not proven**
  Here, companies make statements about sustainability in their communication that are not proven. Motto: "Boldly claimed is half proven!"
- **Vague Statements**
  In communication, unclear terms such as "fair", "sustainable", "regional" and "climate neutral" are used, which are partly not legally protected and/or only pretend sustainability without providing evidence.
- **Empty Labels**
  Companies sometimes use self-designed labels that suggest "sustainability" without being backed by concrete measures or third-party audits (cf. Sect. 5.3.5).
- **Irrelevant Statements**
  Here, statements are made that are true, but have no substance in terms of sustainability. The already quoted "controlled contract cultivation" on the *Bahlsen* packaging is mentioned again here.

- **Highlighting as a Lesser Evil**
  The own offer is compared with an even less environmentally friendly offer in order to make the own offer appear in a better light. The already presented *"Blue Angel"* contributes to this (cf. Sect. 5.3.5).
- **Untruths**
  Here, statements are made in communication that are simply false and therefore must be called a lie.

Companies that lose themselves in non-binding platitudes in brand management and do not follow up their words with actions are quickly punished by the (online) community today. More and more users are engaging as **fact-checkers,** who prefer to quickly lift the "green cloaks"—to expose companies and brands. Therefore, greenwashing is a very dangerous approach. Investors need and many customers increasingly demand honesty and transparency with regard to the impact of corporate actions on planet, people, and profit.

> **Note Box** Purposewashing and greenwashing are increasingly easily and quickly exposed by a critical public today.
> Lies have particularly short legs in the age of social media!

Responsible CEOs, CMOs, managing directors, etc., as well as responsible PR and marketing communication, are characterized by the fact that they do not suggest either ecological, social, or economic sustainability in corporate action if the relevant measures are lacking. Dishonest action increasingly has direct consequences today. In the financial sector, executives have already had to vacate their positions because investments were portrayed as more sustainable than they actually were.

Companies are advised to adhere to the following **rules of honest communication** (see Grimm & Malschinger, 2021, p. 195 f.)

- **Materiality**
  Communication should focus on the most important facts—here on content that really says something about the transparency of the achieved sustainability.
- **Completeness**
  All important topics should be addressed—nothing essential should be left unmentioned. This includes, for example, openly reporting about a climate neutrality that can only be achieved gradually.
- **Balance**
  Honest communication includes not only communicating the positive aspects, but also the difficulties and challenges to be overcome.
- **Comparability**
  The information presented to support a statement is presented in such a way that the results achieved are understandable and thus also comparable.

- **Accuracy/Clarity**
  Communication is based on facts that are presented precisely and do not leave much room for interpretation.
- **Timeliness**
  The information presented should be kept as up-to-date as possible.
- **Reliability**
  Statements and promises must be verifiable by third parties. Ideally, supporting evidence is already included in the communication to support statements.

Every company must critically work out how far it has already progressed in green brand management before **communicating its sustainability position**. When companies and brands embark on the "green journey", the question arises as to how the achievement of goals can be checked. After all, the results achieved must be communicated—understandably—internally and externally. Only then can a "green" image be credibly built. Credibility in sustainable action is also achieved by measuring the executives—ideally all employees—against the achievement of sustainability goals and rewarding them (proportionally) accordingly. This is how words really become actions!

Against the background of honest communication, the following examples are particularly critical to examine. ***Coca-Cola*** has long been trying to polish its image by producing a **plastic bottle made of ocean plastic**. In order to cushion the planned switch away from the refillable glass bottle towards the plastic bottle "green", sample bottles were presented that consisted of approximately 25% recycled plastic collected from beaches. A convincing idea. However, it says on the Coca-Cola website (2023a):

> "Currently, no sale in retail is planned."

Here, therefore, an **innovation in plastic recycling** in bottle production is presented—but not used.

Another example is provided by ***Adidas***. The company advertises on its website **"products made with recycled material".** The recycling materials used are made from waste. This requires less water and energy in production. It continues (Adidas, 2023):

> "We see opportunities to make the best for our athletes also better for our planet. That's why we use recycled materials, develop designs from natural raw materials, and produce products that can become something new after their use. We want the whole world to see new possibilities and fight plastic waste—because only then can the next generation stay in the game."

To what extent such recycled materials are actually used in production remains unclear even with more intensive research. In the *Adidas* product description, it only reads:

> "The design is made with recycled material, which is made from textile waste, scraps, and household waste, and thus has a better ecological balance than newly produced materials."

Is *Adidas* just presenting a "green cloak" to allow customers to have a "clear conscience" when selling and thus cash in on the feel-good factor? Or is the shoe actually made largely of recycled material? You don't find out—and become suspicious.

An example of **Greenwashing** is also provided by the company ***AIDA Cruises.*** Here it says (AIDA, 2023):

> "Blue Angel for AIDAnova—AIDAnova has a good angel—the Blue Angel for environmentally friendly ship design. This makes it the first cruise ship in the world to be awarded the federal government's environmental label. Since September 2019, the Blue Angel has been shining on the stern of the ship—as a widely visible sign of our Green-Cruising course. The use of environmentally friendly liquefied natural gas and many other technical innovations convinced the jury."

The background to this award is that the *AIDAnova* is powered by **low-emission liquefied natural gas** (LNG). This merely means that the use of the previously used and particularly environmentally damaging heavy oil is being avoided. However, the use of liquefied natural gas also causes high environmental pollution during extraction (possibly through fracking), during the liquefaction of the gas with high energy input, during transport (under high pressure and with high energy consumption at −161 °C), during processing (from liquid to gaseous) and during consumption due to CO_2 emissions. If one then considers the energy input required to transport a ship with up to 5200 passengers and a crew of 1500 across the world's oceans, it becomes clear: This is far from being **Green Cruising**.

Companies should refrain from this **Virtue Signaling**—a signaling of virtuousness—if it does not correspond to the core of today's actions or if the company is not at least on the way there (cf. on signaling Sect. 5.3.4). Therefore, companies can safely refrain from the **greening of logos** if the logo is the only thing that turns green. An example of this is provided by ***McDonald's*** from 2009.

However, it can also happen that supposedly "green" alternatives were not accepted by customers. ***Coca-Cola*** provides a suitable example for this. A few years ago, the company failed to present a **green Coke.** With ***Coca-Cola Life*****,** the company wanted to create a greener and healthier image—away from the negative sugar image. The use of the plant-based sweetener Stevia was intended to reduce the sugar content. However, the offer was flatly rejected by customers.

The list of such missteps could be continued indefinitely. Here are just a few particularly serious **examples of Greenwashing:**

- **Bamboo dishware**
  Many single-use plastic products (straws, cotton swabs, to-go cups, cutlery) have been banned from production within the EU since 03.07.2021. This is intended to reduce plastic waste. Companies now offer a supposedly sustainable and environmentally friendly solution: bamboo dishware. After all, bamboo is a fast-growing raw material that—and this is the prerequisite—is biodegradable in its pure form.

However, cups and cutlery are rarely made of 100% pure bamboo. Bamboo dishware often also contains melamine-formaldehyde resins. These plastics act as binders and give many products their stability. This plastic content can—depending on the product—make up between 20 and 80% of the "bamboo" product. The plastic components mean that this bamboo dishware is no longer biodegradable (cf. AOK, 2022).
In addition, various consumer centers have already classified such **bamboo-plastic products** as **harmful to health.** The reason for this is that melamine and formaldehyde are not heat-resistant. Therefore, at temperatures above 70 degrees Celsius, health-threatening amounts of these plastics can be released and enter the food. This can happen with hot drinks.
Whether bamboo products contain plastics is usually not recognizable, as there is **no labeling requirement.** In fact, such bamboo-plastic products should not be sold at all. After all, plastic products that come into contact with food require approval for all added additives (cf. AOK, 2022).

- **Backpack made of 100% ocean plastic**
  The German backpack manufacturer *Got Bag* produces backpacks from ocean plastic. The start-up advertised that it would make these backpacks 100% from ocean plastic. However, research has shown that this statement is not true. The company has now admitted that only the fabric of the products is made entirely of ocean plastic, but not other parts, such as the buckles. The communication, it was admitted in 2022, was not precise and is now being corrected (cf. SWR, 2022).
  Today it is communicated that products are made from ***Ocean Impact Plastic***. How high the respective **proportion of *Ocean Plastic*** is, is not specified for each product. Instead, it is generally stated that as part of the ***Got Bag Clean-up Program*** on the north coast of Java, over 2000 participants collect plastic from coastal waters, mangrove forests, and the delta region of the local rivers. After cleaning and sorting, the plastic is pressed into pellets and processed into robust yarn in China and made into backpacks and bags. The *Got Bags* are then shipped to Europe and the USA (cf. Got Bag, 2023).
- ***Coca-Cola's*** **"A World Without Waste" initiative**
  Coca-Cola announced an **ambitious goal** as early as 2018: By 2030, for every bottle or can sold by *Coca-Cola,* one can or bottle will be collected and recycled. It should not matter where these bottles or cans come from. Every packaging should be given more than one life. The **vision of *Coca-Cola*** is:

▶ A world without waste.

To achieve this goal, **packaging** must first be **recyclable.** For this, *Coca-Cola* is working on packaging that is 100% recyclable. In addition, work is being done to use more **recycled material** to **close the loop.** The goal is a recycling rate of 50% for bottles. However, the packaging must actually be **returned to the cycle** and not disposed of indiscriminately, as is still the case in many countries today. To this end, *Coca-Cola* supports **recycling initiatives** in various countries. Through its international commitment, the company has the potential to become active in (almost) all countries of the world to make recycling easier and more accessible for everyone (cf. Coca-Cola, 2023b).

The need for action—also and especially from *Coca-Cola*—is undisputed. After all, **plastic bottles** are responsible for a **quarter of marine pollution.** The **beverage industry** produces **470 billion plastic bottles** every year. However, most bottles are only used once and then thrown away. The solution to the world's plastic problem, therefore, lies not in increasing the recycling rate—including fishing out *Coca-Cola* bottles from the sea—but in avoiding plastic bottles (see Arte, 2022).

The problem can therefore only be tackled at the root by using **fewer single-use plastic bottles.** The best solution, however, is not pursued by *Coca-Cola:* the **return to refillable bottles.** This would lead to an **internalization of external costs** at *Coca-Cola,* which the company has so far passed on to the general public (see Sect. 1.5). These costs would then be reflected in the calculation and also in the profit of *Coca-Cola.*

Why are refillable bottles the best solution? They reduce the **consumption of raw materials** and at the same time the **amount of plastic to be collected worldwide.** At the same time, national **waste systems are also relieved,** if they exist at all. Moreover, **recycling facilities** do not have to be set up worldwide to process the collected plastic. Even today, the amount of plastic to be processed often exceeds the available capacities. This raises the question of whether *Coca-Cola,* by focusing solely on recycling, can really bring about a turnaround in plastic waste—and whether it wants to.

Food for Thought

The slogan **"A world without waste"** sounds fantastic. However, *Coca-Cola* is not striving for a world without waste at all! *Coca-Cola* wants to continue to see the 470 billion plastic bottles produced annually as waste—but within a circular economy. Has anyone at *Coca-Cola* ever calculated the **billions of investments in collection systems and recycling facilities** that would be necessary in all countries of the world to cope with this flood of waste? After all, there is no recycling without collection. Has it already been determined who should actually make these investments so that *Coca-Cola* and Co. can continue to use their plastic single-use bottles?

In view of these points, this initiative also belongs to the **greenwashing activities.** After all, no honest effort by *Coca-Cola* to solve the real problems is visible. The customers of *Coca-Cola* are merely being blinded with a bit of green sand to continue to reach for plastic single-use bottles seemingly without concern.

The website https://greenwash.com/liefert regularly provides interesting reports about companies that can't stop greenwashing.

A comprehensive study by the consumer organization Foodwatch (2022) comes to the conclusion in an **analysis in the food market:**

> "Behind the climate-neutral label is a huge business from which everyone benefits—except for climate protection. Even manufacturers of beef dishes and water in disposable plastic bottles can easily present themselves as climate protectors without saving a gram of CO_2, and label providers like *Climate Partner* make a lot of money from the mediation of CO_2 credits."

In detail, Foodwatch (2022) has gained the following insights:

- None of the providers of climate seals make **specific requirements for CO_2-reduction** to food manufacturers. None of the seal providers analyzed here (including *Climate Partner* and *Myclimate*) define corresponding requirements. Even manufacturers of unecological products can achieve "climate friendliness"—simply by purchasing CO_2 credits.
- Above all, the providers of climate seals earn millions from the **mediation of questionable CO_2-credits**. *Foodwatch* estimates that *Climate Partner* alone earned approximately 1.2 million € in 2022 from the mediation of CO_2 credits from forest projects to eleven customers. According to *Foodwatch*'s research, *Climate Partner* charges about a 77% markup for the mediation of credits from a Peruvian forest project per credit.
- Many companies advertise with **misleading climate claims.** *Aldi* sells "climate-neutral" milk—without knowing how much CO_2 is emitted during production. *Danone* promotes *Volvic* mineral water as "climate-neutral", which must be packaged in single-use plastic bottles and imported hundreds of kilometers from France—mostly by truck. *Granini* only offsets 7% of total emissions—and receives a "CO_2-neutral" label for fruit juice. *Gustavo Gusto* awards itself the label "first climate-neutral frozen pizza manufacturer in Germany", although the pizzas advertised in this way contain climate-intensive animal ingredients such as salami and cheese. *Hipp* markets baby food with beef as "climate-positive", even though the production of beef causes particularly high emissions. Such production can never be "climate-positive".

Based on these results, *Foodwatch* calls for a **ban on misleading climate advertising** for food at the EU level. A clear **regulation of sustainable advertising promises** is demanded here. After all, terms like "CO_2-neutral" or "climate-positive" do not say anything about whether a product is actually climate-friendly. *Foodwatch* recommends that instead of spending money on misleading climate seals, the respective manufacturers should engage more in climate protection measures along their own supply chain (see Foodwatch, 2022).

> **Food for Thought**
> Companies should focus more on the actual reduction of greenhouse gases in their value chains, rather than engaging in greenwashing by purchasing climate certificates.

Questions you should ask yourself

- What consequences can we draw from the results of the *Edelman Trust Study*?
- Who in our company has already dealt with this study?
- How consistently do we strive to build trust within the company?
- Do we achieve a consistency of words and actions—day by day?
- Do we have a fact-based communication of our company's climate protection strategy?
- Do we take on a protective role for our customers in certain areas?
- Where do our stakeholders, but especially our customers, experience that we take care of them?
- Do we rely on understandable communication—not only, but especially in terms of sustainability?
- How fair is our interaction with our stakeholders—especially our customers and employees?
- Do we have a purpose that truly guides our actions?
- Do we also trust our "trust partners" and expect something from them and demand something from them?
- Does our communication follow the "rules of honest communication"?
- Where do we run the risk of also engaging in purposewashing or greenwashing?
- Do we use seals and labels to document our sustainability in a trustworthy and honest way?
- Where does the responsibility for building trust lie in our company?
- Do we trust those who are supposed to build trust?

References

Adidas. (2023). Produkte mit recyceltem Material. https://www.adidas.de/mit_recycelten_materialien. Accessed 2 Febr 2023.

AIDA. (2023). Green Cruising. AIDAnova: Kreuzfahrt mit dem Blauen Engel. https://aida.de/kreuzfahrt/schiffe/aidanova/umwelt. Accessed 1 Jan 2023.

AOK. (2022). Bambusgeschirr: Nachhaltig oder gesundheitsschädlich? https://www.aok.de/pk/magazin/nachhaltigkeit/muell-vermeiden/bambusgeschirr-nachhaltig-oder-gesundheitsschaedlich/. Accessed 19 Sept 2022.

Arte. (2022). Coca-Cola und das Plastikproblem. https://www.arte.tv/de/videos/098824-000-A/coca-cola-und-das-plastikproblem/. Accessed 19 Sept 2022.

Coca-Cola. (2023a). Getränkeflasche mit Meeresplastik. https://www.coca-cola-deutschland.de/verantwortung/verpackungen/musterflasche-mit-meeresplastik-cr. Accessed 2 Jan 2023.

Coca-Cola. (2023b). Warum eine Welt ohne Müll möglich ist. https://www.coca-cola-deutschland.de/verantwortung/handeln-verandern/unser-ziel-eine-welt-ohne-muell. Accessed 2 Jan 2023.

Edelman. (2022). 2022 Edelman Trust Barometer. https://www.edelman.com/trust/2022-trust-barometer. Accessed 26 July 2022.

Foodwatch. (2022). foodwatch fordert Verbot irreführender Klima-Werbung. https://www.foodwatch.org/de/pressemitteilungen/2022/foodwatch-fordert-verbot-irrefuehrender-klima-werbung/. Accessed 6 Dec 2022.

Got Bag. (2023). Shop. https://got-bag.com/products/rolltop#BLACK. Accessed 1 Febr 2023.

Grimm, A., & Malschinger, A. (2021). *Green Marketing 4.0. Ein Marketing-Guide für Green Davids und Greening Goliaths.* Springer Gabler.

Morgan, R. M., & Hunt, S. D. (1994). The commitment-trust theory of relationship marketing. *Journal of Marketing, 58*(3). https://doi.org/10.2307/1252308.

SWR. (2022). Nach Greenwashing-Vorwurf. Mainzer Rucksack-Hersteller Got Bag verspricht mehr Transparenz. https://www.swr.de/swraktuell/rheinland-pfalz/mainz/greenwashing-vorwurf-gegen-mainzer-meeresplastik-rucksack-hersteller-got-bag-100.html. Accessed 19 Sept 2022.

Development of a "Green Journey" for Your Own Company

8

If it weren't for the last minute, nothing would ever get done.

Mark Twain

Abstract

Every company must embark on a "Green Journey" today. This is required not only by increasing legal requirements, but also by an increasing number of stakeholders. Storytelling plays a significant role in conveying this journey. Which stakeholders to involve in the "Green Journey" can be determined using the Stakeholder-Onion-Model. A stage model for the development and expansion of sustainable corporate management identifies relevant areas of action.

8.1 Relevance of a "Green Journey"

If there is still uncertainty in companies about the **significance of a "Green Journey",** a look at a recent study by Serviceplan (2021, p. 31) can help. For this, 288 Chief Marketing Officers (CMOs) from Germany, Austria, and Switzerland were surveyed. The **marketing managers** cover all sectors and represent both small and large companies. The results can be summarized with the following headlines:

- **Sustainability** is becoming a long-term mega-topic.
- The **brand** is coming back into focus, after e-commerce dominated the agenda last year.
- CMOs expect **creativity** and **digital expertise** as top skills from agencies.

© The Author(s), under exclusive license to Springer Fachmedien Wiesbaden GmbH, part of Springer Nature 2024

R. Kreutzer, *The Path to Sustainable Corporate Management*,
https://doi.org/10.1007/978-3-658-43974-3_8

- The CMO sees himself as a **business driver,** who co-designs new business models and becomes an important driver of the transformation agenda.

The study first asked: "Which three topics will dominate marketing in the coming year?" In response to this open question, which was answered by 234 CMOs, **sustainability and climate change** landed in first place with 39% of the mentions. With a significant gap, **digitalization** follows with 29% (see Fig. 8.1; Serviceplan, 2021, p. 6).

Another question was: "How important are the following marketing topics in 2022?" The answers to this closed question again show that **sustainability** has become the top topic for CMOs. Its importance has even increased from 2021 to 2022. 90% of respondents expect an increase in importance (will "strongly" or "rather increase"). **Digital transformation** is now "only" in third place after **data-driven marketing** was ranked first last year (see Fig. 8.2; Serviceplan, 2021, p. 6).

Regarding the results from Fig. 8.2, it can be assumed that **data-driven marketing** and the **digital transformation** have already found a permanent place on the CMOs' agendas and that intensive work is being done on them. In contrast, there is still a great need to catch up with **sustainability communication**—not least due to the visible weather phenomena caused by climate change.

Another question of the *Serviceplan* study addressed the most important tasks of a CMO. The results show that today, **brand building and brand strengthening** again dominate the CMO agenda (30%). In second place is **digitalization/digital transforma-**

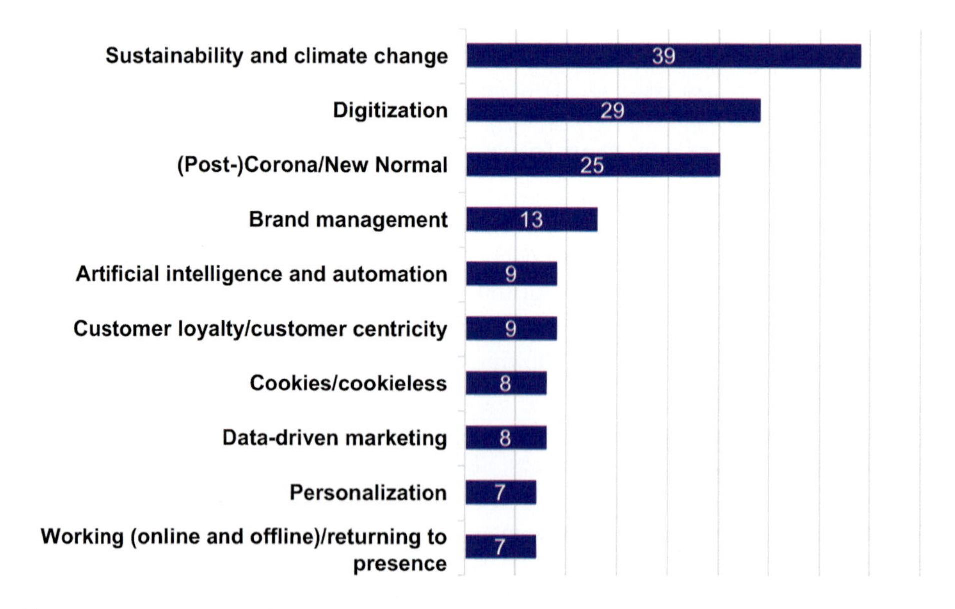

Fig. 8.1 Topics from the CMO's perspective that will determine marketing in the coming years—in %. (Data source: Serviceplan, 2021, p. 6)

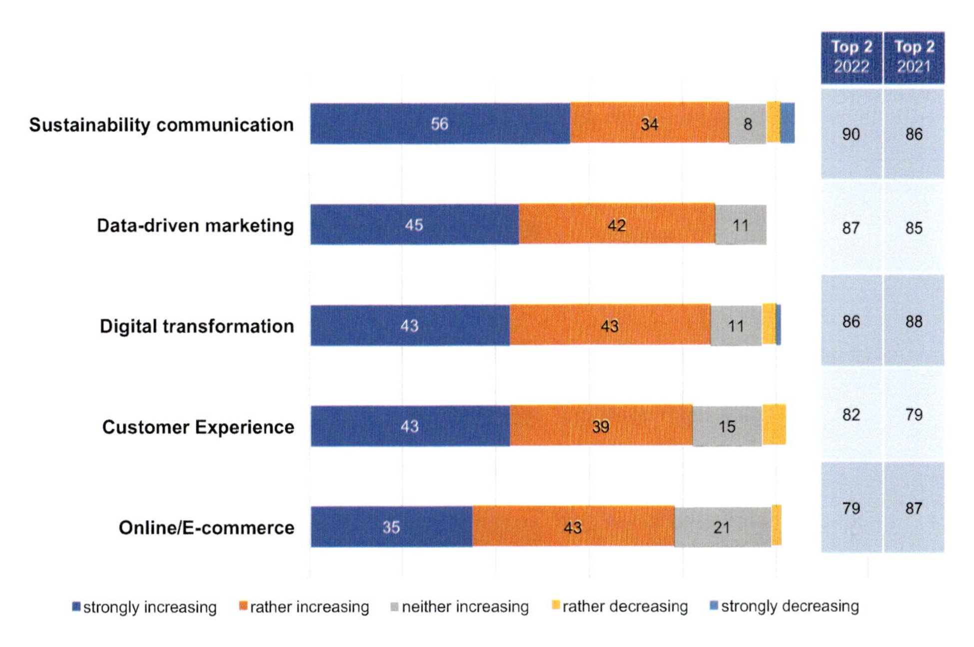

Fig. 8.2 Importance of selected marketing topics in %—2022. (Data source: Serviceplan, 2021, p. 6)

tion (27%). After **employee management** (22%), **sustainability** is in fourth place (15%) among the most important tasks for CMOs—far ahead of **marketing efficiency/ROI** and **innovation/co-designing new business models** (see Fig. 8.3; Serviceplan, 2021, p. 11).

> **Food for Thought** Those who do not take sustainability seriously enough in their business model may only be able to inspire a few **customers** in a few years' time.
>
> Perhaps the **investors** will even lose interest in investing in unsustainable business models beforehand. Then not only is the money from the customer side missing, but also from investors.

The following will show what responsible managers can do to advance the topic of sustainability in their companies. The topic of storytelling is deliberately placed at the forefront of the further explanations—after all, even the best strategies will not achieve success if they do not reach not only the minds but also and above all the hearts of people.

Fig. 8.3 The most important tasks of a CMO—in %. (Data source: Serviceplan, 2021, p. 11)

8.2 The Importance of Storytelling

An indispensable prerequisite for the establishment of sustainable corporate management is a convincing **narrative of storytelling**. With a **narrative**, a **meaningful story** is meant, which simultaneously conveys certain values and emotions that go hand in hand with the topic of sustainability. The narrative is intended to influence the way listeners perceive the world—here primarily the corporate context (cf. in depth Etzold, 2018, pp. 101–165; Friedmann, 2021; Kreutzer, 2021, pp. 70–77).

> **Memory Box** If we want to "touch" people with our stories, we must not stop at **Facts & Figures**. These are indispensable as a **foundation** for a story. However, Facts & Figures must be used in **pictorial narratives** that start films in the listeners and trigger positive emotions. This will rarely be achieved with Excel sheets projected onto the wall. Consequently, **narratives** form the **core of storytelling** (cf. also Storr, 2019; Pyczak, 2021).
>
> The following applies (Shank, 2021):
>
> **"Humans are not ideally set up to understand logic. They are ideally set up to understand stories."**

What can, what should such narratives achieve? Narratives can first of all give **courage** to believe in one's own abilities and possibilities (motto: **The sky is the limit!**). An important aspect in this context is the so-called **self-efficacy.** This refers to a person's expectation of being able to successfully carry out desired actions on the basis of their own abilities and competencies. People who believe that they can make a difference and achieve even high goals have a high **self-efficacy expectation.**

The **self-efficacy expectation** should be systematically promoted through narratives. The person and the organization in which he or she works are then no longer a victim of their environment, but the **result of their own actions.** Not circumstances, luck, chance, and fate determine the course of events, but one's own powerful action. In addition, narratives—well told—can also provide a certain **orientation in stormy times** to tackle the big challenges with vigor.

Storytelling therefore means nothing other than telling stories. It is about conveying explicit and implicit knowledge through certain leitmotifs, symbols, metaphors, etc. Both forms of knowledge are indispensable for a change process. **Explicit knowledge** can be easily conveyed through language. This is also referred to as methodical or specific knowledge, intellectual knowledge or **Embrained Knowledge**. This is about the facts already mentioned. Words and numbers are sufficient for their transmission. Due to its formalization, explicit knowledge can be easily stored on various media, processed in various ways, and passed on.

This possibility of simple storage, processing, and sharing is not given with implicit knowledge. **Implicit knowledge**—also tacit knowledge, experiential knowledge or **Embodied Knowledge**—arises from routines and a skill gained through diverse experiences. It is reflected in memories, beliefs, and values. This implicit knowledge refers to the abilities of a person or even a company. Often, the carriers of this implicit knowledge are not aware of their own knowledge treasure. Implicit knowledge can also be difficult to express linguistically. As a result, implicit knowledge eludes formal linguistic expression. Implicit knowledge is usually considered action-bound. This means that it primarily becomes visible and tangible in action. Simple examples of implicit knowledge are walking, cycling, and swimming. It is extremely easy when you can do it—but difficult to explain to someone who wants to learn it. The rule here is: "Being able to do it, without being able to say exactly how!"

> **Reminder Box** A special **challenge of storytelling** is not only to make explicit knowledge easily understandable and memorable. Rather, it is precisely the implicit knowledge that needs to be conveyed through appropriate stories—and ideally anchored through daily action and offered as a model for imitation learning.

Not least against this background, **top executives as role models** play such a significant role in the goal of sustainable corporate management. The necessary words can be conveyed through **townhall meetings** in which the executives address all employees. In addition, stories about the central projects and the results already achieved belong as a **standard item on the agenda** of all relevant meetings. And then it is necessary to act accordingly.

> **Reminder Box** Listeners retain about 60% of a good story—usually only 5% of numbers and statistics.

Why do these narratives have such great importance in today's times? We are all in a phase of **information overload** (also **Information Overload):** We receive dozens of newsletters with more or less relevant content every day. *Google* searches often yield not just thousands, but even hundreds of thousands of hits. Social media fire hundreds or thousands of posts, status updates, and stories into our field of vision every day that we "must not" miss (keyword FOMO effect, see Sect. 1.3.2).

Every day we could participate in several webinars and listen to podcasts for 90 h a day. *Amazon Prime, Maxdome, Netflix, Spotify* & Co. offer us additionally inexhaustible amounts of music and videos, all of which want to be consumed by us. After all, we often don't have to pay anything more for more intensive use due to a flat rate. Then there are also the traditional media such as newspapers and magazines to inform us comprehensively and seriously through competently prepared content. Can we do that? No.

Nevertheless, we as senders want to penetrate the recipient with our information. This is where storytelling comes into play again. We need to tell a story vividly, clearly, and understandably to gain the **attention** and **concentration** of other people. Moreover, what we say should remain in memory as long as possible to influence thinking and action. The story used can be based on facts and also contain fictional elements. It is exciting that even complex knowledge—as has been proven over millennia—can be passed on through narratives.

For this purpose, storytelling can create a **meta-level** to convey a **superordinate message**. This should indeed evoke stronger emotions such as fear, anger, joy, desire, etc. in the listeners. After all, such emotions contribute to better memorization of the content. Those who are particularly courageous as storytellers can also incorporate fables and anecdotes into their story and let the quintessence of a superordinate wisdom or a punchy punchline flow in. This way, you stay in memory. A **convincing performance** in the presentation of the story is part of it.

Great visions and captivating stories that captivate our attention usually start with "Just suppose ..." or "Imagine ...". This creates an exciting **context** for the listeners. This way, we can create powerful images in the mind. An exciting story also needs turning points—and heroes!

Executives and employees can become heroes of the Green Journey, successfully mastering turning points. For this, it is important that the stories **link causes and effects**—to make clear what is possible if... This significantly increases the **belief in self-efficacy**. It can also be reported which requirements of the supply chain law have already been implemented. In addition, information is provided on what successes have already been achieved in entering the circular economy, what is currently being worked on, and what is coming next. Ideally, reports from neutral media can also be presented, so that the outside world also notices that this company and its team are on the right track.

Through the concepts described in the previous chapters, every company can gain the necessary explicit knowledge to informatively underpin **stories about sustainable corporate management**.

> **Food for Thought** One thing to avoid in storytelling is:
> Dishonesty and lack of sincerity. That would be a disregard for the listeners!
> Therefore, it is also part and parcel of storytelling to clearly point out the expected frictions and painful cuts that can come with sustainable corporate management.
> Here, **clarity** is required!
> After all, everyone on the team expects these to occur.

8.3 Use of the Stakeholder-Onion-Model

Who should be involved in the **process of the Green Journey** when and how intensively? To be able to identify relevant stakeholders with their different goals, expectations, but also with their respective power potential early on, the **Stakeholder-Onion-Model** should be used (see Fig. 8.4).

With the **Stakeholder-Onion-Model** shown in Fig. 8.4, the relationships of stakeholders to a project goal—here sustainable corporate management—can be visualized. To achieve the highest possible **transparency**, further information should be added to describe the respective relationships in the Stakeholder-Onion-Model. For this purpose, the different layers of the Onion-Model are analyzed from the inside out. These layers have given the model its name (Onion for "onion").

Depending on the task, four or five layers can be worked out and filled with corresponding stakeholders. It is important that the relationships of the stakeholders to the project are first analyzed based on the **intensity of their respective involvement**. After all, it is often interest groups with little power and influence (such as employees) who are most comprehensively challenged and affected in the course of a project.

The following are the steps for the **implementation of the Onion Model**:

1. **Step**
   First, a small circle is drawn at the bottom of a sheet. This visualizes the product, service, or topic to be developed. Here, the focus is on **sustainable transformation**.
2. **Step**
   A second circle is placed around this first circle. This is where the interest groups are located, which are directly affected by the products, services, processes to be changed, or the topic of "sustainability" in general. Either the stakeholders to be located here are directly involved in the **process of development** or in the **process**

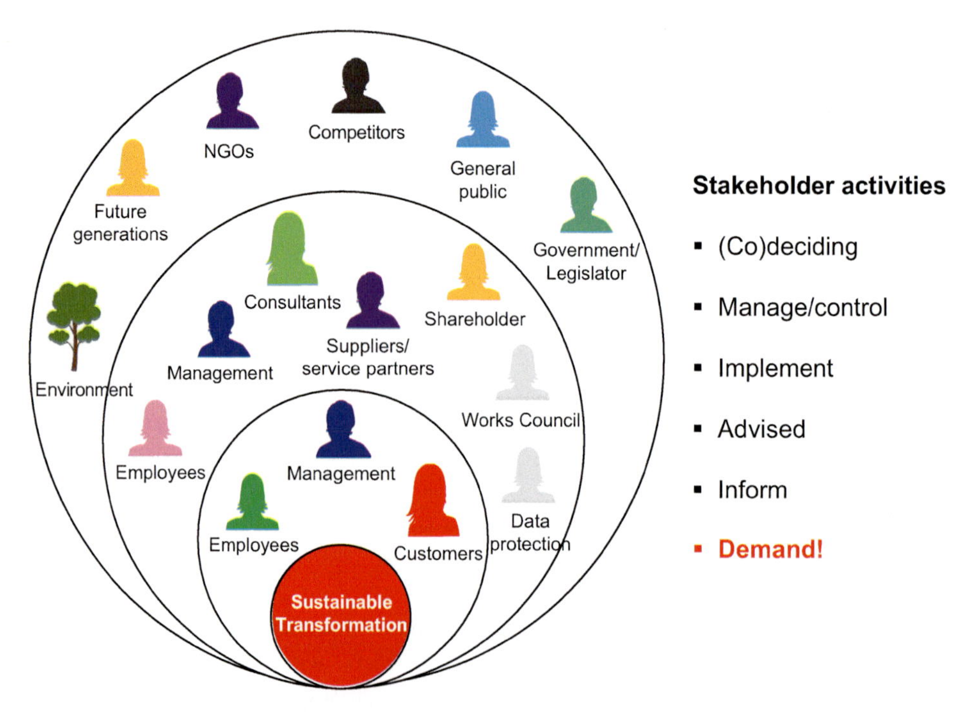

Fig. 8.4 Stakeholder-Onion-Model for the identification of relevant stakeholders in sustainable corporate management

of implementation. In the process of sustainable transformation, these are initially the **employees** and **executives** internally, and primarily the **prospects and customers** externally.

3. **Step**
   In a third circle, the stakeholders who are involved in the development or implementation are positioned. These can initially be **additional employees and executives**. Although they are not directly involved in the process of transformation, they are nevertheless influenced by it in their daily work. Also located here are **service partners** (e.g., consultants, IT service providers) and **suppliers** (e.g., for recyclate, renewable energy, ecological materials), whose input is indispensable for the transformation process. In this circle, the **shareholders** are also to be located. They determine whether a sustainable transformation can succeed by providing or withholding financial resources.
4. **Step**
   In the fourth circle, the stakeholders are to be positioned who are outside the company but still very important for the success of the process. These include, among others, **governments** and the **legislators**. Also, the **general public** and the **competitors** are to be included in this circle. Also, the **future generations** and relevant **NGOs** are to

be integrated here. Here, the **environment** could, should, or must be located as an independent stakeholder!

5. **Step**
   The Onion Model already contains a multitude of information after these four steps. In the 5th step, **relationship arrows** can be added to this model. The direction of the arrow indicates the direction of a relationship or whether it is possibly mutual. The thickness of the arrows can represent the **intensity of the relationship** and the color the **quality of the relationship**. Following a traffic light system, strained relationships can be marked with red, neutral, purely work-related relationships with yellow, and supportive relationships with green.

In a further level of information, the different **roles of the stakeholders** can be highlighted. They can actively manage or control a process. Other roles are responsible for implementation or have an advisory function. Another group of stakeholders is only to be informed about upcoming changes. Shareholders and possibly also the works council must first approve serious measures. Finally, the environment can implicitly demand changes because resources are not or no longer sufficiently available—or because natural disasters are increasing and leading to an internalization of external costs (see Fig. 1.37).

> **Note Box** The additional **stakeholder "Environment"** makes actual demands that must be met to ensure the long-term survival of humanity or the blue planet.

Another interesting level of information can be added by marking the **attitude of the stakeholders towards the project** with different colors. Following the already introduced traffic light method, opponents can be marked with red, neutrals with yellow, and supporters or promoters with green. In addition, the **importance of the stakeholders** can be represented. Particularly important stakeholders are marked in bold. These additional levels of information were not integrated into Fig. 8.4 to avoid overloading this representation here.

The greatest **importance of this stakeholder analysis** is given to the information that is gained in the course of creating the Onion Model. The Onion Model "forces" the users to broaden their perspective and not only consider the "usual suspects" in the stakeholder analysis. This ensures that no important stakeholder perspective is neglected.

If such a comprehensive inventory of the expectations and fears of the various stakeholders is omitted, this can later "trip up" the responsible managers. Then a slowdown or even a stop of the sustainable transformation process is possible. This should be avoided by proactive action.

> **Food for Thought** We should consider the environment as an independent stakeholder in the stakeholder online model.

8.4 Staged Model for the Development and Expansion of Sustainable Corporate Management

To develop or expand sustainable corporate management, you can follow the **staged model of sustainable corporate management**.

- **Implementation of a sustainability team**
  - A decisive and indispensable step towards more sustainability throughout the company is the **installation of a sustainability team**. From the start, this team should be equipped with personnel, financial, spatial, and technical resources. The first task of this team is to determine the **status quo of their own company** in the Green Journey. A look at the different **stages of a Green Journey** in Fig. 8.5 can help with this.
  - The company has the furthest way to go if it falls into the **"not started"** group. Here, the topic of sustainability is (still) not on the company's agenda. In this case, it is essential to establish a high-performance sustainability team. This team's main task is to check whether a disregard for stakeholder expectations in terms of sustainability could lead to a loss of competitiveness in the short or medium term. In addition, many other analyses described so far need to be carried out to identify any possible mandatory (legal) need for action.
  - If a company falls into the **"experimenter"** group, then initial impulses for getting involved in sustainability topics have already been set. Often in this group, only isolated proofs of concept have been carried out, without deriving follow-up measures from the results achieved. Here, the company also remains very much in the status quo. The share of sustainable value creation remains minimal. The task of a sustainability team here is primarily to bring together the various threads and develop a convincing concept for sustainable corporate management.

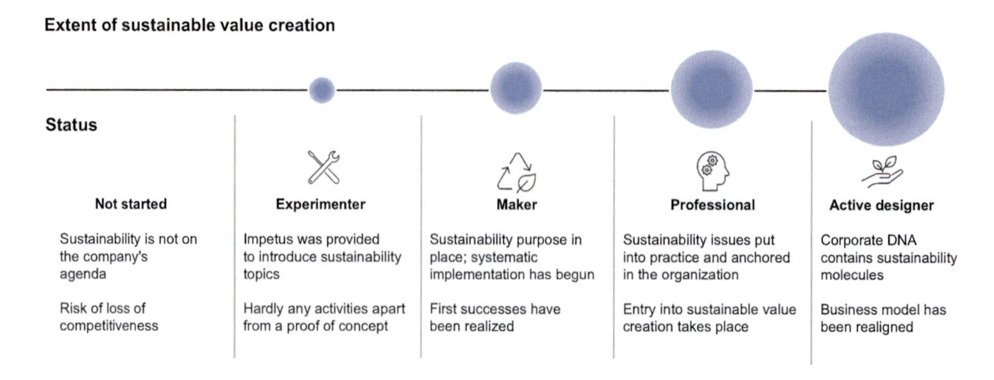

Fig. 8.5 Stages of a Green Journey

- If the company is already in the segment of the **"doers",** then at least a – perhaps still vague – sustainability purpose or a corresponding vision has been developed. A more systematic implementation has been started. Initial successes of sustainable value creation have already been realized. Often, such companies already have precursors of a sustainability team, without having found a more solid organizational structure yet. This structure – equipped with resources – needs to be created now at the latest. The sustainability team here already finds fertile ground on which further steps defined in this work can be initiated.
- If a company has already reached the **professional** status, then a sustainability team has certainly already been installed – perhaps without yet calling it that. This can now be made up for, to set clear signs of sustainable corporate management both internally and externally. After all, at this stage, several sustainability topics have already been implemented and anchored organizationally. As a result, the share of sustainable value creation has already been significantly increased. The further steps towards sustainable corporate management can now be consistently driven forward by the sustainability team.
- If the company already belongs to the **active designers,** then the question of a sustainability team has certainly already been resolved long ago. The company's DNA already contains many sustainability molecules to permanently align the company towards sustainable corporate management. For this purpose, the business model has been realigned in some areas or completely. This path must be consistently continued.

If a company is serious about sustainable corporate management, then a **sustainability sponsor** should be established at the highest hierarchy level right from the start of the Green Journey. This person acts as a sparring partner on the journey towards sustainability. On the other hand, the sustainability sponsor is also needed to support the foreseeable need to overcome organizational inertia and resistance to necessary changes in the company.

> **Note Box** A sustainability team that does not receive back-up from a sustainability sponsor must live with a high risk of failure.

To expand the sphere of influence of the sustainability team, so-called **sustainability ambassadors** can be deployed. These can be employees from various business areas and departments who are specifically trained for their role as sustainability ambassadors. By **establishing sustainability ambassadors on site,** their knowledge of sustainable solutions can directly influence daily business operations.

- **Appointment of a Chief Sustainability Officer**
  A significant step further is the appointment of a Chief Sustainability Officer (CSuO). A Chief Sustainability Officer is positioned at the highest level of hierarchy. Their

explicit task is to align all of the company's activities towards sustainability. So far, Chief Sustainability Officers are primarily found in the financial services sector. According to a study by Deloitte (2022), more than three quarters of the surveyed financial service providers already have a dedicated CSuO or an equivalent position.
The **core tasks of a Chief Sustainability Officer** include the analysis of (legal) requirements regarding sustainable corporate governance and the derivation of strategic consequences for their own company. Crucial for the implementation of a sustainability strategy in the company is the personnel and financial resources to initiate and monitor transformation processes. This also includes meeting the necessary reporting requirements. A further task is to significantly contribute to the further development or reorientation of the business model. A sustainability team would report directly to the Chief Sustainability Officer.
The position of a Chief Sustainability Officer can be temporary. Once the desired sustainable transformation has been achieved and all offers and processes are ideally "climate neutral", the Chief Sustainability Officer can retire. Until then, a lot of water will flow down the Rhine! For the foreseeable future, however, the importance of the CSuO in most companies will continue to grow significantly. After all, more and more companies are realizing that there is no alternative to a sustainable orientation of corporate governance.

- **Clarification of legal framework conditions**
  A first task of the sustainability team or the Chief Sustainability Officer is to check which legal requirements already exist for their own company and their own industry and which further ones are foreseeable. It is also necessary to check which obligations their own business partners have to fulfill. After all, they have to access information from the companies they are networked with in order to fulfill their own obligations. At this stage, support from their own **legal department** and/or **specialized service providers** is often indispensable.
- **Conducting an (initial) impact analysis (Impact Analysis)**
  Depending on the industry, the size of their own company as well as the integration into networks of other companies, the legal requirements presented in Chap. 2 have different relevance for each company. The **impact analysis** shown in Fig. 8.6 helps to set the right priorities for their own measures. It should be noted again that legal requirements for their own business partners can also affect their own company. This includes comprehensive obligations to provide information and evidence that large companies need from their business partners in order to fulfill their own legal requirements. Based on such an analysis, the relevant **priorities** can be derived.

> **Note Box** If a company does not fall into the defined classes for the validity of various legal obligations (such as in terms of company size or industry), such requirements can still affect their own company if direct business partners have to fulfill corresponding obligations. This is not only the case with the supply chain law.

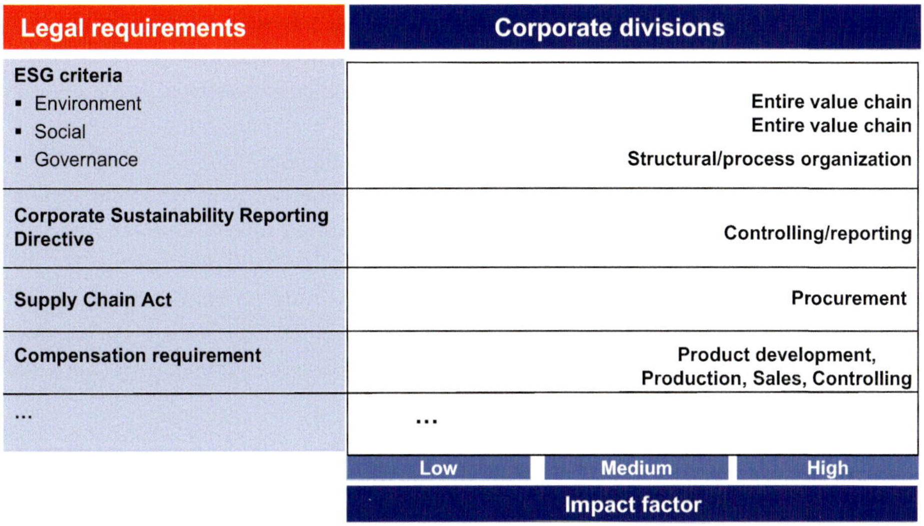

Legal requirements	Corporate divisions
ESG criteria ▪ Environment ▪ Social ▪ Governance	**Entire value chain** **Entire value chain** **Structural/process organization**
Corporate Sustainability Reporting Directive	**Controlling/reporting**
Supply Chain Act	**Procurement**
Compensation requirement	**Product development, Production, Sales, Controlling**
...	...
	Low – Medium – High Impact factor

Fig. 8.6 Impact analysis

- **Derivation of quantified objectives for the defined priorities**
  For the processing of the defined priorities, qualified **project management** is required. This begins with **quantified objectives,** to describe the relevant scope of action to the company and its managers. Only such objectives can unfold the necessary **control and motivation function**. Only quantified objectives can be checked at the end of reporting periods to fulfill the **control function** of objectives. In addition, only then can the defined **reporting obligations** be fulfilled. For this purpose, the concepts presented in Sect. 6.2 can be used.
  The quantitative objectives must be worked out by the sustainability team or the Chief Sustainable Officer together with the relevant company departments. Which areas need to be particularly involved is determined by the presented analysis of those affected.
- **Development of action packages**
  The biggest task now is to develop relevant **measures**. For this purpose, a milestones plan needs to be developed. This should include binding reporting dates to top management and quarterly result checks. In addition, continuous analyses of those affected should be initiated to quickly pick up on new developments. The various company departments should be extensively involved in all these work steps.
- Motto: **Turn those affected into participants!**
  A **sustainability strategy** is most successfully implemented when methods for emission reduction are gradually incorporated into the strategies. It is crucial that, in addition to the necessary data and technologies, the necessary budgets and the necessary know-how are also provided.

In addition, it should be checked which **cooperation partners** need to be involved in order to achieve certain sustainability goals more quickly, comprehensively and/or cost-effectively. Various potential partners have already been presented in the relevant chapters.

- **Complementary training for all employees and managers**
  Parallel to the development and implementation of measures to increase the sustainability of corporate activities, internal or external **training measures** should be initiated. Only in this way can the entire workforce be taken on the Green Journey.
  Such trainings are offered by various institutions. To build up the necessary knowledge for CO_2 reduction in your own company, companies like *Climate Partner* also offer online events on climate protection (keyword *"Climate Partner Academy";* see Climate Partner, 2023).
- **Use of storytelling**
  An ecological transformation will not succeed without convincing storytelling—internally and externally. This requires comprehensive coordination of the activities of the sustainability team or the Chief Sustainable Officer with the **communications department.** In addition, the **company management** needs to be extensively involved in this **communication process** in order for the ecological transformation to succeed (see Sect. 8.1).
- **Annual sustainability review**
  Once a year, a large sustainability review can be conducted to—ideally—celebrate the already achieved goals. In addition, possible **target deviations** need to be analyzed in order to initiate countermeasures. At the same time, further goals can already be formulated to give the Green Journey a new direction.

Where do companies see the biggest challenges associated with the ever more comprehensive **compliance**? An international study by BCG (2022, pp. 4, 19–21) provides exciting answers to this question (see Fig. 8.7). For this purpose, 250 compliance professionals from companies in various industries around the world were asked to name the most important problems and challenges.

In summary, it is noted that compliance requirements are becoming increasingly comprehensive. A compliance team today must bring more diverse skills than before. Therefore, the **search for qualified compliance staff** leads the ranking of challenges ("recruitment of talents" 68%). After all, the success of a company's compliance management depends largely on the recruitment of new and the further qualification of existing employees. If the potential for compliance as a competitive advantage is to be used, highly qualified employees must be recruited for this. In larger companies, it may even be necessary to appoint a **Chief Sustainability Officer**.

Another challenge is seen by 43% in the **organizational complexity**. To meet compliance requirements—also and especially in terms of sustainability—the various areas of the company must work together very intensively. Information and process silos must be overcome for this. These processes are to be supported by a powerful IT infrastructure (38%), which is not yet installed in all companies.

Companies also see themselves challenged by the **costs of complying with regulations** (37%) and by the **increasing regulatory control** (36%). This leads to **staff capacities for implementation** (33%) becoming a bottleneck. Every third company (32%) still sees itself confronted with the need to develop a **compliance strategy**.

Further challenges are seen in the further development of corporate culture, the budget, and the use of internal and external knowledge. The integration of business goals and digitization goals is also not yet solved in some cases. A quarter of those surveyed also still see the "sense of urgency on the part of management" as a bottleneck.

> **Note Box** The **Green Journey**—once started—will run for the vast majority of companies for many years. Products and services often have to be redesigned, processes and the necessary production facilities and logistics concepts have to be revised. At the same time, suppliers must be held more accountable if the company itself has committed to sustainability goals. And customers and their behavior also need to be further developed in order to achieve important sustainability goals.
>
> Finally, it should be noted: Sustainable corporate management does not stop at the company's exit gate. Sustainable corporate management must think and act holistically in order not to further endanger our beautiful blue planet.

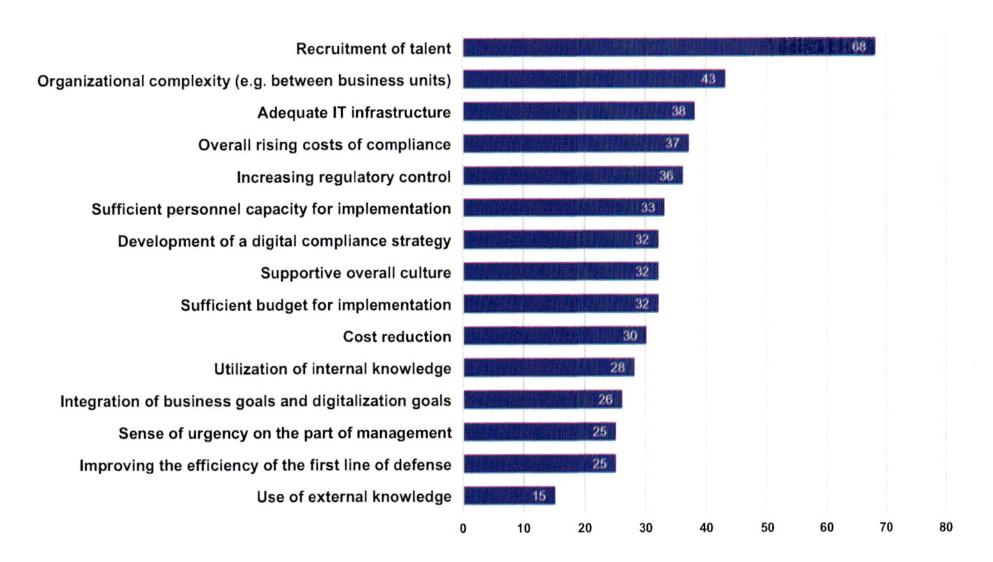

Fig. 8.7 Central challenges for the organization of compliance in the future. (Data source: BCG, 2022, pp. 4, 19–21)

8.5 Novamondo—Our Journey towards More Sustainability

Guest contribution by Andreas Huthwelker and Christian Schlimok

For more than 20 years, ***Novamondo*** has been accompanying numerous international organizations and companies in their brand development and supporting them in their digital and analog communication. In addition to strategic consulting, the creation of high-quality communication, UX and UI design, as well as programming and content production are among the core competencies of our team. Since our clients are largely public organizations, foundations, and companies with a high demand for sustainability, this was a topic at Novamondo that was almost self-evident and permanently present.

"Sustainability? We do that anyway!", was a typical statement in internal exchange. Or: "We have inspiring clients who themselves have a claim to be sustainable." And: "We use recycled paper in print production!" As the term gained more and more importance at the societal level, but also in tenders and inquiries from potential clients, we began to look at it more closely about five years ago.

Formulate and reflect values

The initial moment was an **internal project for further development** of our brand. With the entire team, we reflected on **company goals, vision, and mission** as well as our four **central values. Sustainability,** we decided together, should definitely be included. But to be able to communicate this, we first had to reflect on what exactly it means:

- What does sustainability mean in general understanding—as well as economically, socially, and ecologically?
- What does sustainability mean for our employees—and for our claim to lead responsibly?
- What ideas do we have as a team to implement sustainability internally—for example in our office routine?
- And how can we realize sustainability in projects with our clients?

Connecting Marketing and Innovation

We initiated a **series of events,** in which we brought external experts and customers into exchange with our team. Together, the aim was to explore the possibilities for even more sustainability in the areas of brand communication and innovation consulting. For example, we dealt with new **forms of print production** and the **co-creative design of analogue products from waste.** We tested **measures to save CO_2** in the design of digital applications and explored new methods for sustainability communication. The newly acquired knowledge flowed directly into our **project work** at Novamondo. In addition, we established **quality standards** as the content basis of our work, which we continuously develop further.

Designing Brand Identity Sustainably

An important **field of application for sustainability** is, for example, the **design of brand experiences** around our **own brand identity.** In our brand communication, we now almost completely do without print products. When we produce analogue formats, they must deliver high added value. Thus, our **artistic edition,** which we co-creatively designed with the entire team, is used as a gift for partners and customers in the long term—and not just for a special occasion. And when we produced a **Christmas card** again after a long time, we did so to explore the use of algae-based printing inks.

In addition, we have been using timeless core elements such as our logo in our **brand design** for almost 20 years. This way, many of our communication products can continue to be used. For our event series, we also experimented with the **multiple use and circularity of posters.** The changing event dates are applied with stickers so that the poster does not have to be printed again and again, saving resources. In addition, we integrated an additional **augmented reality layer** that can be newly played out from event to event.

Finding Goals and Developing a Strategy

In 2021, we developed our own **sustainability strategy** at Novamondo. In addition to our overarching corporate goals, it is extensively reflected here how we can realize the value of sustainability even more holistically within the framework of our activities. Following the Sustainable Development Goals (SDG) Compass of the UN Global Compact (see SDG Compass, 2022; see these goals Fig. 1.3), the world's largest and most important initiative for sustainable and responsible corporate governance, we identified our significant sustainability topics in a **materiality analysis.**

In comparison with external and internal stakeholders such as customers and employees, we analyzed the requirements that they consider particularly important for the realization of sustainability in the corporate governance of Novamondo. Subsequently, we selected from the total of **17 Sustainable Development Goals** all those that best fit the activities of Novamondo.

Thus, we contribute to **equal opportunity and high-quality education** (SDG 4) by advising organizations in the education sector on their brand communication and digitization. We also realize our own products, such as the learning app that we develop as part of a start-up participation. In addition, we try to research and gain new insights in educational formats for kindergartens and schools as well as in cooperation with universities. For example, we are concerned with the question of how sustainability can be reflected in exchange with children and young people and what impulses for future educational formats for adults can be derived from this.

Further examples of lived sustainability goals in our everyday life are the focus on topics such as **Smart and Circular City** in our projects (SDG 11 for sustainable cities and communities). Through the **consultation of public and political institutions** (SDG 16 for peace, justice and strong institutions) and our **commitment to the ecological transformation of companies and circular economy** (see Circular Applied Research Lab, 2022), among other things in teaching at universities (SDG 9 and 12 for industry, innovation and infrastructure and sustainable consumption and production methods), we contribute to further fields of sustainability.

Reporting and Getting Certified

Our sustainability strategy and the associated projects form the content basis for our **sustainability report** (see Novamondo, 2022). In addition, there are data that we systematically record as part of our **CO_2 balance.** To reflect our sustainability from the outside, we have had ourselves certified as a **B-Corp** in a one-year process (see B Corporations, 2022). We had previously dealt extensively with various certifications in the field of sustainability and had also sought advice on this. What we liked about B-Corp was the holistic view of corporate governance from both an ecological and social perspective, as well as the integration into an international community of companies that want to combine economic success with sustainability.

Creating Structures and Staying Committed

For the mentioned initiatives to have a long-term effect, **structures** are necessary that promote the continuous exchange of new ideas, quality standards, projects, and products, and involve our entire team. We are also working on this at Novamondo and are trying to solve this through **clear roles and responsibilities**. Sustainability is written as an overarching value in the formal goals at Novamondo.

In the course of the **certification as a B-Corp**, we have also adapted our articles of association as a GmbH accordingly. But paper is known to be patient, the planet is not:

Therefore, it is essential that our entire team continuously exchanges ideas about the understanding of sustainability and its realization, and together continues to take small steps towards a societal future that preserves our earth as a habitat for future generations.

Finding New Ideas, Rethinking and Moving Forward

The demand for sustainability poses systemic challenges to agencies and consulting firms in the field of marketing and brand communication. The possible **transition to a circular economy** is further promoted and legally anchored by the EU taxonomy. This development calls into question many areas and measures that still shape our actions around marketing and brand communication today. In addition, the **principles of Circular Design,** which go far beyond recyclability in the design of sustainable analog products, provide the entry into completely new ways of thinking and acting.

With a view to concrete environmental problems and resource scarcity, we must ask ourselves to what extent marketing in its previous forms is still meaningful and socially accepted. The following questions arise, among others:

- Are **merchandising items,** which consume natural resources and energy, still justifiable in an oversaturated consumer society? This question remains relevant even if these items were produced "responsibly" by today's standards.

- How do you design successful **corporate publishing** given the fact that in Germany alone about 1.2 billion kg of mailbox advertising end up unread in the trash (cf. Venugopal, 2022)?
- How can **posters** and the corresponding **logistics** be designed in such a way that the environment is less affected?
- What new measures can be used to stay in touch with people as a **brand**?
- The digital world causes about 2 to 4% of global greenhouse gas emissions—and the trend is rising (cf. Freitag et al., 2021). Can the **digitization of communication** with its massive energy consumption show a sustainable alternative?
- Is a **shift of communication to the digital space** purposeful—considering the high social consequences that come with too intensive use of digital devices?
- Which **forms of brand communication** create such a high added value that the "social costs" for it can be accepted?
- What role does **design** play in the future if we do not want to continue using it primarily to seduce consumption and resource use?

As this brief outline shows, many **new ideas** and a **passionate and long-term commitment** are required to meet the demand for sustainability in a new world. We are working on this at *Novamondo*. A guiding principle from our brand communication accompanies us (see Fig. 8.8):

> **Stay curious, inspire the world!**

Questions you should ask yourself

- How is the relevance of a Green Journey seen in my company?
- Where are the biggest skeptics still—and how can they be "brought on board"?
- How good is our storytelling?
- Do we know enough about the success factors of storytelling and—even more crucial—do we also take these into account in internal and external communication?
- Who is responsible for sustainability storytelling in our company?
- Is the Stakeholder-Onion-Model already being used in our company?
- How can its standard use—not only in sustainable management—be ensured?
- What is our stance on the idea of including the "environment" as its own stakeholder in our considerations?
- Does our company have a sustainability team?
- Are we already using sustainability ambassadors in the various departments and areas to incorporate sustainability thinking into everyday life?
- Should we establish a Chief Sustainability Manager in the company?
- Are we using a convincing step-by-step concept for the development or expansion of our sustainable corporate management?

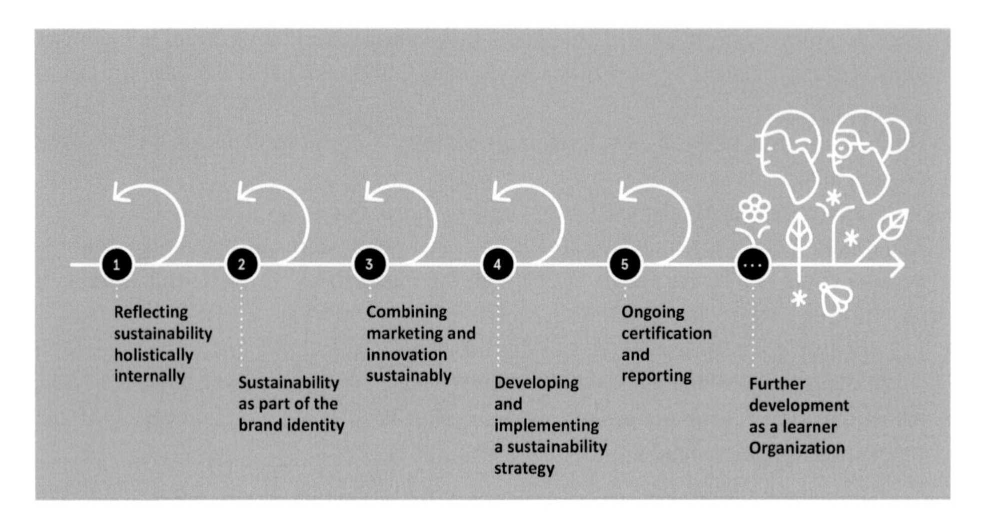

Fig. 8.8 Process for sustainable corporate management. (Source: © Novamondo)

- Is there a comprehensive training agenda on the topic of sustainability?
- Who is responsible for such a training agenda?
- Have we already managed to make (almost) all employees and managers eager for a journey towards sustainability?
- What could accelerate this journey and really involve everyone?

References

B Corporations. (2022). Unternehmertum verpflichtet. https://www.bcorporation.de. Accessed 21 Dec 2022.

BCG. (2022). 2022 Global ESG, Compliance & Risk Report, Value creation amid rising global uncertainty. Retrieved December 2, 2022, from https://media-publications.bcg.com. Accessed 2 Dec 2022.

Circular Applied Research Lab. (2022). Die Plattform für ökologische Transformation von Unternehmen. https://www.circular-applied-research.de. Accessed 21 Dec 2022.

Climate Partner. (2023). Klimaschutz im Unternehmen – jetzt starten. https://www.climatepartner.com/de. Accessed 2 Jan 2023.

Deloitte. (2022). Die Rolle des Chief Sustainability Officers im Finanzsektor im Wandel. https://www2.deloitte.com/de/de/pages/trends/rolle-des-chief-sustainability-officers.html. Accessed 16 Sept 2022.

Etzold, V. (2018). *Strategie, Planen – Erklären – Umsetzen.* Gabal.

Freitag, C., Berners-Lee, M., Widdicks, K., Knowles, B., Blair, G. S., & Friday, A. (2021). The real climate and transformative impact of ICT: A critique of estimates, trends, and regulations. *Patterns,* 2021. https://doi.org/10.1016/j.patter.2021.100340

Friedmann, J. (2021). *Storytelling for media.* UVK.

Kreutzer, R. T. (2021). *Toolbox für Digital Business. Leadership, Geschäftsmodelle, Technologien und Change-Management für das digitale Zeitalter*. Springer Gabler.

Novamondo. (2022). Nachhaltige Markenkommunikation und Innovationsberatung. https://novamondo.de/designagentur-fuer-nachhaltige-markenkommunikation-und-innovationsberatung. Accessed 21 Dec 2022.

Pyczak, T. (2021). *Tell me! Wie Sie mit Storytelling überzeugen. Inkl. Praxisbeispiele. Für alle, die erfolgreich sein wollen in Beruf, PR und Online-Marketing*. Rheinwerk.

SDG Compass. (2022). Learn more about the SDGs. https://www.sdgcompass.org/sdgs/. Accessed 21 Dec 2022.

Serviceplan. (2021). CMO Barometer 2022. https://www.serviceplan.com/de/news/cmo-barometer-2022.html. Accessed 2 Aug 2022.

Shank, R. C. (2021). *Effective storytelling techniques for social media*. https://masterclasses.marketing-interactive.com/virtual-masterclasses/effective-storytelling-techniques-for-social-media/. Accessed 5 Jan 2021.

Storr, W. (2019). *Science of storytelling*. William Collins.

Venugopal, T. (2022). *8 Fakten, wie Briefkastenwerbung unsere Umwelt zerstört*. https://www.cleanupnetwork.com/news/nachhaltigkeit/briefkasten-keine-werbung-papiermuell/. Accessed 21 Dec 2022.

Printed by Printforce, the Netherlands